WARD'S

World Motor Vehicle Data 2013

Price per copy: US $375.00
Also available in electronic format or CD-ROM
Add sales tax for some states
Add $25 for overseas destinations

ISBN Number 978-0-910589-66-6
ISSN Number 1553-8176

WardsAuto Group
3000 Town Center, Suite 2750
Southfield, MI 48075 USA
Phone: (248) 357-0800 FAX: (248) 357-0810
www.WardsAuto.com

Foreword

WardsAuto Group is pleased to present the 2013 edition of *Ward's World Motor Vehicle Data*™. Our hope is that this publication will serve as a key reference tool for those who work in the auto industry and for the many groups with an avid interest in the business — including economists, government officials, investors and the academic community.

A few notes about the contents:

- Careful attempts were made to eliminate double-counting of production in multiple countries. Double-counting occurs when a vehicle is substantially produced in one country, and the subassembly sets are shipped to another country, which completes the final assembly. Sometimes, both countries report the vehicle in their totals. Every effort was made to identify these situations, and correct for them, with output credited in the regional and world summaries to the country of final assembly.
- Production typically is attributed to the selling company, in those cases where one manufacturer produces the vehicle under the badge of another.
- Sales are a mix of manufacturer-reported sales and new government registrations, depending on local practice.
- The detail level of the data also is governed by the local market. In some cases, light commercial vehicles (typically trucks) are broken out from heavy CVs and buses. In other markets, that distinction isn't available. When provided, these distinctions are provided in the country tables. But because this detail isn't universally available, the world and regional data show only a car/cv split.
- Local practice determines the classification of cars vs. commercial vehicles as well. Sport-utility vehicles and vans are considered trucks in North America, but often are considered cars in other markets. Conversely, cars used as taxi-cabs are often counted as commercial vehicles outside N. America.

WardsAuto Group has served the automotive industry's information needs for decades, through our comprehensive website, *WardsAuto.com*, the industry's weekly newsletter, *Ward's Automotive Reports*®, along with other services including online data products, magazines and digital newsletters.

Acknowledgements

WardsAuto Group is grateful for the cooperation of many organizations in providing information included in the 2013 edition of *Ward's World Motor Vehicle Data*™.

Information sources included industry associations and/or government agencies in the following countries: Argentina, Australia, Austria, Belgium, Brazil, Canada, Chile, China, Colombia, Czech Republic, Denmark, Egypt, Finland, France, Germany, Greece, Hungary, India, Indonesia, Ireland, Israel, Italy, Japan, Lithuania, Luxembourg, Malaysia, Mexico, Netherlands, New Zealand, Norway, Pakistan, Poland, Portugal, Romania, Russia, Serbia, Singapore, Slovakia, South Africa, South Korea, Spain, Sweden, Switzerland, Taiwan, Thailand, Turkey, Ukraine, Uruguay, United Kingdom, Venezuela and Vietnam. Other data sources included the International Road Federation, World Bank, and International Monetary Fund.

James W. Bush: Managing Director, Ward's Automotive Group
John Sousanis: Director, Information Content **Paul Zajac:** Manager, Industry Data
Heather Rowe: Research Associate **Erin Berger:** Research Assistant
Rebecca Norris: Data Operations Manager

Barbara Liske: Customer Service
Phone: (248) 799-2645

Table of Contents

World Markets See Sign of Recovery in 2012

Full-year global vehicle sales reached an estimated 81.8 million units in 2012 thanks to double- digit growth in North America and the Asia-Pacific region, which more than compensated for the downturn in Europe and slower growth in South America.

North American sales rose 12.4% to 17.5 million units, led by U.S. sales of nearly 14.8 million units, a 13.4% increase over 2011 as consumers took advantage of low interest rates and generous incentives despite a sluggish labor market. This was the fourth straight year of increased sales for the U.S. since the recession struck the global economy. Canada saw moderate sales growth as deliveries increased 5.9% and total unit sales hit their highest level since 2002. In Mexico, supported by a growing economy and strong fleet sales, deliveries increased 9.4% to slightly over 1 million units, the highest level since 2008.

Sales in the Asia-Pacific region rose 10.3%, as many markets posted solid gains over 2011, when vehicle supply was hampered by natural disasters in Japan and Thailand.

The region was home to the three fastest growing markets last year: Japan (+27.5%), Indonesia (+24.8%), and Thailand, where a growing economy and favorable year-over-year comparisons with a flood-plagued 2011 combined for a 80.7% surge in sales to a record 1.4 million units.

China, the world's largest market, saw sales rise to a record 19.3 million vehicles. The growth rate of 4.6% was higher than 2011's 2.6%, but the country still felt the impact of the EU economic crisis as it is China's biggest export market. Market stabilization after years of massive gains also affected the growth rate as China's share of world sales fell slightly to 23.6% from prior-year's 23.8%.

Japan remained the world's third-largest market, with it's 5.3 million deliveries accounting for 6.6% of global volume, compared with 5.4% in the prior year. Collectively, the Asia-Pacific region accounted for 43.6% of 2012 global sales, up two share points.

As the meltdown in Europe continued, sales volume fell 5.9%. Russia (+10.5%) and the U.K (+3.8%) were the only major markets to post gains. Deliveries in Russia rose to an all time record of 3.1 million units as global automakers flocked to one of the few healthy markets in Europe. Currently the second largest market in Europe, some analysts predict that Russia will soon overtake Germany as the region's No.1 car market.

Brazil sales for the year hit a record 3.8 million units as the government approved a stimulus package aimed at boosting auto loans. However, rising inflation, interest rates and widespread public protests have raised concerns over the long- term outlook for sales. In Argentina, South America's second largest market, vehicle deliveries fell 6% to 830,000 units. Overall, South American sales rose a modest 2.3% to 5.9 million vehicles which gave the region 7.2% of the world market.

In 2012, global production hit an all time record of 84.1 million vehicles, up over 5% from 2011 levels and, except for Western Europe, seemed to have fully recovered from the global collapse of 2009.

To keep up with demand some manufacturers operated 3 shifts and were at near capacity as North American production hit 15.8 million vehicles in 2012, up 17.2% over 2011 and a staggering 80% since bottoming out in 2009 when production was a meager 8.8 million units. Led by the U.S. with a 19.3% year over year increase, North America grew its share of world production to nearly 19% in 2012.

In the Asia-Pacific region, China maintained its position as the world's no.1 producer as total vehicle production reached 19.3 million vehicles, a growth rate of 4.6% over 2011, a much slower pace than prior years.

Japan manufacturing was restored to full capacity following the natural disasters of 2011; total unit production hit 9.9 million, 18.4% higher than the earthquake and tsunami plagued prior year, securing their spot as the no. 3 global producer.

Overall, the Asia-Pacific region accounted for over 50% of global production.

As the European economic crisis continued to unfold in 2012, total production in the European Union was substantially reduced by nearly 8% due to work stoppages as a result of weak demand. There were a few bright spots in non- EU countries such as the United Kingdom, where production hit 1.6 million vehicles, up 7.6% over prior-year levels. Russia also showed signs of recovery as 2.2 million vehicles were built, 12.2% above 2011 and setting an all-time record. ❑

Ward's World Motor Vehicle Data 2013

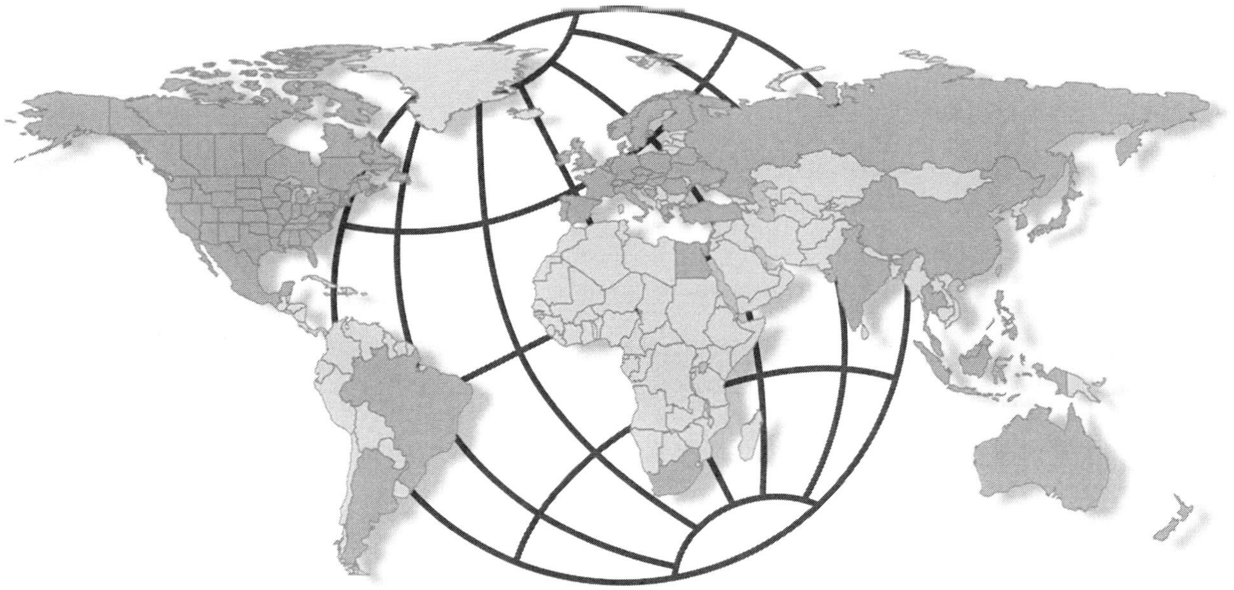

Asia/Pacific Data

AUSTRALIA

ECONOMIC SUMMARY

Common Language	English
Currency	Australian Dollar
Steering	Right Hand Drive

2012 Vehicles in Operation

Passenger Cars	12,714,235
Commercial Vehicles	3,265,521
Total	15,979,756
Persons per Car	1.8
Persons per Vehicle	1.4

2012 GDP (billions of US $)

Nominal GDP	1,541.2
Real GDP	1,529.0
GDP % annual growth	3.6
GDP per capita	67,159

Current Account Balance (billions of US $)

2012: -56.4
2011: -33.8

Current Account Balance in % of GDP

2012: -3.7
2011: -2.3

Total Number of Fatal Road Accidents (most recent data)

2010: 1,367

Capital	Canberra
Population	22,262,501
Population Density	7.5
Surface Area sq km/miles	7,682,300 / 2,966,152

Road Infrastructure

Paved Road Surface %	38.7
Total Road Length in km/miles	913,000 / 567,312
Road Traffic (Vehicles per mile)	28.2

2012 Total Vehicle Percent Market Share by Company

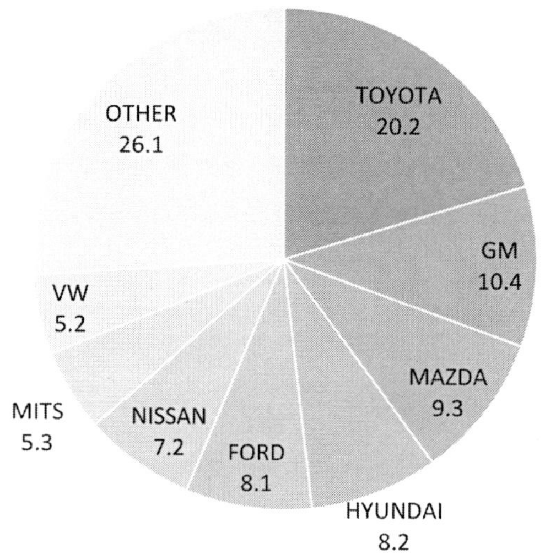

Total Vehicle Production and Sales 2002-2012

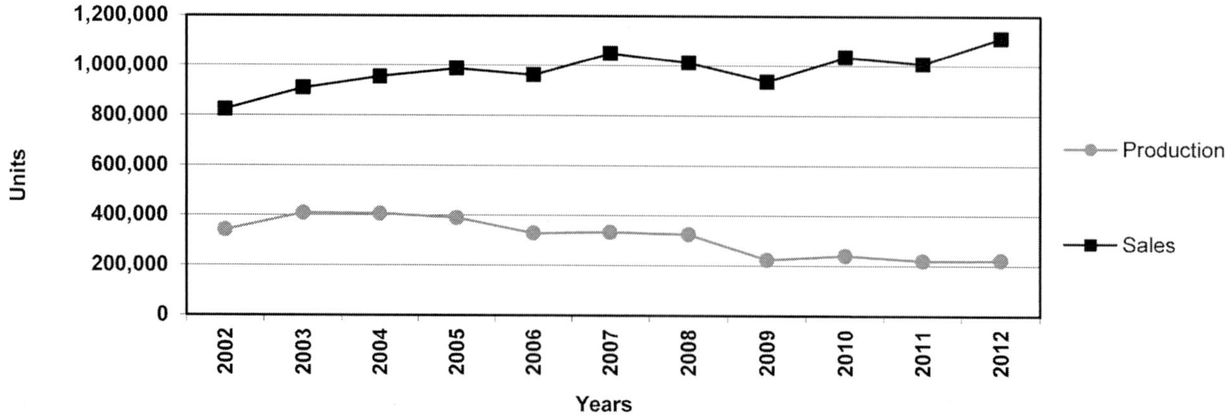

AUSTRALIA

PRODUCTION

Year	Cars	Commercial Vehicles	Total
2012	189,949	31,275	221,224
2011	189,503	29,873	219,376
2010	205,334	34,109	239,443
2009	188,158	35,196	223,354
2008	285,590	38,528	324,118
2007	306,416	25,823	332,239
2006	297,986	29,750	327,736
2005	320,258	68,727	388,985
2004	340,864	64,450	405,314
2003	365,611	42,341	407,952
2002	306,867	33,599	340,466
2001	313,762	30,108	343,870
2000	323,649	21,273	344,922
1999	302,925	17,940	320,865
1998	337,500	17,069	354,569
1997	325,957	16,983	342,940
1996	328,827	16,056	344,883
1995	314,142	29,709	343,851
1994	328,909	9,159	338,068
1993	294,070	7,015	301,085
1992	269,496	26,326	295,822
1991	310,660	22,110	332,770
1990	386,031	26,645	412,676
1989	333,170	22,050	355,220
1988	314,875	22,860	337,735
1987	302,057	20,462	322,519
1986	364,754	25,912	390,666
1985	375,812	28,749	404,561
1984	342,316	25,993	368,309
1983	334,685	21,921	356,606
1982	384,228	38,014	422,242

AUSTRALIA

PRODUCTION BY MANUFACTURER

	2012	% Chg.	2011		2012	% Chg	2011
Cars				Commercial Vehicles			
Ford	15,035	-13.6	17,407	Ford	22,667	11.2	20,387
GM (Holden)	73,490	-6.4	78,481	GM (Holden)	8,608	-9.3	9,486
Toyota	101,424	8.3	93,615	**Total Comm. Vehs.**	**31,275**	**4.7**	**29,873**
Total Cars	**189,949**	**0.2**	**189,503**	**Total Vehicles**	**221,224**	**0.8**	**219,376**

AUSTRALIA

SALES

Year	Cars	Commercial Vehicles	Total
2012	576,855	535,177	1,112,032
2011	559,314	449,123	1,008,437
2010	592,122	443,452	1,035,574
2009	540,562	396,766	937,328
2008	596,765	415,399	1,012,164
2007	637,019	412,963	1,049,982
2006	598,394	364,272	962,666
2005	608,804	379,465	988,269
2004	589,985	365,244	955,229
2003	588,511	321,300	909,811
2002	540,240	284,069	824,309
2001	529,452	243,229	772,681
2000	553,673	233,427	787,100
1999	547,575	239,270	786,845
1998	584,360	223,309	807,669
1997	544,129	182,074	726,203
1996	492,058	157,991	650,049
1995	488,372	154,185	642,557
1994	460,698	155,588	616,286
1993	414,397	140,909	555,306
1992	453,215	89,853	543,068
1991	416,783	81,322	498,105
1990	487,742	126,496	614,238
1989	464,322	133,015	597,337
1988	418,699	101,833	520,532
1987	369,335	81,487	450,822
1986	410,878	115,262	526,140
1985	530,209	159,153	689,362
1984	482,102	155,861	637,963
1983	433,514	132,089	565,603
1982	480,073	146,544	626,617

AUSTRALIA

SALES BY COMPANY

	2012	% Chg.	2011
Cars			
Aston Martin	94	-6.9	101
Audi	8,981	-13.4	10,367
Bentley	65	6.6	61
BMW	*11,256*	*12.3*	*10,026*
Mini	*1,878*	*0.8*	*1,863*
BMW Total	13,134	10.5	11,889
Caterham	3	–	–
Chery	551	-22.2	708
Chrysler	*1,359*	*134.3*	*580*
Dodge	*1,838*	*-19.1*	*2,273*
Chrysler Total	3,197	12.1	2,853
Mercedes-Benz	*14,876*	*5.9*	*14,047*
Smart	*142*	*-39.8*	*236*
Daimler Total	15,018	5.1	14,283
Ferrari	98	-26.9	134
Alfa Romeo	*906*	*-17.0*	*1,091*
Fiat	*513*	*-3.6*	*532*
Fiat Total	1,419	-12.6	1,623
Ford	48,238	-4.5	50,516
Holden	*74,884*	*-16.1*	*89,255*
Opel	*541*	–	–
GM Total	75,425	-15.5	89,255
Honda	31,079	27.6	24,359
Hyundai	69,427	7.7	64,469
Jaguar Land Rover	786	13.4	693
Kia	23,083	25.3	18,428
Lamborghini	41	28.1	32
Maserati	124	-11.4	140
Mazda	68,430	5.1	65,117
McLaren	23	–	–
Mitsubishi	15,856	-20.0	19,816
Morgan	16	-20.0	20
Infiniti	*36*	–	–
Nissan	*15,174*	*-3.5*	*15,724*

	2012	% Chg.	2011
Nissan Total	15,210	-3.3	15,724
Citroen	*1,416*	*16.4*	*1,217*
Peugeot	*3,914*	*-14.7*	*4,587*
Peugeot Citroen Total	5,330	-8.2	5,804
Porsche	516	-4.4	540
Lotus	*69*	*15.0*	*60*
Proton	*856*	*-25.3*	*1,146*
Proton Total	925	-23.3	1,206
Renault	2,676	10.9	2,413
Rolls-Royce	22	10.0	20
Saab	27	-82.0	150
Ssangyong	37	-41.3	63
Subaru	13,014	-13.0	14,962
Suzuki	19,120	-4.2	19,958
Lexus	*4,489*	*0.5*	*4,466*
Toyota	*101,492*	*21.9*	*83,248*
Toyota Total	105,981	20.8	87,714
Skoda	*2,330*	*-14.0*	*2,709*
Volkswagen	*34,944*	*12.4*	*31,077*
Volkswagen Total	37,274	10.3	33,786
Volvo	1,635	-22.5	2,110
Total Cars	**576,855**	**3.1**	**559,314**
Light Trucks			
Audi	5,554	34.0	4,144
BMW	*7,157*	*-4.3*	*7,482*
Mini	*516*	*20.6*	*428*
BMW Total	7,673	-3.0	7,910
Chery	582	-47.8	1,114
Chrysler	*190*	*-55.8*	*430*
Jeep	*18,014*	*108.3*	*8,648*
Chrysler Total	18,204	100.5	9,078
Mercedes-Benz	*4,375*	*12.7*	*3,883*
Mitsubishi Fuso	*591*	–	–
Daimler Total	4,966	27.9	3,883

AUSTRALIA

SALES BY COMPANY

	2012	% Chg.	2011
Fiat	115	-3.4	119
Ford	41,910	2.9	40,727
Holden	*39,781*	*8.0*	*36,840*
Hummer	–	–	*3*
GM Total	39,781	8.0	36,843
Great Wall	11,006	27.0	8,665
Honda	4,733	-17.7	5,748
Hyundai	22,109	-1.9	22,539
Isuzu	7,782	21.7	6,397
Kia	7,675	14.6	6,700
Jaguar Land Rover	7,929	36.0	5,832
Mazda	35,456	52.7	23,216
Mitsubishi	43,012	4.2	41,292
Infiniti	*49*	–	–
Nissan	*64,573*	*23.7*	*52,202*
Nissan Total	64,622	23.8	52,202
Citroen	*286*	*44.4*	*198*
Peugeot	*1,157*	*82.8*	*633*
Peugeot Citroen Total	1,443	73.6	831
Porsche	857	6.7	803
Proton	149	-62.4	396
Renault	1,905	67.0	1,141
Ssangyong	1,553	0.6	1,543
Subaru	27,175	42.7	19,049
Suzuki	4,937	29.2	3,820
Lexus	*2,350*	*24.9*	*1,881*
Toyota	*116,684*	*19.0*	*98,055*
Toyota Total	119,034	19.1	99,936
Skoda	*1,172*	*48.0*	*792*
Volkswagen	*19,053*	*44.4*	*13,194*
Volkswagen Total	20,225	44.6	13,986
Volvo	3,740	18.3	3,162
Total Light Trucks	**504,127**	**19.7**	**421,076**

	2012	% Chg.	2011
Heavy Trucks			
Freightliner	*798*	*10.5*	*722*
Mercedes-Benz	*3,146*	*-3.2*	*3,250*
Mitsubishi Fuso	*2,908*	*-22.4*	*3,749*
Western Star	*1,003*	*14.8*	*874*
Daimler Total	7,855	-8.6	8,595
Dennis Eagle	135	237.5	40
Fiat	804	-10.0	893
Ford	260	–	–
Foton	111	32.1	84
Hino	4,216	26.5	3,334
Hyundai	65	-51.9	135
International	18	-75.0	72
Isuzu	7,190	8.7	6,617
Iveco	2,086	23.5	1,689
MAN	*428*	*59.7*	*268*
Volkswagen Truck & Bus	*838*	*78.7*	*469*
MAN Total	1,266	71.8	737
DAF	*318*	*89.3*	*168*
Kenworth	*2,360*	*33.6*	*1,766*
PACCAR Total	2,678	38.5	1,934
Scania	584	28.6	454
Toyota	–	–	321
Mack Truck	*958*	*16.1*	*825*
Renault Truck	*430*	*532.4*	*68*
UD Trucks	*734*	*-16.4*	*878*
Volvo Truck	*1,351*	*11.2*	*1,215*
Volvo Truck Total	3,473	16.3	2,986
Other	309	98.1	156
Total Heavy Trucks	**31,050**	**10.7**	**28,047**
Total Comm. Vehs.	**535,177**	**19.2**	**449,123**
Total Vehicles	**1,112,032**	**10.3**	**1,008,437**

CHINA

ECONOMIC SUMMARY

Common Language	Chinese/Mandarin
Currency	Yuan
Steering	Left Hand Drive

2012 Vehicles in Operation

Passenger Cars	52,165,000
Commercial Vehicles	57,275,000
Total	109,440,000
Persons per Car	25.9
Persons per Vehicle	12.3

2012 GDP (billions of US $)

Nominal GDP	8,227.0
Real GDP	2,637.9
GDP % annual growth	7.8
GDP per capita	1,948

Current Account Balance (billions of US $)

2012: 213.7
2011: 201.7

Current Account Balance in % of GDP

2012: 2.6
2011: 2.8

Total Number of Fatal Road Accidents (most recent data)

2011: 62,387

Capital	Beijing
Population	1,349,585,838
Population Density	365.3
Surface Area sq km/miles	9,596,901 / 3,694,959

Road Infrastructure

Paved Road Surface %	84.1
Total Road Length in km/miles	4,106,387 / 2,551,591
Road Traffic (Vehicles per mile)	42.9

2012 Total Vehicle Percent Market Share by Company

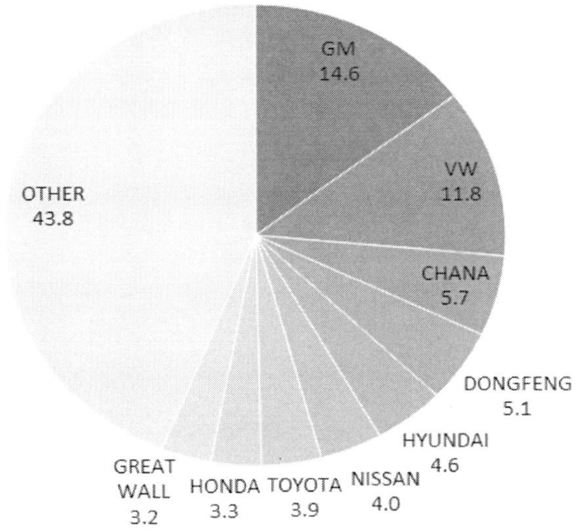

Total Vehicle Production 2002-2012

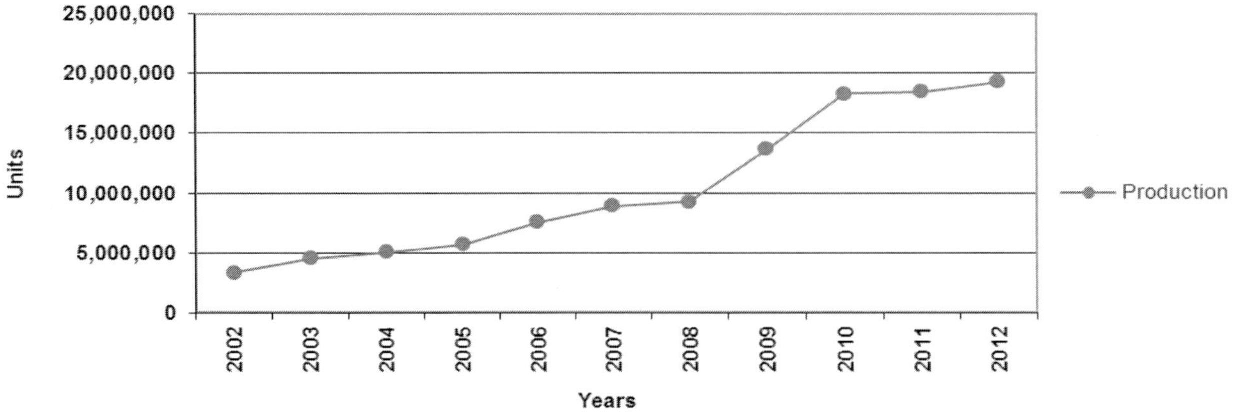

CHINA

PRODUCTION

Year	Cars	Trucks	Buses	Total
2012	10,704,962	8,397,753	169,093	19,271,808
2011	10,052,854	8,201,683	164,339	18,418,876
2010	9,494,018	8,609,046	161,603	18,264,667
2009	7,322,018	6,184,086	142,449	13,648,553
2008	4,970,674	4,131,477	131,139	9,233,290
2007	5,184,971	3,524,451	176,039	8,885,461
2006	5,160,874	2,280,619	124,740	7,566,233
2005	3,585,877	2,016,133	66,153	5,668,163
2004	2,316,262	1,514,734	1,239,531	5,070,527
2003	2,124,350	1,229,601	1,195,210	4,549,161
2002	1,175,198	1,096,197	1,064,208	3,335,603
2001	703,521	802,353	828,566	2,334,440
2000	604,677	764,005	700,387	2,069,069
1999	565,366	756,277	508,680	1,830,323
1998	507,103	661,701	459,025	1,627,829
1997	487,695	659,318	435,615	1,582,628
1996	381,510	684,369	390,321	1,456,200
1995	320,578	737,035	377,175	1,434,788
1994	247,639	781,266	322,085	1,350,990
1993	229,697	623,184	443,897	1,296,778
1992	162,725	460,274	438,722	1,061,721
1991	81,055	361,310	266,455	708,820
1990	42,409	269,098	197,735	509,242
1989	28,820	342,835	215,281	586,936
1988	36,798	364,000	246,153	646,951
1987	20,865	299,356	152,317	472,538
1986	12,297	218,863	141,593	372,753
1985	5,207	236,934	201,236	443,377
1984	6,010	179,846	130,511	316,367
1983	6,046	137,100	96,740	239,886
1982	4,030	121,789	–	125,819

CHINA

PRODUCTION BY MANUFACTURER

	2012	% Chg	2011
Cars			
Beijing AIC	22,913	–	–
Beijing Benz (Mercedes-Benz)	70,598	-11.4	79,670
Beijing Hyundai	683,521	15.5	591,657
Brilliance BMW	128,871	31.5	98,001
Brilliance China (Brilliance)	118,342	-4.0	123,326
BYD	366,201	-5.3	386,777
Chana	162,565	-1.7	165,401
Ford	*274,350*	*-13.0*	*315,309*
Mazda	*211,529*	*151.4*	*84,142*
Volvo	*5,529*	*-67.0*	*16,769*
Chana Ford Total	491,408	18.1	416,220
Chana Suzuki	170,797	-22.3	219,889
Changhe	34,517	41.8	24,337
Chery	408,265	-11.1	459,227
Dongfeng	36,788	103.8	18,051
Dongfeng Honda	105,518	10.7	95,344
Dongfeng Kia	371,777	16.0	320,502
Dongfeng	*45,141*	–	–
Nissan	*586,771*	*-11.7*	*664,660*
Dongfeng Nissan Total	631,912	-4.9	664,660
Citroen	*226,109*	*-2.5*	*231,833*
Peugeot	*215,983*	*24.1*	*174,102*
Dongfeng PSA Total	442,092	8.9	405,935
Dongfeng Yulon (Luxgen)	32,103	353.5	7,079
FAW	*8,643*	–	–
Hongqi	*69,894*	*-36.8*	*110,674*
Mazda	*100,524*	*-17.4*	*121,696*
FAW Total	179,061	-22.9	232,370
FAW Haima	61,919	-24.9	82,427
FAW Toyota	377,475	-5.4	398,918
Audi	*244,087*	*22.2*	*199,780*
Volkswagen	*1,009,451*	*33.0*	*759,246*
FAW Volkswagen Total	1,253,538	30.7	959,026
FAW Xiali	181,305	-28.5	253,631
Emgrand	*160,996*	*40.8*	*114,324*
Englon	*22,405*	–	*854*
Geely	*228,247*	*-17.0*	*275,022*
Gleagle	*50,877*	*12.4*	*45,276*
Geely Total	462,525	6.2	435,476
Great Wall	181,870	14.4	158,996
Guangzhou AC	32,555	94.7	16,723
Guangzhou Changfeng (Liebao)	1	-99.8	626
Guangzhou Fiat	13,505	–	–
Everus	*23,881*	*-6.6*	*25,570*
Honda	*228,658*	*-26.4*	*310,843*
Guangzhou Honda Total	252,539	-24.9	336,413
Guangzhou Toyota	158,468	-1.7	161,204
Chana	*24,901*	–	–
Hafei	*6,865*	*6.1*	*6,468*
Hafei Total	31,766	391.1	6,468
Haima Zhengzhou (Zhengzhou)	11,083	-21.0	14,034
Honda	28,730	18.7	24,202
Huatai	162	-82.6	931
Jianghuai Auto Co.	137,300	-6.2	146,405
Jiangnan	27,332	-16.0	32,536
Lifan	157,894	56.8	100,701
MG	*13,616*	*867.7*	*1,407*
Roewe	*69,975*	*13.6*	*61,602*
Nanjing Total	83,591	32.7	63,009
MG	*63,617*	*28.0*	*49,714*
Roewe	*41,605*	*4.9*	*39,672*
Shanghai AIC Total	105,222	17.7	89,386
Buick	*606,172*	*3.6*	*584,926*
Cadillac	*6,546*	*-9.7*	*7,249*
Chevrolet	*639,520*	*17.5*	*544,308*
Shanghai GM Total	1,252,238	10.2	1,136,483
Baojun	*86,661*	*294.5*	*21,968*
Chevrolet	*28,771*	*-56.9*	*66,739*
Shanghai GM Wuling Total	115,432	30.1	88,707
Skoda	*48,702*	*6.1*	*45,896*
Volkswagen	*1,028,041*	*6.3*	*967,146*
Shanghai Volkswagen Total	1,076,743	6.3	1,013,042
Southeast (Mitsubishi)	101,879	3.4	98,488
Young Man (Proton)	56,194	68.1	33,426
Zhejiang (Feidie)	4,970	-13.1	5,717
Zotye	81,477	-6.8	87,433
Total Cars	**10,704,962**	**6.5**	**10,052,854**
Light Trucks			
Beijing AIC	*152,948*	*-1.4*	*155,101*
Weiwang	*43,855*	*213.1*	*14,008*
Beijing AIC Total	196,803	16.4	169,109
Beijing Benz (Mercedes-Benz)	26,241	–	502
Beijing Hyundai	171,786	12.8	152,231
Brilliance BMW	21,181	–	227
Brilliance	*112,141*	*74.4*	*64,308*
Jinbei	*1,044*	*48.7*	*702*
Brilliance China Total	113,185	74.1	65,010
BYD	89,243	42.5	62,648

CHINA

PRODUCTION BY MANUFACTURER

	2012	% Chg.	2011
Chana	533,491	3.0	518,042
Chana Ford	2,132	-35.4	3,300
Changhe	91,383	-23.2	118,939
Changzheng	–	–	733
Chery	155,686	-12.6	178,196
China National	6,311	310.9	1,536
China Yituo	3,846	-21.1	4,872
Dayun	8,168	23.3	6,626
Dongfeng	698,256	-3.7	724,795
Dongfeng Honda	179,905	13.5	158,510
Dongfeng Kia	113,491	2.5	110,763
Dongfeng Nissan	125,721	-15.9	149,463
Dongfeng PSA (Peugeot)	1,371	–	–
FAW	*70,481*	*5.3*	*66,926*
Hongta	*28*	*-78.1*	*128*
Jilin	*91,114*	*-11.6*	*103,084*
Mazda	*1,897*	*-72.0*	*6,771*
FAW Total	163,520	-7.6	176,909
Haima	*46,107*	*-7.3*	*49,763*
Mazda	*21,433*	*8.1*	*19,828*
FAW Haima Total	67,540	-2.9	69,591
FAW Toyota	123,726	-4.3	129,352
FAW Volkswagen (Audi)	89,346	52.0	58,776
Fire Fighting Vehicles	–	–	336
Foton	409,372	-0.8	412,612
Geely (Gleagle)	32,123	–	–
Golden Dragon	15,916	9.9	14,483
GoNow	36,617	55.1	23,602
Great Wall	442,556	35.1	327,566
Guangzhou	*437*	*-43.5*	*773*
Liebao	*6,169*	–	–
Guangzhou AC Total	6,606	754.6	773
Changfeng	–	–	*3*
Liebao	–	–	*23,822*
Mitsubishi	–	–	*8,419*
Guangzhou Changfeng Total	–	–	32,244
Guangzhou Hino	323	37.4	235
Guangzhou Honda	56,008	70.9	32,763
Mitsubishi	*13,359*	–	–
Liebao	*5,133*	–	–
Guangzhou Mitsubishi Total	18,492	–	–
Guangzhou Toyota	92,801	-16.6	111,322
Guilin	2,464	44.3	1,708
Guizhou	4,736	34.4	3,524
Hafei	55,363	-52.2	115,772
Haima Zhengzhou	39,222	-32.8	58,369
Heibao	–	–	36,331
Hengtong	20	33.3	15
Huanghai	*35,044*	*-4.0*	*36,491*
Oldsmobile	–	–	*22*
Huanghai Total	35,044	-4.0	36,513
Huatai	*7,137*	*986.3*	*657*
Hyundai	*27,418*	*61.7*	*16,957*
Huatai Total	34,555	96.2	17,614
Jialing	142	–	–
Jianghuai Auto Co.	284,594	4.1	273,491
Jinbei	245,906	-4.0	256,273
JMC (Jiangling)	217,653	7.6	202,209
Kaima	80,933	11.1	72,822
King Long	20,535	37.8	14,905
Lifan	85,132	12.8	75,459
Linghe	1,995	15.3	1,731
Mei Ya	2,430	-46.0	4,497
Chana	*140,067*	–	–
Nanjing	*130,190*	*-54.0*	*283,049*
Roewe	*5,381*	–	–
Nanjing Total	275,638	-2.6	283,049
Nanjun	37,308	-17.6	45,275
New Forta	7,482	1.8	7,353
New Longma	1,188	-25.5	1,595
Isuzu	*1,735*	*29.0*	*1,345*
Qingling	*63,117*	*-0.2*	*63,249*
Qingling Total	64,852	0.4	64,594
Sanjing	42	-51.7	87
Shaanxi	9,319	-18.8	11,472
Shandong	3,300	31.2	2,516
Roewe	*1,481*	*-69.2*	*4,811*
Shanghai AIC	*7,582*	*91.7*	*3,956*
Shanghai AIC Total	9,063	3.4	8,767
Buick	*71,443*	*0.7*	*70,962*
Chevrolet	*19,764*	–	–
Shanghai GM Total	91,207	28.5	70,962
Shanghai GM Wuling	1,384,736	16.1	1,192,411
Shanghai Volkswagen	205,214	24.1	165,379
Sichuan	6,867	-42.2	11,875
Mercedes-Benz	*7,516*	*-28.8*	*10,563*
Mitsubishi	*777*	*-89.1*	*7,126*
Southeast	*4,362*	*-9.7*	*4,828*
Southeast Total	12,655	-43.8	22,517
Sunwin	105	-58.2	251

PRODUCTION BY MANUFACTURER

	2012	% Chg.	2011
Tangjun	52,425	7.1	48,935
Tri-Ring Special	606	-42.4	1,052
Yaxing	629	-44.3	1,129
Yutong	4,261	41.8	3,004
Feidie	*27,917*	*21.7*	*22,942*
Zhejiang	*645*	*491.7*	*109*
Zhejiang Total	28,562	23.9	23,051
Zhengzhou Nissan	46,914	-9.4	51,799
Zoomlion	268	-39.6	444
Zotye	8,242	-54.1	17,968
ZX Auto	63,221	2.4	61,755
Total Light Trucks	**7,517,974**	**6.1**	**7,084,539**
Medium/Heavy Trucks			
Changzheng	71	-97.9	3,309
China National	120,705	-19.4	149,748
China Yituo	4,119	-19.3	5,104
Dayun	12,159	-17.0	14,644
Dongfeng	193,413	-23.9	254,087
FAW	151,295	-21.8	193,586
Fire Fighting Vehicles	–	–	8
Foton	87,685	-16.1	104,570
Guangzhou Hino	5,174	37.6	3,761
Hualing Heavy Duty	14,091	-39.9	23,429
Huanghai	413	325.8	97
Jianghuai Auto Co.	39,059	-27.4	53,786
Jinbei	12,337	12.9	10,928
Kaima	4,166	-55.9	9,440
Lifan	29,631	-0.2	29,686
Nanjing	2,469	-15.7	2,928
Nanjun	22,590	-16.6	27,095
New Longma	1,220	0.7	1,212
North Benz (Mercedes-Benz)	22,500	-35.3	34,758
Qingling	24,106	-18.0	29,412
Sanjing	318	622.7	44
Shaanxi	76,699	-19.0	94,741
Shandong	–	–	99
Shanghai AIC (Iveco)	15,614	-42.2	27,005
Taiyuan	309	-90.0	3,096
Tangjun	16,742	21.1	13,823
Tri-Ring Special	7,688	-16.6	9,213
Feidie	*12,553*	*-19.7*	*15,629*
Zhejiang	*2,653*	*39.2*	*1,906*
Zhejiang Total	15,206	-13.3	17,535
Total Med./Hvy. Trucks	**879,779**	**-21.2**	**1,117,144**

	2012	% Chg.	2011
Buses			
Ankai	5,683	14.7	4,956
BYD	262	-18.9	323
Chana	1,682	-12.6	1,925
China National	776	78.8	434
Dadi	–	–	116
Dayun	–	–	15
Dongfeng	14,919	-6.0	15,870
FAW	8,744	-27.9	12,132
Foton	4,055	21.3	3,344
Golden Dragon	8,138	-2.3	8,328
Guangzhou AC	1,138	57.0	725
Guangzhou Hino	313	148.4	126
Guilin	2,048	2.8	1,992
Hengtong	1,646	318.8	393
Huanghai	5,085	17.1	4,343
Jialing	182	–	–
Jianghuai Auto Co.	9,720	34.7	7,218
JMC (Jiangling)	62	–	–
King Long	33,637	0.6	33,451
Lions	126	-40.8	213
Nanjing	1,497	-45.1	2,729
Nanjun	845	-45.0	1,535
Neoplan	1,150	34.5	855
New Forta	–	–	511
New Longma	203	45.0	140
Sanjing	1	-66.7	3
Shaanxi	265	-74.2	1,026
Shaolin	2,600	8.0	2,407
Sichuan	–	–	485
Sunwin	3,145	8.4	2,902
Xian Silver	254	12.4	226
Yanjing	919	-15.7	1,090
Yaxing	3,423	38.6	2,469
Young Man (Neoplan)	3,408	4.6	3,257
Yutong	47,174	10.0	42,891
Zhejiang	–	–	18
Zhongtong	5,984	5.6	5,667
Zoomlion	9	-96.0	224
Total Buses	**169,093**	**2.9**	**164,339**
Total Comm. Vehs.	**8,566,846**	**2.4**	**8,366,022**
Total Vehicles	**19,271,808**	**4.6**	**18,418,876**

CHINA

SALES

Year	Cars	Trucks	Buses	Total
2012	10,683,427	8,453,259	169,749	19,306,435
2011	10,040,167	8,299,796	165,389	18,505,352
2010	9,409,855	8,493,385	158,696	18,061,936
2009	7,322,906	6,152,604	142,585	13,618,095
2008	4,983,100	4,213,015	134,251	9,330,366
2007	5,146,160	3,498,475	174,498	8,819,133
2006	4,646,622	2,311,382	118,813	7,076,817
2005	3,734,823	1,995,886	83,567	5,814,276
2004	3,068,580	1,923,630	78,851	5,071,061
2003	2,667,035	1,651,651	72,120	4,390,806
2002	1,757,121	1,409,506	81,431	3,248,058
2001	1,210,013	1,094,315	59,337	2,363,665
2000	1,021,634	1,027,528	42,143	2,091,305
1999	868,826	931,108	36,701	1,836,635
1998	737,902	829,779	29,521	1,597,202
1997	690,069	855,041	20,794	1,565,904
1996	555,959	880,608	22,099	1,458,666
1995	476,176	939,454	26,149	1,441,779
1994	353,414	960,048	23,839	1,337,301

CHINA

SALES BY COMPANY

	2012	% Chg.	2011
Cars			
Audi	238,543	19.8	199,068
Beijing AIC	20,652	–	–
BMW	126,733	32.8	95,439
Brilliance	118,549	-11.5	133,895
BYD	368,071	-4.5	385,451
Chana	182,295	12.4	162,253
Changhe	37,489	64.6	22,778
Chery	405,831	-11.9	460,399
Daimler (Mercedes-Benz)	67,735	-12.6	77,536
Dongfeng	75,084	329.8	17,470
FAW	6,545	–	–
Feidie	4,657	-20.2	5,834
Fiat	11,288	–	–
Ford	274,522	-13.5	317,462
Baojun	80,323	267.5	21,854
Buick	*623,425*	*8.8*	*572,817*
Cadillac	*7,007*	–	*7,008*
Chevrolet	*669,468*	*9.9*	*609,106*
GM Total	1,380,223	14.0	1,210,785
Great Wall	180,592	15.3	156,648
Guangzhou AC	32,611	91.8	17,006
Hafei	7,226	10.7	6,527
Haima	*61,812*	*-25.0*	*82,381*
Zhengzhou	*8,921*	*-34.2*	*13,565*
Haima Total	70,733	-26.3	95,946
Everus	*24,576*	*-0.5*	*24,696*
Honda	*372,250*	*-12.4*	*424,994*
Honda Total	396,826	-11.8	449,690
Hongqi	72,579	-35.7	112,818
Huatai	131	-82.3	741
Hyundai	688,137	17.5	585,589
Jianghuai Auto Co.	138,320	-5.3	145,996
Kia	363,301	12.3	323,422
Lifan	156,345	58.3	98,793
Mazda	312,052	51.9	205,435
MG Motors	77,774	57.0	49,522
Mitsubishi	99,972	0.8	99,177
Nissan	600,375	-9.8	665,449
Citroen	*223,801*	*-2.7*	*230,002*
Peugeot	*216,227*	*24.2*	*174,137*
Peugeot Citroen Total	440,028	8.9	404,139
Proton	45,421	28.5	35,335
Shanghai AIC (Roewe)	115,148	5.8	108,815
Suzuki	170,037	-22.7	220,008
Toyota	531,027	-5.6	562,716
Skoda	*48,380*	*-0.8*	*48,759*
Volkswagen	*2,022,710*	*16.7*	*1,733,843*
Volkswagen Total	2,071,090	16.2	1,782,602
Volvo	7,453	-49.9	14,884
Xiali	185,018	-26.9	253,035
Yulon (Luxgen)	31,106	340.7	7,058
Emgrand	*160,517*	*41.8*	*113,219*
Englon	*21,090*	–	*941*
Geely	*228,345*	*-16.4*	*273,182*
Gleagle	*50,195*	*10.5*	*45,410*
Zhejiang Geely Total	460,147	6.3	432,752
Jiangnan	*29,535*	*-8.9*	*32,416*
Zotye	*82,226*	*-3.6*	*85,278*
Zotye Total	111,761	-5.0	117,694
Total Cars	**10,683,427**	**6.4**	**10,040,167**
Light Trucks			
Audi	90,157	53.8	58,632
Beijing AIC	*158,688*	*-0.3*	*159,242*
Weiwang	*46,368*	*362.9*	*10,016*
Beijing AIC Total	205,056	21.1	169,258
BMW	20,641	–	5
Brilliance	*107,929*	*68.6*	*64,009*
Jinbei	*249,277*	*-6.2*	*265,638*
Brilliance Total	357,206	8.4	329,647
BYD	87,985	39.6	63,033
Chana	*695,487*	*34.4*	*517,465*
Jiangling	*214,711*	*2.9*	*208,606*
Chana Total	910,198	25.4	726,071
Changhe	92,671	-24.1	122,141
Changzheng	–	–	733
Chery	157,474	-13.1	181,316
China National	6,080	303.5	1,507
China Yituo	3,858	-20.3	4,843
Daimler (Mercedes-Benz)	33,395	197.7	11,219
Dayun	8,534	33.8	6,378
Dongfeng	685,888	-4.2	716,232
FAW	72,033	-4.1	75,096
FAW Jinlin (Jilin)	90,942	-12.3	103,645
Feidie	28,200	23.6	22,815
Fire Fighting Vehicles	–	–	353
Ford	1,800	-43.7	3,195
Foton	412,294	-3.5	427,216
Buick	*74,016*	*10.6*	*66,912*
Chevrolet	*19,601*	–	–
Oldsmobile	–	–	*156*
Wuling	*1,347,880*	*11.5*	*1,209,006*
GM Total	1,441,497	13.0	1,276,074
GoNow	36,808	47.3	24,984
Great Wall	444,010	34.5	330,163
Changfeng	*35*	*337.5*	*8*
Guangzhou AC	*462*	*-37.8*	*743*
Liebao	*21,806*	*-6.5*	*23,322*
Guangzhou AC Total	22,303	-7.4	24,073
Guilin	2,165	26.2	1,715
Guizhou	3,916	-22.8	5,071
Hafei	62,720	-46.8	117,921
Haima	80,861	-23.8	106,092
Heibao	–	–	36,121
Hengtong	21	50.0	14
Hino	220	-46.3	410
Honda	230,784	20.0	192,321
Hongta	94	-46.0	174
Huanghai	34,902	-6.3	37,246
Huatai	6,675	–	510
Hyundai	198,723	15.4	172,187
Isuzu	1,865	52.4	1,224
Jialing	97	–	–

CHINA

SALES BY COMPANY

	2012	% Chg.	2011
Jianghuai Auto Co.	293,414	5.0	279,509
Kaima	80,323	11.3	72,147
Kia	117,142	7.4	109,096
Golden Dragon	*15,993*	*10.0*	*14,541*
King Long	*20,593*	*38.2*	*14,898*
King Long Total	36,586	24.3	29,439
Lifan	84,946	13.5	74,830
Linghe	1,995	15.3	1,731
Mazda	24,032	-7.2	25,898
Mei Ya	2,585	-41.2	4,396
Mitsubishi	7,821	-45.5	14,351
Nanjing	130,447	-53.8	282,259
Nanjun	37,423	-16.9	45,020
New Forta	7,529	1.3	7,431
New Longma	1,321	-11.5	1,492
Nissan	178,418	-8.3	194,648
Qingling	62,593	-1.5	63,535
Sanjing	42	-51.7	87
Shaanxi	9,657	-20.6	12,155
Shandong	3,300	30.6	2,526
Roewe	*7,396*	*82.1*	*4,062*
Shanghai AIC	*7,076*	*149.8*	*2,833*
Sunwin	*105*	*-58.2*	*251*
Shanghai AIC Total	14,577	104.0	7,146
Sichuan	7,229	-33.5	10,864
Southeast	6,013	33.6	4,500
Tangjun	52,354	7.3	48,796
Toyota	214,538	-10.9	240,747
Tri-Ring Special	899	4.9	857
Volkswagen	209,106	30.4	160,413
Yaxing	640	-40.2	1,070
Yutong	4,214	38.0	3,054
Zhejiang	557	411.0	109
Zhejiang Geely (Gleagle)	31,297	–	–
Zoomlion	254	-44.3	456
Zotye	10,433	-39.4	17,226
ZX Auto	63,231	2.4	61,773
Total Light Trucks	**7,526,989**	**5.6**	**7,127,196**
Medium/Heavy Trucks			
Brilliance (Jinbei)	14,850	90.1	7,813
Changzheng	89	-97.3	3,333
China National	131,013	-16.7	157,211
China Yituo	4,107	-20.3	5,156
Daimler (Mercedes-Benz)	24,270	-39.4	40,022
Dayun	11,946	-22.3	15,366
Dongfeng	205,228	-23.6	268,545
FAW	151,757	-24.2	200,202
Feidie	12,589	-19.6	15,659
Fire Fighting Vehicles	–	–	8
Foton	95,517	-12.6	109,248
Hino	5,218	30.2	4,008
Hualing Heavy Duty	15,069	-41.4	25,695
Huanghai	3	–	–
Iveco	17,008	-46.0	31,500
Jianghuai Auto Co.	39,446	-31.0	57,185
Kaima	4,211	-53.7	9,100
Lifan	29,012	1.7	28,521

	2012	% Chg.	2011
Nanjing	2,041	[illegible]	2,677
Nanjun	22,513	-16.6	26,989
New Longma	1,132	-10.7	1,267
Qingling	24,414	-16.6	29,256
Sanjing	318	622.7	44
Shaanxi	84,618	-19.8	105,468
Shandong	–	–	99
Taiyuan	730	-76.2	3,062
Tangjun	16,562	19.4	13,869
Tri-Ring Special	9,401	-1.0	9,494
Zhejiang	2,608	37.0	1,903
Total Med./Hvy. Trucks	**926,270**	**-21.0**	**1,172,600**
Buses			
Ankai	5,947	17.9	5,043
BYD	262	-18.9	323
Chana	*1,630*	*-16.3*	*1,947*
Jiangling	*36*	–	–
Chana Total	1,666	-14.4	1,947
China National	709	79.9	394
Dadi	–	–	116
Dayun	–	–	17
Dongfeng	15,219	-3.8	15,813
FAW	8,756	-28.2	12,187
Foton	4,339	43.7	3,019
Guangzhou AC	1,141	19.4	956
Guilin	2,053	6.6	1,926
Hengtong	1,657	274.0	443
Hino	314	145.3	128
Huanghai	4,775	7.5	4,443
Jialing	172	–	–
Jianghuai Auto Co.	9,633	35.9	7,089
Golden Dragon	*7,798*	*-9.1*	*8,574*
King Long	*33,924*	*0.5*	*33,742*
King Long Total	41,722	-1.4	42,316
Lions	124	-41.8	213
MAN (Neoplan)	4,556	12.4	4,053
Nanjing	1,566	-41.5	2,679
Nanjun	831	-48.0	1,599
New Forta	–	–	511
New Longma	203	45.0	140
Sanjing	1	-66.7	3
Shaanxi	263	-74.2	1,018
Shanghai AIC (Sunwin)	3,145	8.4	2,901
Shaolin	2,570	7.4	2,394
Sichuan	–	–	482
Xian Silver	254	12.4	226
Yanjing	923	-15.6	1,094
Yaxing	3,380	38.9	2,433
Yutong	47,474	8.8	43,634
Zhejiang	–	–	18
Zhongtong	6,085	8.5	5,607
Zoomlion	9	-96.0	224
Total Buses	**169,749**	**2.6**	**165,389**
Total Comm. Vehs.	**8,623,008**	**1.9**	**8,465,185**
Total Vehicles	**19,306,435**	**4.3**	**18,505,352**

INDIA

ECONOMIC SUMMARY

Common Language	Hindi/English
Currency	Indian Rupee
Steering	Right Hand Drive

Capital	New Delhi
Population	1,220,800,359
Population Density	1063.5
Surface Area sq km/miles	2,973,193 / 1,147,956

2012 Vehicles in Operation

Passenger Cars	18,796,000
Commercial Vehicles	10,558,000
Total	29,354,000
Persons per Car	65.0
Persons per Vehicle	41.6

Road Infrastructure

Paved Road Surface %	47.4
Total Road Length in km/miles	3,320,410 / 2,063,207
Road Traffic (Vehicles per mile)	14.2

2012 GDP (billions of US $)

Nominal GDP	1,824.8
Real GDP	1,050.6
GDP % annual growth	4.0
GDP per capita	859

Current Account Balance (billions of US $)

2012:	-93.3
2011:	-62.8

Current Account Balance in % of GDP

2012:	-5.1
2011:	-3.4

Total Number of Fatal Road Accidents (most recent data)

2010:	133,938

2012 Total Vehicle Percent Market Share by Company

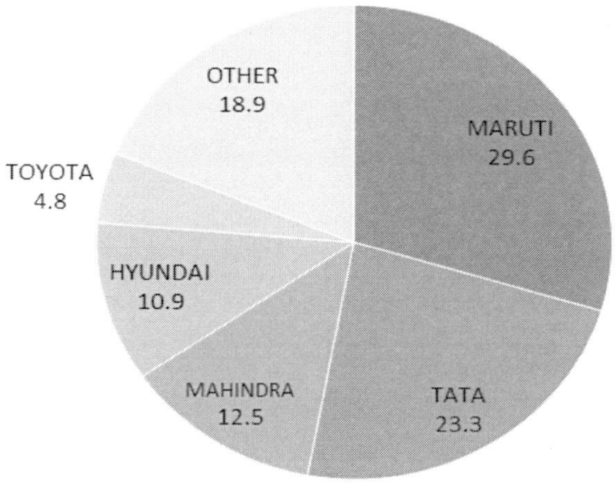

Total Vehicle Production and Sales 2002-2012

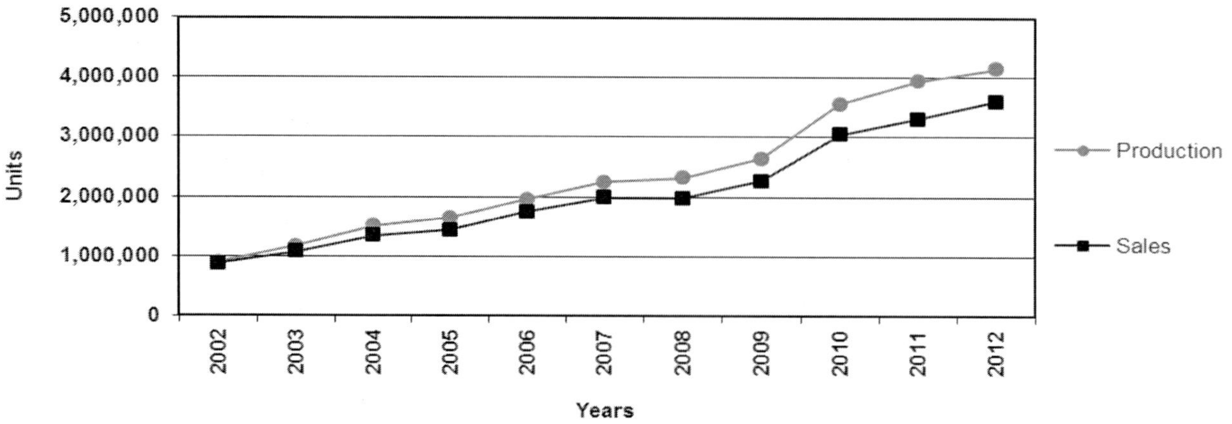

INDIA

PRODUCTION

Year	Cars	Trucks	Buses	Total
2012	2,520,261	1,575,460	53,248	4,148,969
2011	2,479,131	1,411,131	50,098	3,940,360
2010	2,316,931	1,182,263	54,609	3,553,803
2009	1,781,399	819,086	42,017	2,642,502
2008	1,506,539	764,992	44,101	2,315,632
2007	1,376,932	828,399	44,429	2,249,760
2006	1,185,544	741,912	30,299	1,957,755
2005	998,951	612,759	30,347	1,642,057
2004	939,679	542,052	29,307	1,511,038
2003	712,072	422,845	26,556	1,161,473
2002	546,050	326,516	20,128	892,694
2001	573,808	253,446	23,075	850,329
2000	604,970	254,707	28,696	888,373
1999	533,177	259,916	24,641	817,734
1998	383,798	218,987	24,894	627,679
1997	409,896	293,701	32,419	736,016
1996	396,456	339,592	26,074	762,122
1995	329,676	278,887	27,453	636,016
1994	237,280	210,383	27,487	475,150
1993	199,571	150,167	21,892	371,630
1992	153,867	143,164	23,133	320,164
1991	178,927	154,151	21,955	355,033
1990	176,821	167,251	20,321	364,393
1989	177,190	141,502	18,433	337,125
1988	159,941	132,359	20,032	312,332
1987	148,495	117,649	21,569	287,713
1986	116,004	102,140	20,694	238,838
1985	102,456	106,234	21,870	230,560
1984	64,013	95,440	21,466	180,919
1983	44,613	90,634	18,391	153,638
1982	42,674	94,042	15,808	152,524

INDIA

PRODUCTION BY MANUFACTURER

	2012	% Chg.	2011
Cars			
BMW	6,140	-2.1	6,270
Fiat	9,992	-45.7	18,395
Ford	115,297	-0.8	116,261
GM	70,320	-22.8	91,139
Hindustan Motors	2,707	-20.9	3,422
Honda	75,201	68.7	44,585
Hyundai	637,960	3.1	618,511
Mahindra & Mahindra	16,468	1.0	16,305
Maruti (Suzuki)	996,450	3.0	967,489
Mercedes-Benz	6,113	-18.0	7,451
Nissan	*141,455*	*13.1*	*125,106*
Renault	*11,252*	*827.6*	*1,213*
Nissan Total	152,707	841	126,319
Land Rover	*516*	–	–
Tata	*229,458*	*-17.2*	*277,154*
Tata Total	229,974	-17	277,154
Toyota	98,868	33.9	73,849
Skoda	*34,729*	*15.3*	*30,122*
Volkswagen	*67,335*	*-17.7*	*81,859*
Volkswagen Total	102,064	-2	111,981
Total Cars	**2,520,261**	**1.7**	**2,479,131**
Light Trucks			
Ashok Leyland	31,028	651.1	4,131
BMW	2,030	-48.0	3,907
Eicher Motors	11,000	-1.2	11,130
Force Motors	27,759	-8.6	30,383
Ford	1,609	-44.8	2,913
GM	20,892	-0.4	20,974
Hindustan Motors	1,987	-15.2	2,342
Hyundai	805	-33.1	1,204
International Cars & Motors	372	-13.7	431
Mahindra	*457,809*	*28.0*	*357,581*
Mahindra Navistar	*9,006*	*-11.9*	*10,224*
Mahindra & Mahindra Total	466,815	16	367,805
Maruti (Suzuki)	186,590	19.3	156,405
Mazda	4,308	-19.6	5,355
Nissan	*1,365*	–	–
Renault	*24,243*	–	*252*
Nissan Total	25,608	–	252
Piaggio	4,502	-59.1	11,009
Tata	427,610	5.5	405,159
Toyota	91,453	46.0	62,630
Volkswagen (Skoda)	826	-70.6	2,807
Total Light Trucks	**1,305,194**	**19.9**	**1,088,837**
Medium/Heavy Trucks			
Ashok Leyland	61,799	-10.7	69,167
Asia Motor Works	7,378	-20.8	9,311
Eicher Motors	32,235	-5.3	34,025
Kamaz	–	–	120
Mahindra & Mahindra	3,461	71.3	2,020
Mazda	4,962	-5.1	5,230
Mercedes-Benz	–	–	485
Tata	159,890	-20.6	201,288
Volvo Truck	541	-16.5	648
Total Med./Hvy. Trks.	**270,266**	**-16.1**	**322,294**
Buses			
Ashok Leyland	25,056	9.5	22,872
Eicher Motors	4,819	32.1	3,648
JCBL	–	–	54
Mahindra & Mahindra (Navistar)	620	429.9	117
Mazda	4,427	42.6	3,105
Tata	17,602	-10.3	19,632
Volvo Truck	724	8.1	670
Total Buses	**53,248**	**6.3**	**50,098**
Total Comm. Vehs.	**1,628,708**	**11.5**	**1,461,229**
Total Vehicles	**4,148,969**	**5.3**	**3,940,360**

Note: Includes Ward's estimates

INDIA

SALES

Year	Cars	Trucks	Buses	Total
2012	2,015,519	1,527,893	49,804	3,593,216
2011	1,950,363	1,303,097	45,979	3,299,439
2010	1,871,041	1,118,946	49,533	3,039,520
2009	1,425,933	799,859	37,888	2,263,680
2008	1,201,178	742,953	35,996	1,980,127
2007	1,178,027	774,860	37,337	1,990,224
2006	1,023,559	698,294	28,924	1,750,777
2005	852,974	560,066	26,573	1,439,613
2004	802,800	516,941	24,597	1,344,338
2003	644,048	408,936	23,541	1,076,525
2002	544,411	316,083	18,639	879,133
2001	601,132	251,395	22,066	874,593
2000	603,065	254,583	28,650	886,298
1999	574,338	258,057	24,551	856,946
1998	403,326	222,338	25,899	651,563
1997	431,483	289,518	31,112	752,113
1996	389,659	335,177	25,528	750,364
1995	330,500	280,374	27,920	638,794
1994	232,798	236,671	*	469,469
1993	199,969	181,229	*	381,198
1992	164,581	165,409	*	329,990
1991	172,485	172,462	*	344,947
1990	174,633	168,206	14,313	357,152
1989	177,409	140,109	18,606	336,124
1988	159,277	201,928	20,366	381,571
1987	150,080	177,866	20,228	348,174
1986	109,919	156,837	21,770	288,526
1985	102,365	148,510	20,924	271,799
1984	62,220	135,167	20,141	217,528
1983	45,419	138,266	11,398	195,083
1982	43,769	129,692	9,544	183,005

* Included with Trucks.

INDIA

SALES BY COMPANY

	2012	% Chg	2011
Cars			
Audi	5,791	26.0	4,596
BMW	3,886	-39.1	6,382
Daimler (Mercedes-Benz)	4,826	-29.4	6,833
Fiat	11,145	-33.4	16,733
Ford	86,090	-7.7	93,268
GM	70,904	-20.0	88,657
Hindustan Motors	2,720	-23.3	3,545
Honda	72,840	54.3	47,222
Hyundai	390,274	4.8	372,259
Jaguar Land Rover	1,597	–	–
Mahindra & Mahindra	16,521	2.0	16,204
Maruti	878,272	4.1	844,001
Nissan	44,439	96.8	22,585
Porsche	220	–	–
Renault	11,015	868.8	1,137
Tata	233,783	-5.3	246,874
Toyota	81,025	10.9	73,094
Skoda	*33,343*	*16.7*	*28,569*
Volkswagen	*66,828*	*-14.8*	*78,404*
Volkswagen Total	100,171	-6.4	106,973
Total Cars	**2,015,519**	**3.3**	**1,950,363**
Light Trucks			
Ashok Leyland	29,217	817.6	3,184
Audi	3,070	132.2	1,322
BMW	5,130	42.5	3,600
Daimler (Mercedes-Benz)	1,280	57.8	811
Eicher Motors	9,388	0.7	9,321
Force Motors	28,041	-1.0	28,336
Ford	1,510	-49.7	3,003
GM	21,155	-5.6	22,399
Hindustan Motors	2,009	-14.3	2,345
Honda	294	-9.8	326
Hyundai	1,002	-30.9	1,450
International Cars & Motors	380	-16.5	455
Mahindra	*419,781*	*26.2*	*332,655*
Mahindra Navistar	*8,596*	*-12.1*	*9,780*
Mahindra & Mahindra Total	428,377	25.1	342,435
Maruti	185,327	20.9	153,280
Mazda	3,941	-14.1	4,587
Nissan	800	159.7	308
Piaggio	4,018	-64.1	11,188
Renault	24,142	–	264
Tata	433,843	23.5	351,165
Toyota	91,216	44.7	63,054
Skoda	*1,335*	*-38.5*	*2,172*
Volkswagen	*45*	*650.0*	*6*
Volkswagen Total	1,380	-36.6	2,178
Total Light Trucks	**1,275,520**	**26.9**	**1,005,011**
Medium/Heavy Trucks			
Ashok Leyland	55,588	-10.6	62,211
Asia Motor Works	7,586	-20.0	9,478
Daimler (Mercedes-Benz)	–	–	178
Eicher Motors	31,870	-2.4	32,649
Mahindra & Mahindra (Navistar)	3,546	40.6	2,522
Mazda	4,408	-8.0	4,793
Tata	148,806	-19.9	185,711
Volvo Truck	569	4.6	544
Total Med./Hvy. Trks.	**252,373**	**-15.3**	**298,086**
Buses			
Ashok Leyland	20,055	3.8	19,325
Eicher Motors	4,503	31.6	3,421
JCBL	–	–	54
Mahindra & Mahindra (Navistar)	705	–	6
Mazda	3,820	12.3	3,403
Tata	20,016	4.7	19,116
Volvo Truck	705	7.8	654
Total Buses	**49,804**	**8.3**	**45,979**
Total Comm. Vehs.	**1,577,697**	**16.9**	**1,349,076**
Total Vehicles	**3,593,216**	**8.9**	**3,299,439**

Note: Includes Ward's estimates

INDONESIA

ECONOMIC SUMMARY

Common Language	Bahasa Indonesia
Currency	Rupiah
Steering	Left Hand Drive

2012 Vehicles in Operation

Passenger Cars	10,494,000
Commercial Vehicles	7,510,000
Total	18,004,000
Persons per Car	23.9
Persons per Vehicle	14.0

2012 GDP (billions of US $)

Nominal GDP	878.2
Real GDP	272.4
GDP % annual growth	6.2
GDP per capita	1,114

Current Account Balance (billions of US $)

2012: -24.2
2011: 1.7

Current Account Balance in % of GDP

2012: -2.8
2011: 0.2

Total Number of Fatal Road Accidents (most recent data)

2012: 31,195

Capital	Jakarta
Population	251,160,124
Population Density	359.1
Surface Area sq km/miles	1,811,569 / 699,450

Road Infrastructure

Paved Road Surface %	57.0
Total Road Length in km/miles	487,314 / 302,803
Road Traffic (Vehicles per mile)	59.5

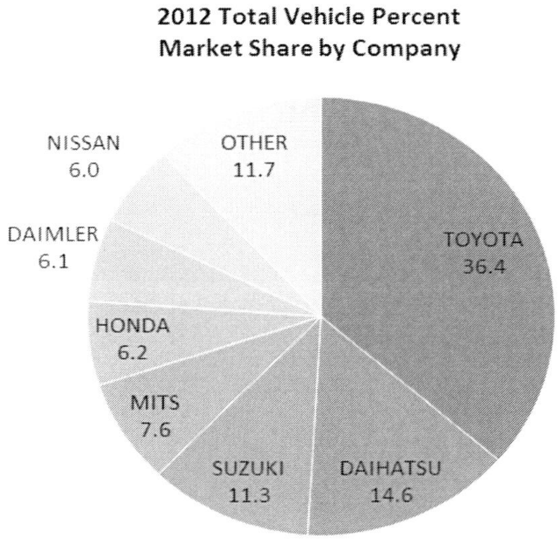

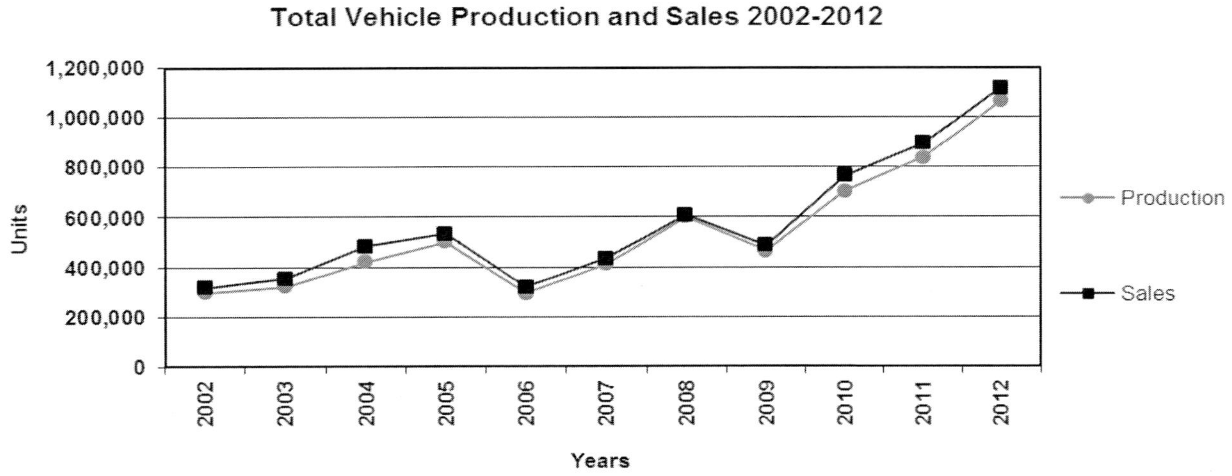

INDONESIA

PRODUCTION

Year	Cars	Commercial Vehicles	Total
2012	743,501	322,056	1,065,557
2011	561,863	276,085	837,948
2010	496,524	205,984	702,508
2009	352,172	112,644	464,816
2008	431,423	169,205	600,628
2007	309,208	102,430	411,638
2006	206,321	89,687	296,008
2005	332,590	168,120	500,710
2004	257,461	164,638	422,099
2003	203,196	118,839	322,035
2002	23,890	272,619	296,509
2001	32,237	295,493	327,730
2000	37,317	307,959	345,276
1999	5,974	117,270	123,244
1998	8,401	49,678	58,079
1997	55,102	334,177	389,279
1996	35,304	290,191	325,495
1995	39,839	347,702	387,541

Note: Car and Commercial Vehicle classification changed in 2003.

INDONESIA

PRODUCTION BY MANUFACTURER

	2012	% Chg.	2011
Cars			
BMW	1,682	99.3	844
Chery	–	–	165
Daihatsu	119,320	11.7	106,866
Geely	727	99.7	364
Honda	55,828	36.7	40,844
Hyundai	3,540	6.4	3,327
Isuzu	5,100	-4.0	5,310
Mercedes-Benz	2,922	51.3	1,931
Mitsubishi	6,327	174.1	2,308
Nissan	65,900	20.1	54,863
Suzuki	61,655	43.3	43,033
Toyota	419,716	39.2	301,514
Audi	*226*	–	–
Volkswagen	*558*	*13.0*	*494*
Volkswagen Total	784	58.7	494
Total Cars	**743,501**	**32.3**	**561,863**
Light Trucks			
Daihatsu	37,947	27.2	29,837
Hyundai	1,335	-34.5	2,039
Isuzu	4,950	-9.3	5,460
Mitsubishi	53,679	4.7	51,288
Suzuki	71,784	14.3	62,797
Volkswagen	52	–	–
Total Light Trucks	**169,747**	**12.1**	**151,421**
Medium/Heavy Trucks			
Hino	36,694	51.9	24,161
Isuzu	21,595	25.0	17,277
Mercedes-Benz	833	-18.9	1,027
Mitsubishi	54,084	-1.5	54,929
Mitsubishi Fuso	7,392	22.9	6,017
Toyota	28,492	57.1	18,133
Volvo Truck (UD Trucks)	3,219	3.2	3,120
Total Med./Hvy. Trucks	**152,309**	**22.2**	**124,664**
Total Comm. Vehs.	**322,056**	**16.7**	**276,085**
Total Vehicles	**1,065,557**	**27.2**	**837,948**

INDONESIA

SALES

Year	Cars	Commercial Vehicles	Total
2012	780,785	335,445	1,116,230
2011	614,999	279,184	894,183
2010	541,475	223,235	764,710
2009	361,907	124,181	486,088
2008	429,294	178,507	607,801
2007	313,663	120,810	434,473
2006	221,423	97,481	318,904
2005	365,565	168,357	533,922
2004	333,282	149,887	483,169
2003	233,912	120,497	354,409
2002	199,191	118,600	317,791
2001	35,297	264,276	299,573
2000	46,888	254,073	300,961
1999	11,017	82,797	93,814
1998	13,237	46,139	59,376
1997	73,215	313,552	386,767
1996	43,914	287,595	331,509
1995	37,921	346,528	384,449
1994	40,412	286,059	326,471
1993	32,686	181,611	214,297
1992	30,341	141,557	171,898
1991	45,774	217,299	263,073
1990	56,510	219,014	275,524
1989	32,179	146,613	178,792

Note: Car and Commercial Vehicle classification changed in 2002.

INDONESIA

SALES BY COMPANY

	2012	% Chg.	2011
Cars			
Audi	432	29.3	334
BMW	2,188	41.1	1,551
Chery	–	–	436
Chrysler	*153*	*163.8*	*58*
Dodge	*100*	*156.4*	*39*
Jeep	*509*	*26.3*	*403*
Chrysler Total	762	52.4	500
Daihatsu	110,149	11.5	98,779
Mercedes-Benz	*3,856*	*-6.5*	*4,123*
Smart	*235*	*-15.2*	*277*
Daimler Total	4,091	-7.0	4,400
FAW	680	–	–
Ford	7,174	-23.0	9,311
GM (Chevrolet)	5,643	21.1	4,658
Honda	69,320	52.6	45,416
Hyundai	5,746	20.3	4,776
Isuzu	11,277	10.1	10,244
Jaguar	*75*	*29.3*	*58*
Land Rover	*256*	*271.0*	*69*
Jaguar Land Rover Total	331	160.6	127
Kia	13,651	50.3	9,081
Mazda	8,054	36.3	5,910
Mitsubishi	51,830	23.3	42,029
Nissan	67,179	19.6	56,192
Peugeot Citroen	333	89.2	176
Proton	1,017	-47.2	1,926
Subaru	204	-6.4	218
Suzuki	51,573	64.5	31,354
Lexus	*356*	*33.8*	*266*
Toyota	*364,706*	*27.8*	*285,329*
Toyota Total	365,062	27.8	285,595

	2012	% Chg.	2011
Volkswagen	1,017	42.6	713
Volvo	23	64.3	14
Zhejiang Geely	1,017	-4.3	1,063
Other	2,032	936.7	196
Total Cars	**780,785**	**27.0**	**614,999**
Light Trucks			
Daihatsu	52,593	29.0	40,766
Daimler (Mitsubishi Fuso)	54,011	-0.1	54,078
Ford	4,784	-24.8	6,359
Hino	13,590	9.1	12,454
Hyundai	638	-18.8	786
Isuzu	18,240	10.7	16,482
Mazda	4,337	43.5	3,023
Mitsubishi	33,546	3.7	32,361
Suzuki	75,004	18.6	63,215
Toyota	40,708	60.6	25,345
Total Light Trucks	**297,451**	**16.7**	**254,869**
Medium/Heavy Trucks			
Mercedes-Benz	*752*	*41.9*	*530*
Mitsubishi Fuso	*9,531*	*60.2*	*5,948*
Daimler Total	10,283	58.7	6,478
Foton	–	–	113
Hino	20,882	71.2	12,198
Isuzu	3,648	80.6	2,020
MAN	256	-44.5	461
Volvo Truck (UD Trucks)	2,925	-3.9	3,045
Total Med./Hvy. Trks.	**37,994**	**56.3**	**24,315**
Total Comm. Vehs.	**335,445**	**20.2**	**279,184**
Total Vehicles	**1,116,230**	**24.8**	**894,183**

JAPAN

ECONOMIC SUMMARY

Common Language	Japanese
Currency	Yen
Steering	Right Hand Drive

2012 Vehicles in Operation

Passenger Cars	59,421,009
Commercial Vehicles	15,061,199
Total	74,482,208
Persons per Car	2.1
Persons per Vehicle	1.7

2012 GDP (billions of US $)

Nominal GDP	5,964.0
Real GDP	6,049.1
GDP % annual growth	2.0
GDP per capita	47,402

Current Account Balance (billions of US $)

2012:	59.0
2011:	119.2

Current Account Balance in % of GDP

2012:	1.0
2011:	2.0

Total Number of Fatal Road Accidents (most recent data)

2012:	4,411

Capital	Tokyo
Population	127,253,075
Population Density	1,215.1
Surface Area sq km/miles	364,485 / 140,728

Road Infrastructure

Paved Road Surface %	80.4
Total Road Length in km/miles	1,210,251 / 752,015
Road Traffic (Vehicles per mile)	99.0

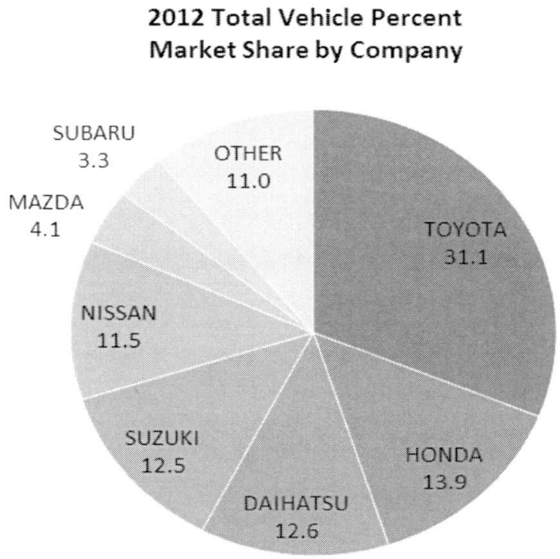

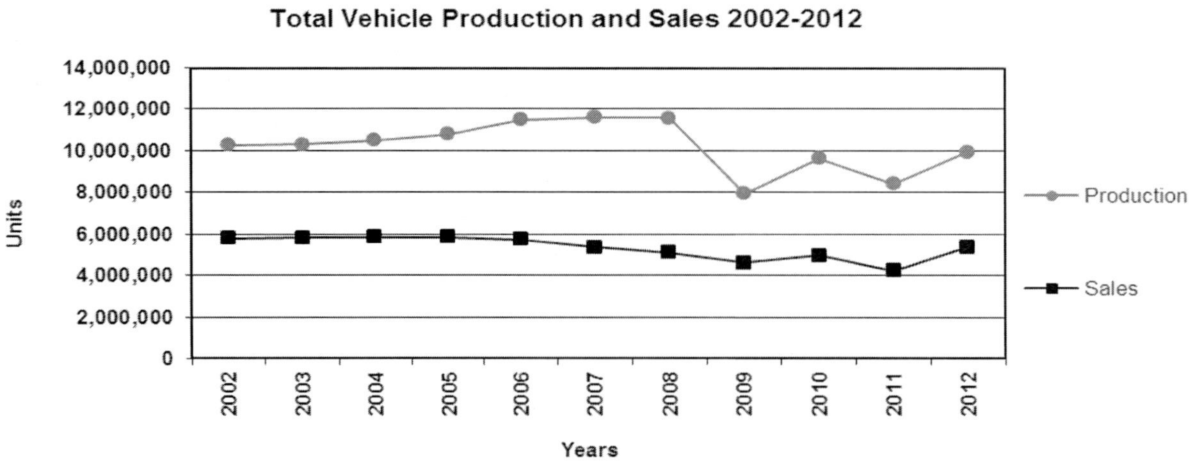

JAPAN

PRODUCTION

Year	Cars	Trucks	Buses	Total
2012	8,554,219	1,266,272	122,220	9,942,711
2011	7,158,525	1,136,020	104,109	8,398,654
2010	8,307,382	1,209,224	109,334	9,625,940
2009	6,862,161	985,100	87,255	7,934,516
2008	9,916,149	1,508,378	139,102	11,563,629
2007	9,944,637	1,538,020	113,670	11,596,327
2006	9,756,515	1,639,081	88,637	11,484,233
2005	9,016,735	1,706,611	76,313	10,799,659
2004	8,720,385	1,730,691	60,442	10,511,518
2003	8,478,328	1,746,916	61,074	10,286,318
2002	8,618,728	1,572,041	66,321	10,257,090
2001	8,117,071	1,601,099	58,092	9,776,262
2000	8,363,471	1,726,472	54,544	10,144,487
1999	8,100,169	1,746,912	48,395	9,895,476
1998	8,055,763	1,937,076	56,953	10,049,792
1997	8,491,480	2,421,373	62,234	10,975,087
1996	7,864,676	2,428,897	53,126	10,346,699
1995	7,610,533	2,537,737	47,266	10,195,536
1994	7,802,037	2,702,970	49,112	10,554,119
1993	8,493,943	2,685,528	48,074	11,227,545
1992	9,378,694	3,068,585	52,005	12,499,284
1991	9,753,069	3,447,914	44,449	13,245,432
1990	9,947,972	3,498,639	40,185	13,486,796
1989	9,052,406	3,931,255	42,074	13,025,735
1988	8,198,400	4,443,994	57,413	12,699,807
1987	7,891,087	4,308,100	49,987	12,249,174
1986	7,809,809	4,407,666	42,342	12,259,817
1985	7,646,816	4,544,688	79,591	12,271,095
1984	7,073,173	4,319,538	72,209	11,464,920
1983	7,151,888	3,903,823	55,948	11,111,659
1982	6,881,586	3,783,218	66,990	10,731,794

PRODUCTION BY MANUFACTURER

	2012	% Chg.	2011
Cars			
Daihatsu	561,256	28.6	436,362
Subaru	32,862	16.3	28,265
Toyota	39,769	159.4	15,329
Daihatsu Total	633,887	32.1	479,956
Acura	254	-55.7	574
Honda	996,578	45.0	687,374
Honda Total	996,832	44.9	687,948
Mazda	821,666	4.6	785,415
Nissan	8,628	-31.8	12,645
Mazda Total	830,294	4.0	798,060
Citroen	5,579	-67.7	17,299
Mitsubishi	433,630	-14.0	504,267
Nissan	7,761	-30.6	11,191
Peugeot	1,628	-51.9	3,385
Mitsubishi Total	448,598	-16.3	536,142
Infiniti	48,950	-23.3	63,792
Nissan	986,776	4.9	940,874
Nissan Total	1,035,726	3.1	1,004,666
Subaru	505,190	37.8	366,518
Toyota	46,622	–	–
Subaru Total	551,812	50.6	366,518
Mazda	43,509	38.5	31,425
Mitsubishi	12,888	-13.3	14,869
Nissan	116,887	8.9	107,377
Suzuki	713,497	8.4	658,018
Suzuki Total	886,781	9.3	811,689
Lexus	65,618	-22.9	85,061
Scion	24,496	39.4	17,568
Subaru	4,512	-63.0	12,184
Toyota	3,075,663	30.4	2,358,733
Toyota Total	3,170,289	28.2	2,473,546
Total Cars	**8,554,219**	**19.5**	**7,158,525**
Light Trucks			
Daihatsu	119,818	-4.3	125,219
Subaru	11,297	514.6	1,838
Toyota	9,404	255.7	2,644
Daihatsu Total	140,519	8.3	129,701
Hino	58,076	49.8	38,760
Honda	32,481	43.3	22,673
Isuzu	97,522	8.0	90,305
Mazda	2,204	4.7	2,106
Nissan	1,052	29.2	814
UD Trucks	255	83.5	139
Isuzu Total	101,033	8.2	93,364
Mazda	9,942	4.2	9,537
Mitsubishi	–	–	836
Nissan	5,314	9.1	4,869
Mazda Total	15,256	0.1	15,242
Mitsubishi	66,558	-1.3	67,452
Suzuki	12	–	–
Mitsubishi Total	66,570	-1.3	67,452
Mitsubishi Fuso	20,978	15.0	18,238
Nissan	106,588	3.1	103,360
Subaru	18,361	-64.7	52,027
Mazda	11,185	-0.9	11,292
Suzuki	163,897	29.2	126,818
Suzuki Total	175,082	26.8	138,110
Toyota	221,265	10.3	200,679
Volvo Truck (UD Trucks)	800	-40.1	1,336
Total Light Trucks	**957,009**	**8.6**	**880,942**
Medium/Heavy Trucks			
Hino	88,561	27.0	69,743
Isuzu	137,483	29.9	105,841
Mazda	75	38.9	54
Nissan	62	19.2	52
UD Trucks	50	233.3	15
Isuzu Total	137,670	29.9	105,962
Mitsubishi Fuso	60,185	8.6	55,431
Volvo Truck (UD Trucks)	20,814	-4.6	21,808
Other	2,033	-4.7	2,134
Total Med./Hvy Trks	**309,263**	**21.2**	**255,078**
Buses			
Hino	6,202	17.4	5,285
Isuzu	2,544	7.1	2,375
Mitsubishi Fuso	5,923	8.0	5,485
Nissan	5,951	19.8	4,969
Toyota	101,359	18.1	85,803
Volvo Truck (UD Trucks)	241	25.5	192
Total Buses	**122,220**	**17.4**	**104,109**
Total Comm. Vehs.	**1,388,492**	**12.0**	**1,240,129**
Total Vehicles	**9,942,711**	**18.4**	**8,398,654**

JAPAN

SALES

Year	Cars	Trucks	Buses	Total
2012	4,572,333	785,450	11,938	5,369,721
2011	3,524,789	674,780	10,651	4,210,220
2010	4,212,266	731,094	12,775	4,956,135
2009	3,923,740	672,943	12,572	4,609,255
2008	4,227,643	839,259	15,333	5,082,235
2007	4,400,297	937,731	15,617	5,353,645
2006	4,641,732	1,080,174	17,600	5,739,506
2005	4,748,409	1,085,904	17,754	5,852,067
2004	4,768,131	1,067,104	18,147	5,853,382
2003	4,460,084	1,342,577	21,256	5,823,917
2002	4,441,493	1,333,760	16,359	5,791,612
2001	4,290,056	1,601,113	15,932	5,907,101
2000	4,260,414	1,687,905	16,571	5,964,890
1999	4,154,084	1,692,654	14,478	5,861,216
1998	4,093,148	1,772,136	14,141	5,879,425
1997	4,492,006	2,217,257	15,763	6,725,026
1996	4,668,728	2,391,790	17,227	7,077,745
1995	4,443,906	2,403,825	17,303	6,865,034
1994	4,210,168	2,298,685	17,843	6,526,696
1993	4,199,451	2,248,803	19,025	6,467,279
1992	4,454,012	2,483,484	21,577	6,959,073
1991	4,868,233	2,632,730	23,796	7,524,759
1990	5,102,659	2,649,909	24,925	7,777,493
1989	4,403,749	2,829,394	23,530	7,256,673
1988	3,717,359	2,980,103	23,542	6,721,004
1987	3,274,800	2,721,581	22,018	6,018,399
1986	3,146,023	2,540,174	21,617	5,707,814
1985	3,104,083	2,431,178	21,573	5,556,834
1984	3,095,554	2,320,938	20,267	5,436,759
1983	3,135,611	2,226,396	20,310	5,382,317
1982	3,038,272	2,201,366	21,793	5,261,431

SALES BY COMPANY

	2012	% Chg.	2011
Cars			
Daihatsu	551,281	28.3	429,669
GM (Chevrolet)	1	-99.8	400
Honda	712,350	49.7	475,747
Hyundai	–	–	1
Mazda	192,463	16.3	165,485
Mitsubishi	93,798	-14.0	109,097
Nissan	518,509	11.7	464,265
Subaru	141,065	31.2	107,522
Suzuki	529,212	21.9	434,001
Lexus	*43,657*	*3.0*	*42,365*
Toyota	*1,489,384*	*43.8*	*1,035,511*
Toyota Total	1,533,041	42.2	1,077,876
Other	300,613	15.3	260,726
Total Cars	**4,572,333**	**29.7**	**3,524,789**
Light Trucks			
Daihatsu	125,890	6.1	118,610
Daimler (Mitsubishi Fuso)	23,917	24.2	19,259
Hino	28,629	26.5	22,638
Honda	32,668	21.7	26,840
Isuzu	47,770	40.7	33,942
Mazda	25,898	6.0	24,440
Mitsubishi	33,928	-11.7	38,422
Nissan	97,392	30.8	74,465
Subaru	36,656	-28.4	51,179
Suzuki	142,898	23.4	115,811
Toyota	134,901	30.8	103,138
Volvo Truck (UD Trucks)	3,124	0.6	3,105
Other	15,116	2.9	14,695
Total Light Trucks	**748,787**	**15.8**	**646,544**
Medium/Heavy Trucks			
Daimler (Mitsubishi Fuso)	8,734	44.4	6,048
Hino	11,189	18.0	9,483
Isuzu	10,512	46.9	7,156
Volvo Truck (UD Trucks)	5,977	12.0	5,338
Other	251	19.0	211
Total Med./Heavy Trks	**36,663**	**29.8**	**28,236**
Buses			
Daimler (Mitsubishi Fuso)	2,064	19.7	1,725
Hino	2,645	24.9	2,117
Isuzu	1,523	52.8	997
Nissan	1,532	-35.4	2,371
Toyota	4,099	22.1	3,357
Volvo Truck (UD Trucks)	3	-88.5	26
Other	72	24.1	58
Total Buses	**11,938**	**12.1**	**10,651**
Total Comm. Vehs.	**797,388**	**16.3**	**685,431**
Total Vehicles	**5,369,721**	**27.5**	**4,210,220**

MALAYSIA

ECONOMIC SUMMARY

Common Language	Malay
Currency	Ringgit
Steering	Right Hand Drive

2012 Vehicles in Operation

Passenger Cars	9,833,000
Commercial Vehicles	1,171,000
Total	11,004,000
Persons per Car	3.0
Persons per Vehicle	2.7

2012 GDP (billions of US $)

Nominal GDP	303.5
Real GDP	240.7
GDP % annual growth	5.6
GDP per capita	8,172

Current Account Balance (billions of US $)

2012:	19.4
2011:	31.7

Current Account Balance in % of GDP

2012:	6.4
2011:	11.0

Total Number of Fatal Road Accidents (most recent data)

2010:	6,872

Capital	Kuala Lumpur
Population	29,628,392
Population Density	233.5
Surface Area sq km/miles	328,657 / 126,895

Road Infrastructure

Paved Road Surface %	80.4
Total Road Length in km/miles	144,403 / 89,728
Road Traffic (Vehicles per mile)	122.6

2012 Total Vehicle Percent Market Share by Company

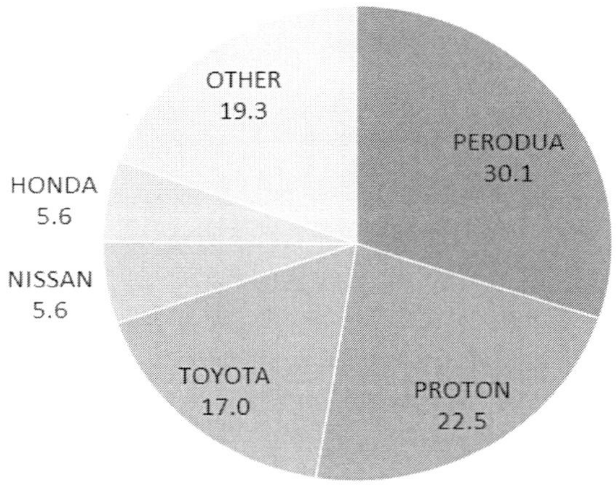

Total Vehicle Production and Sales 2002-2012

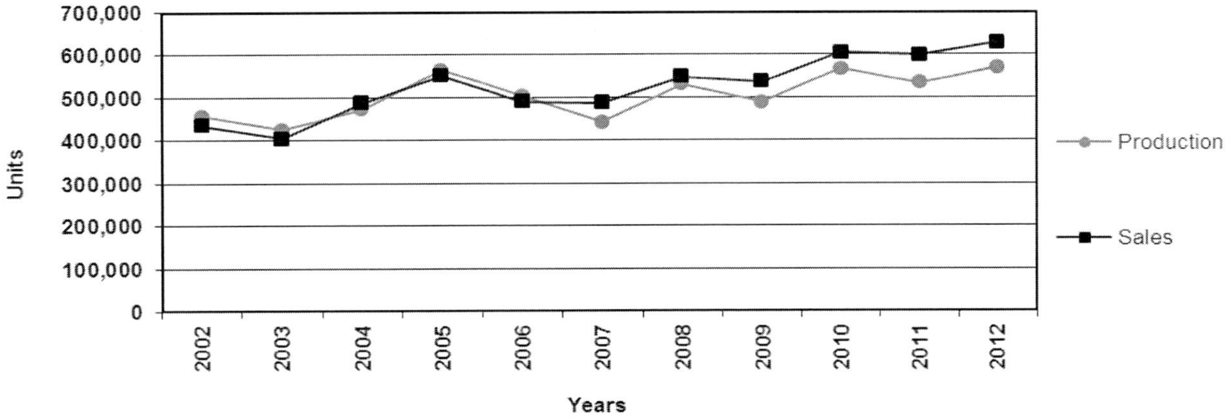

MALAYSIA

PRODUCTION

Year	Cars	Commercial Vehicles	Total
2012	399,913	169,707	569,620
2011	384,935	148,760	533,695
2010	403,474	164,241	567,715
2009	372,873	116,396	489,269
2008	419,963	110,847	530,810
2007	347,971	93,707	441,678
2006	377,952	125,096	503,048
2005	422,065	141,445	563,510
2004	364,852	107,123	471,975
2003	324,911	99,196	424,107
2002	380,050	76,772	456,822
2001	355,863	72,838	428,701
2000	295,318	63,877	359,195
1999	257,607	46,372	303,979
1998	143,756	17,700	161,456
1997	337,717	100,976	438,693
1996	280,222	84,899	365,121
1995	231,280	57,058	288,338
1994	147,351	43,834	191,185
1993	120,864	34,929	155,793
1992	116,965	35,054	152,019
1991	135,479	81,099	216,578
1990	116,526	75,054	191,580
1989	86,148	48,772	134,920
1988	74,144	23,787	97,931
1987	33,878	15,305	49,183
1986	41,896	19,942	61,838
1985	69,769	37,261	107,030
1984	96,357	27,680	124,037
1983	100,226	18,238	118,464
1982	85,321	15,650	100,971

MALAYSIA

PRODUCTION BY MANUFACTURER

	2012	% Chg.	2011
Cars			
BMW	4,918	18.4	4,152
Honda	26,177	30.7	20,032
Hyundai	5,270	53.2	3,439
Kia	6,155	5.4	5,839
Mazda	2,599	64.5	1,580
Mercedes-Benz	5,163	0.8	5,120
Nissan	15,070	76.8	8,524
Perodua	152,995	-0.4	153,683
Peugeot Citroen	2,036	-40.1	3,399
Proton	130,701	-8.3	142,585
Suzuki	5,750	16.4	4,939
Toyota	38,904	30.1	29,903
Volkswagen	1,662	–	–
Volvo	2,513	44.4	1,740
Total Cars	**399,913**	**3.9**	**384,935**
Light Trucks			
BMW	310	-41.5	530
Chery	568	-81.1	3,009
Hino	7,031	9.4	6,425
Honda	3,241	-6.1	3,450
Hyundai	559	-58.9	1,361
Isuzu	9,895	36.3	7,258
Jinbei	453	20.5	376
Kia	3,501	38.7	2,524
Mitsubishi Fuso	2,683	56.7	1,712
Nissan	25,590	54.8	16,536
Daihatsu	–	–	*355*
Perodua	*41,990*	*9.3*	*38,432*
Perodua Total	41,990	8.3	38,787
Peugeot Citroen	959	530.9	152
Proton	30,711	29.9	23,644
Toyota	34,874	2.1	34,141
Volvo Truck (UD Trucks)	1,227	-9.8	1,360
Other	332	-6.5	355
Total Light Trucks	**163,924**	**15.7**	**141,620**
Medium/Heavy Trucks			
Foton	377	-39.8	626
Hicom	1,506	-45.7	2,773
Hino	642	32.6	484
Hyundai	1,386	-4.3	1,448
Mercedes-Benz	96	-46.7	180
Scania	359	-31.6	525
Volvo Truck	417	108.5	200
Other	1,000	10.6	904
Total Med./Hvy. Trks.	**5,783**	**-19.0**	**7,140**
Total Comm. Vehs.	**169,707**	**-19.0**	**148,760**
Total Vehicles	**569,620**	**6.7**	**533,695**

MALAYSIA

SALES

Year	Cars	Commercial Vehicles	Total
2012	427,611	200,142	627,753
2011	417,990	181,770	599,760
2010	414,597	190,414	605,011
2009	406,742	129,827	536,569
2008	425,955	122,237	548,192
2007	379,543	107,633	487,176
2006	366,738	124,030	490,768
2005	400,835	150,207	551,042
2004	380,568	107,037	487,605
2003	319,885	85,163	405,048
2002	359,934	75,020	434,954
2001	327,497	68,972	396,469
2000	282,103	61,070	343,173
1999	239,647	48,900	288,547
1998	137,691	26,160	163,851
1997	307,907	96,930	404,837
1996	275,615	89,173	364,788
1995	224,991	60,801	285,792
1994	155,765	44,670	200,435
1993	128,600	39,328	167,928
1992	109,432	35,652	145,084
1991	121,660	60,217	181,877
1990	106,454	59,407	165,861
1989	73,793	35,564	109,357
1988	53,532	18,060	71,592
1987	35,265	13,731	48,996
1986	47,028	20,819	67,847
1985	63,857	31,142	94,999
1984	86,810	23,105	109,915
1983	90,251	18,063	108,314
1982	86,245	16,202	102,447

MALAYSIA

SALES BY COMPANY

	2012	% Chg.	2011
Cars			
Audi	1,061	50.9	703
BMW	*5,686*	*25.7*	*4,524*
Mini	*341*	*13.3*	*301*
BMW Total	6,027	24.9	4,825
Chery	82	-26.1	111
Daimler (Mercedes-Benz)	5,492	3.3	5,315
Ford	4,052	-3.7	4,207
GM (Chevrolet)	952	52.6	624
Honda	30,872	8.8	28,373
Hyundai	6,921	2.6	6,746
Kia	8,470	8.5	7,810
Mazda	4,052	0.7	4,023
Mitsubishi	1,216	-45.6	2,234
Nissan	14,363	95.1	7,362
Perodua	147,752	3.6	142,587
Peugeot Citroen	5,197	11.6	4,657
Porsche	237	-5.2	250
Proton	117,004	-13.1	134,594
Renault	9	-18.2	11
Subaru	53	211.8	17
Suzuki	6,874	2.3	6,721
Toyota	54,136	9.7	49,362
Volkswagen	11,252	62.6	6,918
Volvo	526	-2.6	540
Other	1,011	–	–
Total Cars	**427,611**	**2.3**	**417,990**
Light Trucks			
Audi	353	57.6	224
BMW	632	32.8	476
Chery	1,554	-46.2	2,886
Daihatsu	1,452	-31.4	2,118
Daimler (Mercedes-Benz)	413	233.1	124
Ford	3,056	2.5	2,981
GM (Chevrolet)	1,074	65.7	648
Honda	4,078	-0.7	4,107
Hyundai	5,017	-0.8	5,060
Isuzu	4,973	-16.6	5,961
Jaguar Land Rover	643	192.3	220
Kia	3,857	17.7	3,278
Mahindra & Mahindra	1	-90.0	10
Mazda	2,280	13.7	2,005
Mitsubishi	10,436	6.3	9,819
Nissan	20,603	-11.9	23,383
Perodua	41,385	10.6	37,402
Peugeot Citroen	917	33.3	688
Porsche	158	-4.2	165
Proton	24,117	0.2	24,063
Renault	81	-5.8	86
Ssangyong	238	-7.4	257
Suzuki	1,213	106.6	587
Toyota	52,486	39.6	37,589
Volkswagen	1,751	305.3	432
Volvo	271	3.8	261
Total Light Trucks	**183,039**	**11.0**	**164,830**
Medium/Heavy Trucks			
Mercedes-Benz	–	–	*100*
Mitsubishi Fuso	*2,180*	*41.6*	*1,540*
Daimler Total	2,180	32.9	1,640
Dongfeng	–	–	144
Hicom	129	-96.8	4,009
Hino	6,433	15.2	5,585
Isuzu	5,700	70.8	3,338
MAN	168	23.5	136
Scania	466	16.2	401
UD Trucks	*1,305*	*-16.1*	*1,555*
Volvo Truck	*140*	*6.1*	*132*
Volvo Truck Total	1,445	-14.3	1,687
Other	582	–	–
Total Med./Hvy. Trks.	**17,103**	**1.0**	**16,940**
Total Comm. Vehs.	**200,142**	**10.1**	**181,770**
Total Vehicles	**627,753**	**4.7**	**599,760**

NEW ZEALAND

ECONOMIC SUMMARY

Common Language	English
Currency	New Zealand Dollar
Steering	Right Hand Drive

2012 Vehicles in Operation

Passenger Cars	2,633,378
Commercial Vehicles	510,284
Total	3,143,662
Persons per Car	1.7
Persons per Vehicle	1.4

2012 GDP (billions of US $)

Nominal GDP	169.7
Real GDP	119.2
GDP % annual growth	2.5
GDP per capita	26,840

Current Account Balance (billions of US $)

2012:	-8.5
2011:	-6.6

Current Account Balance in % of GDP

2012:	-5.0
2011:	-4.1

Total Number of Fatal Road Accidents (most recent data)

2012:	308

Capital	Wellington
Population	4,365,113
Population Density	42.2
Surface Area sq km/miles	267,710 / 103,363

Road Infrastructure

Paved Road Surface %	66.1
Total Road Length in km/miles	94,094 / 58,467
Road Traffic (Vehicles per mile)	53.8

2012 Total Vehicle Percent Market Share by Company

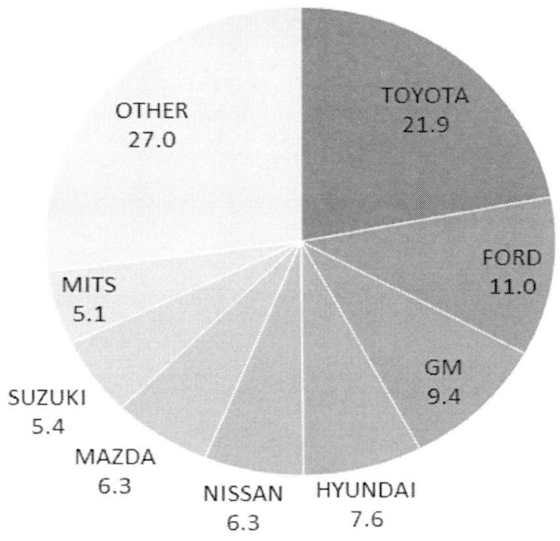

Total Vehicle Sales 2002-2012

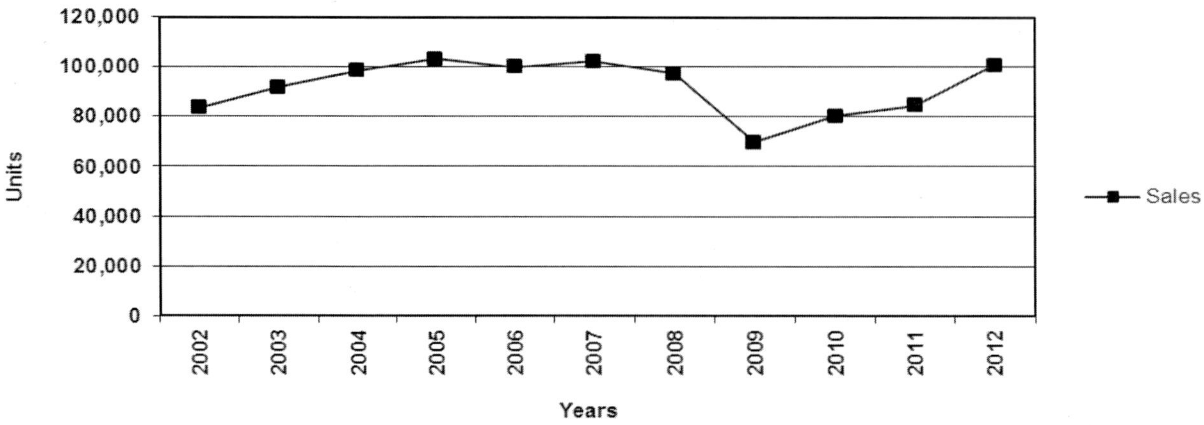

NEW ZEALAND

SALES

Year	Cars	Commercial Vehicles	Total
2012	50,381	50,203	100,584
2011	45,062	39,422	84,484
2010	45,565	34,770	80,335
2009	42,974	26,971	69,945
2008	58,082	39,103	97,185
2007	59,021	43,274	102,295
2006	60,618	39,226	99,844
2005	61,963	41,268	103,231
2004	74,693	23,762	98,455
2003	70,455	21,345	91,800
2002	63,576	20,045	83,621
2001	57,887	17,283	75,170
2000	57,762	16,937	74,699
1999	58,073	14,028	72,101
1998	54,154	11,828	65,982
1997	58,558	12,934	71,492
1996	176,178	29,844	206,022
1995	146,656	29,598	176,254
1994	123,853	30,534	154,387
1993	97,663	27,402	125,065
1992	92,110	23,772	115,882
1991	102,966	24,985	127,951
1990	159,746	29,445	189,191
1989	134,828	23,275	158,103
1988	88,589	16,652	105,241
1987	89,628	16,119	105,747
1986	80,021	18,583	98,604
1985	84,434	23,284	107,718
1984	98,437	27,035	125,472
1983	75,851	25,024	100,875
1982	85,479	30,712	116,191

Note: Prior to 2005 SUV data was in Car.

NEW ZEALAND

SALES BY COMPANY

	2012	% Chg.	2011
Cars			
Aston Martin	15	-44.4	27
Audi	1,022	3.3	989
Bentley	15	36.4	11
BMW	*973*	*42.3*	*684*
Mini	*302*	*29.1*	*234*
BMW Total	1,275	38.9	918
Chery	165	52.8	108
Chrysler	*114*	*418.2*	*22*
Dodge	*519*	*16.4*	*446*
Chrysler Total	633	35.3	468
Daihatsu	180	283.0	47
Daimler (Mercedes-Benz)	837	12.2	746
Ferrari	16	45.5	11
Alfa Romeo	*191*	*17.9*	*162*
Fiat	*79*	*-4.8*	*83*
Fiat Total	270	10.2	245
Ford	5,974	25.9	4,744
GM (Holden)	5,383	8.8	4,949
Great Wall	1	–	–
Honda	2,380	-0.5	2,393
Hyundai	4,297	7.6	3,994
Jaguar	113	189.7	39
Kia	1,748	37.9	1,268
Lamborghini	7	40.0	5
Maserati	18	-10.0	20
Mazda	3,333	-16.3	3,980
Mitsubishi	1,500	47.9	1,014
Nissan	989	-49.7	1,966
Citroen	*225*	*8.7*	*207*
Peugeot	*809*	*34.4*	*602*
Peugeot Citroen Total	1,034	27.8	809
Porsche	61	27.1	48
Proton (Lotus)	1	–	1
Renault	65	14.0	57
Rolls-Royce	1	–	–
Ssangyong	2	-50.0	4
Subaru	561	-18.5	688
Suzuki	4,795	12.4	4,267
Lexus	*292*	*-21.5*	*372*
Toyota	*10,616*	*31.5*	*8,075*
Toyota Total	10,908	29.1	8,447
Skoda	*412*	*-9.1*	*453*
Volkswagen	*2,108*	*9.3*	*1,929*
Volkswagen Total	2,520	5.8	2,382
Volvo	89	-19.8	111
Zhejiang Geely	8	-85.7	56
Other	165	-34.0	250
Total Cars	**50,381**	**11.8**	**45,062**
Light Trucks			
Audi	478	44.8	330
BMW	*665*	*25.7*	*529*
Mini	*128*	*70.7*	*75*
BMW Total	793	31.3	604
Chery	61	-4.7	64
Chrysler	–	–	*8*
Dodge	*115*	–	*115*
Jeep	*660*	*73.2*	*381*
Chrysler Total	775	53.8	504
Daihatsu	141	354.8	31
Daimler (Mercedes-Benz)	571	-28.3	796
Fiat	48	-53.4	103
Ford	5,077	31.8	3,851
GM (Holden)	4,042	31.9	3,065
Great Wall	995	84.9	538
Hino	–	–	1
Honda	623	192.5	213
Hyundai	3,260	20.9	2,696
Isuzu	298	36.7	218
Kia	1,196	3.2	1,159
Land Rover	455	7.6	423
Mazda	2,965	55.2	1,911
Mitsubishi	3,674	-3.2	3,795
Nissan	5,368	67.1	3,213
Citroen	*28*	–	*1*
Peugeot	*204*	*20.7*	*169*
Peugeot Citroen Total	232	36.5	170

NEW ZEALAND

SALES BY COMPANY

	2012	% Chg.	2011
Porsche	99	1.0	98
Renault	42	23.5	34
Ssangyong	754	71.4	440
Subaru	1,125	44.4	779
Suzuki	617	9.2	565
Lexus	*197*	*49.2*	*132*
Toyota	*10,961*	*16.1*	*9,441*
Toyota Total	11,158	16.6	9,573
Skoda	*169*	*191.4*	*58*
Volkswagen	*1,789*	*17.9*	*1,518*
Volkswagen Total	1,958	24.2	1,576
Volvo	154	113.9	72
Other	348	104.7	170
Total Light Trucks	**47,307**	**27.9**	**36,992**
Medium/Heavy Trucks			
Freightliner	*124*	*11.7*	*111*
Mercedes-Benz	*78*	*-23.5*	*102*
Mitsubishi Fuso	*509*	*43.4*	*355*
Sterling	–	–	*2*
Western Star	*61*	*-35.1*	*94*
Daimler Total	772	16.3	664
GM (GMC)	1	–	–
Hino	428	61.5	265
Hyundai	44	100.0	22
International	11	22.2	9
Isuzu	587	35.9	432
Iveco	81	11.0	73
Jianghuai Auto Co.	27	200.0	9
MAN	*81*	*-2.4*	*83*
Volkswagen Truck & Bus	*1*	–	–
MAN Total	82	-1.2	83
Nissan	2	-60.0	5
DAF	*122*	*84.8*	*66*
Kenworth	*86*	*-34.4*	*131*
Peterbilt	*1*	–	–
PACCAR Total	209	6.1	197
Scania	153	-27.8	212
Mack Truck	*59*	*63.9*	*36*
Renault Truck	*6*	*100.0*	*3*
UD Trucks	*107*	*16.3*	*92*
Volvo Truck	*168*	*2.4*	*164*
Volvo Truck Total	340	15.3	295
Other	159	-3.0	164
Total Med./Hvy. Trks.	**2,896**	**19.2**	**2,430**
Total Comm. Vehs.	**50,203**	**27.3**	**39,422**
Total Vehicles	**100,584**	**19.1**	**84,484**

PAKISTAN

ECONOMIC SUMMARY

Common Language	Punjabi/Urdu
Currency	Rupee
Steering	Left Hand Drive

2012 Vehicles in Operation

Passenger Cars	1,997,000
Commercial Vehicles	588,000
Total	2,585,000
Persons per Car	96.8
Persons per Vehicle	74.8

2012 GDP (billions of US $)

Nominal GDP	231.9
Real GDP	61.8
GDP % annual growth	3.7
GDP per capita	346

Current Account Balance (billions of US $)

2012: -4.6
2011: 0.2

Current Account Balance in % of GDP

2012: -2.0
2011: 0.1

Total Number of Fatal Road Accidents (most recent data)

2011: 5,271

Capital	Islamabad
Population	193,238,868
Population Density	649.2
Surface Area sq km/miles	770,875 / 297,635

Road Infrastructure

Paved Road Surface %	69.4
Total Road Length in km/miles	260,760 / 162,029
Road Traffic (Vehicles per mile)	16.0

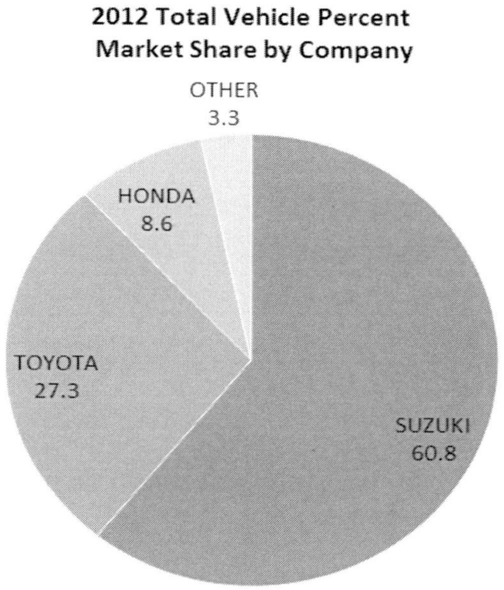

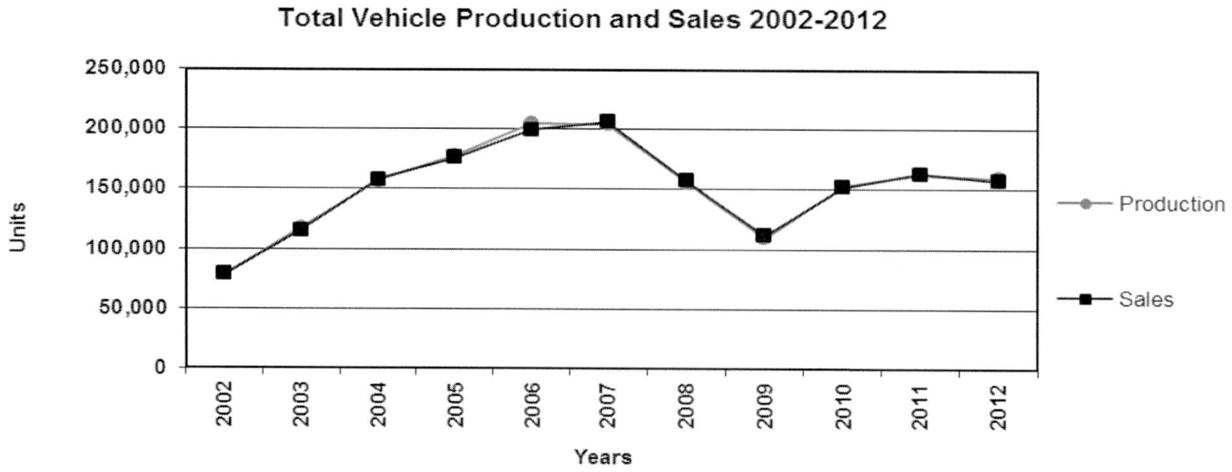

PAKISTAN

PRODUCTION

Year	Cars	Trucks	Buses	Total
2012	117,698	41,199	640	159,537
2011	124,267	37,805	466	162,538
2010	117,643	34,777	550	152,970
2009	84,677	24,068	688	109,433
2008	111,628	43,447	898	155,973
2007	158,305	44,749	1,073	204,127
2006	164,505	39,355	940	204,800
2005	142,975	33,650	839	177,464
2004	133,722	21,476	1,762	156,960
2003	103,662	12,519	1,380	117,561
2002	66,432	10,938	1,296	78,666
2001	42,679	8,111	1,086	51,876
2000	41,556	6,353	1,326	49,235
1999	35,332	5,078	1,460	41,870
1998	42,927	5,539	1,124	49,590
1997	38,676	7,183	591	46,450
1996	37,032	8,262	456	45,750
1995	33,419	7,950	474	41,843

PRODUCTION BY MANUFACTURER

	2012	% Chg.	2011
Cars			
Honda	14,835	1.5	14,612
Suzuki	62,397	0.4	62,153
Daihatsu	*1,575*	*-70.2*	*5,289*
Toyota	*38,891*	*-7.9*	*42,213*
Toyota Total	40,466	-14.8	47,502
Total Cars	**117,698**	**-5.3**	**124,267**
Light Trucks			
Master	39	34.5	29
Sigma	321	-53.4	689
Suzuki	33,973	11.8	30,376
Toyota	4,566	11.6	4,091
Total Light Trucks	**38,899**	**10.6**	**35,185**

	2012	% Chg	2011
Medium/Heavy Trucks			
Hino	1,168	17.3	996
Isuzu	354	-12.2	403
Master	561	-37.8	902
Volvo Truck (UD Trucks)	217	-32.0	319
Total Med./Hvy. Trks.	**2,300**	**-12.2**	**2,620**
Buses			
Hino	514	30.5	394
Isuzu	96	39.1	69
Master	24	700.0	3
Volvo Truck (UD Trucks)	6	–	–
Total Buses	**640**	**37.3**	**466**
Total Comm. Vehs.	**41,839**	**9.3**	**38,271**
Total Vehicles	**159,537**	**-1.8**	**162,538**

PAKISTAN

SALES

Year	Cars	Trucks	Buses	Total
2012	116,769	40,138	674	157,581
2011	124,508	38,290	473	163,271
2010	117,337	34,401	611	152,349
2009	86,445	24,569	682	111,696
2008	113,562	42,698	922	157,182
2007	159,457	45,470	1,089	206,016
2006	160,256	38,494	928	199,678
2005	141,567	33,381	1,068	176,016
2004	134,550	21,585	1,605	157,740
2003	101,848	12,271	1,363	115,482
2002	66,315	10,720	1,332	78,367
2001	44,378	8,204	1,065	53,647
2000	40,310	6,398	1,384	48,092
1999	34,911	5,259	1,362	41,532
1998	41,337	5,337	1,060	47,734

PAKISTAN

SALES BY COMPANY

	2012	% Chg.	2011
Cars			
Daihatsu	2,044	-58.6	4,932
Honda	13,551	-12.4	15,472
Suzuki	62,558	1.4	61,690
Toyota	38,616	-9.0	42,414
Total Cars	**116,769**	**-6.2**	**124,508**
Light Trucks			
Hyundai	–	–	113
Master	36	9.1	33
Sigma	267	-62.5	712
Suzuki	33,205	8.3	30,649
Toyota	4,450	2.8	4,327
Total Light Trucks	**37,958**	**5.9**	**35,834**

	2012	% Chg	2011
Medium/Heavy Trucks			
Hino	1,025	0.5	1,020
Isuzu	350	18.2	296
Master	557	-33.8	842
Volvo Truck (UD Trucks)	248	-16.8	298
Total Med./Hvy. Trks.	**2,180**	**-11.2**	**2,456**
Buses			
Hino	537	34.9	398
Isuzu	108	47.9	73
Master	23	–	2
Volvo Truck (UD Trucks)	6	–	–
Total Buses	**674**	**42.5**	**473**
Total Comm. Vehs.	**40,812**	**5.3**	**38,763**
Total Vehicles	**157,581**	**-3.5**	**163,271**

PHILIPPINES

ECONOMIC SUMMARY

Common Language	Filipino/English
Currency	Philippine Peso
Steering	Left Hand Drive

2012 Vehicles in Operation	
Passenger Cars	488,600
Commercial Vehicles	1,424,800
Total	1,913,400
Persons per Car	216.4
Persons per Vehicle	55.3

2012 GDP (billions of US $)	
Nominal GDP	250.4
Real GDP	153.5
GDP % annual growth	6.6
GDP per capita	1,603

Current Account Balance (billions of US $)	
2012:	7.2
2011:	7.1

Current Account Balance in % of GDP	
2012:	2.9
2011:	3.2

Total Number of Fatal Road Accidents (most recent data)	
2011:	1,171

Capital	Manila
Population	105,720, 644
Population Density	918.3
Surface Area sq km/miles	298,170 / 115,124

Road Infrastructure	
Paved Road Surface %	25.6
Total Road Length in km/miles	213,151 / 132,446
Road Traffic (Vehicles per mile)	14.4

2012 Total Vehicle Percent Market Share by Company

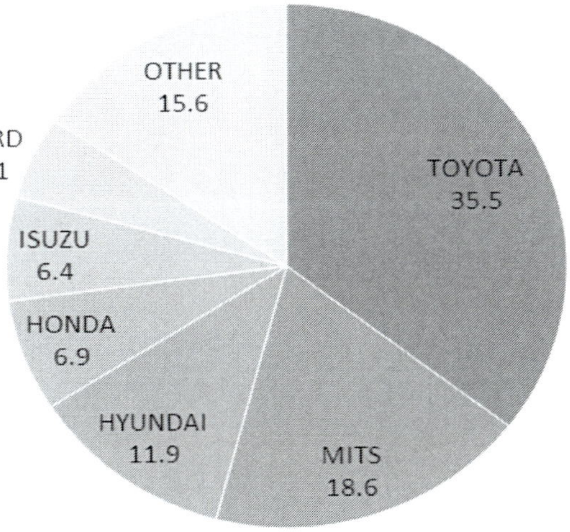

Total Vehicle Sales 2002-2012

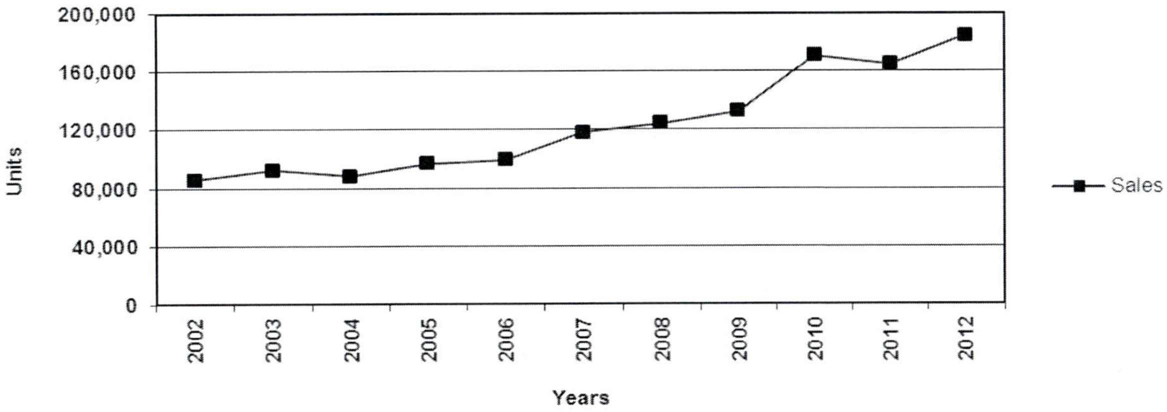

PHILIPPINES

SALES

Year	Cars	Commercial Vehicles	Total
2012	67,144	117,104	184,248
2011	57,576	107,345	164,921
2010	59,698	110,957	170,655
2009	46,228	86,216	132,444
2008	44,426	80,023	124,449
2007	40,958	76,945	117,903
2006	38,479	61,062	99,541
2005	35,631	61,432	97,063
2004	33,277	54,798	88,075
2003	24,317	68,019	92,336
2002	21,728	63,859	85,587
2001	23,684	52,986	76,670
2000	28,826	55,123	83,949
1999	27,580	46,834	74,414
1998	34,688	45,543	80,231
1997	75,760	68,675	144,435
1996	88,977	73,110	162,087
1995	71,195	56,967	128,162
1994	58,322	44,970	103,292
1993	50,000	32,612	82,612
1992	35,100	25,208	60,308
1991	27,800	20,151	47,951
1990	34,005	23,854	57,859

PHILIPPINES

SALES BY COMPANY

	2012	% Chg.	2011
Cars			
Audi	123	-35.6	191
BMW	*464*	*-15.3*	*548*
Mini	*166*	–	–
BMW Total	630	15.0	548
Daimler (Mercedes-Benz)	600	-19.1	742
Ford	4,416	-2.6	4,535
GM (Chevrolet)	1,731	21.0	1,431
Honda	10,882	9.5	9,940
Hyundai	9,458	8.3	8,736
Kia	3,384	57.5	2,149
Mazda	45	-58.3	108
Mitsubishi	2,742	2.5	2,674
Nissan	833	-22.7	1,077
Porsche	92	-31.3	134
Subaru	1,752	–	–
Suzuki	2,881	13.9	2,529
Toyota	27,465	21.2	22,656
Volvo	110	-12.7	126
Total Cars	**67,144**	**16.6**	**57,576**
Light Trucks			
BMW	209	-14.3	244
Mercedes-Benz	*201*	*-20.2*	*252*
Mitsubishi Fuso	*514*	*54.4*	*333*
Daimler Total	715	22.2	585
Ford	4,982	-5.0	5,243
Foton	26	-16.1	31
GM (Chevrolet)	1,597	19.9	1,332
Hino	48	9.1	44
Honda	1,771	6.0	1,671
Hyundai	12,538	8.5	11,561
Isuzu	11,172	17.0	9,550
Kia	4,143	57.3	2,633
Mazda	1,459	-51.8	3,030
Mitsubishi	31,530	7.1	29,452
Nissan	5,571	-22.0	7,144
Porsche	78	-35.5	121
Suzuki	1,620	12.6	1,439
Toyota	37,931	18.8	31,937
Volvo	102	-14.3	119
Yuejin	114	-12.3	130
Total Light Trucks	**115,606**	**8.8**	**106,266**
Medium/Heavy Trucks			
Daimler (Mitsubishi Fuso)	129	-10.4	144
Hino	432	-7.3	466
Isuzu	587	117.4	270
MAN	80	-32.2	118
Volvo Truck (UD Trucks)	270	233.3	81
Total Med./Hvy. Trks.	**1,498**	**38.8**	**1,079**
Total Comm. Vehs.	**117,104**	**9.1**	**107,345**
Total Vehicles	**184,248**	**11.7**	**164,921**

SINGAPORE

ECONOMIC SUMMARY

Common Language	Mandarin/English
Currency	Singapore Dollar
Steering	Right Hand Drive

Capital	Singapore
Population	5,460,302
Population Density	20,604.9
Surface Area sq km/miles	687 / 265

2012 Vehicles in Operation

Passenger Cars	605,149
Commercial Vehicles	205,789
Total	810,938
Persons per Car	9.0
Persons per Vehicle	6.7

Road Infrastructure

Paved Road Surface %	100.0
Total Road Length in km/miles	3,356 / 2,085
Road Traffic (Vehicles per mile)	388.9

2012 GDP (billions of US $)

Current GDP	276.5
Real GDP	249.3
GDP % annual growth	1.3
GDP per capita	46,132

Current Account Balance (billions of US $)

2012:	51.4
2011:	65.3

Current Account Balance in % of GDP

2012:	18.6
2011:	24.6

Total Number of Fatal Road Accidents (most recent data)

2011:	197

2012 Total Vehicle Percent Market Share by Company

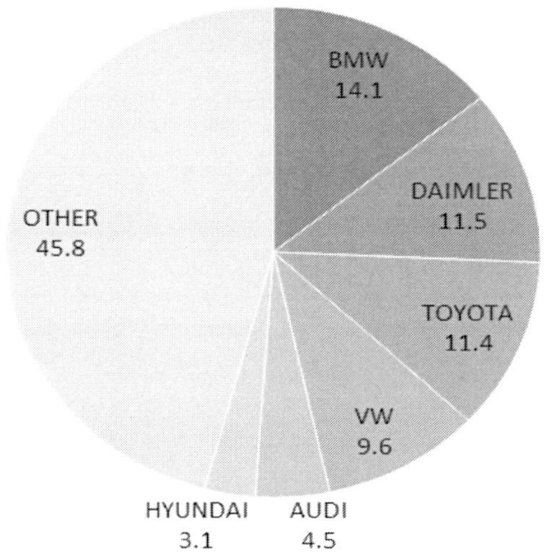

Total Vehicle Sales 2002-2012

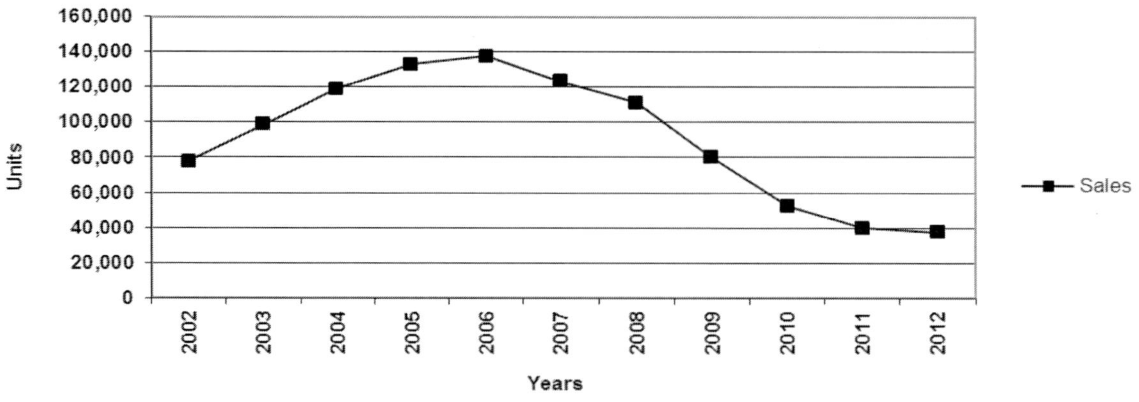

SINGAPORE

SALES

Year	Light Vehicles	Medium/Heavy Trucks	Total
2012	33,291	4,341	37,632
2011	33,732	6,077	39,809
2010	48,119	4,618	52,737
2009	73,735	6,320	80,055
2008	101,312	9,595	110,907
2007	111,722	11,440	123,162
2006	123,451	14,146	137,597
2005	118,091	14,821	132,912
2004	103,609	15,531	119,140
2003	84,471	14,365	98,836
2002	67,073	10,873	77,946
2001	70,462	14,638	85,100
2000	61,928	23,778	85,706
1999	41,667	11,915	53,582
1998	32,802	9,879	42,681
1997	30,799	8,955	39,754
1996	33,148	8,151	41,299
1995	37,870	8,025	45,895
1994	37,778	6,768	44,546
1993	43,594	7,835	51,429

SINGAPORE

SALES BY COMPANY

	2012	% Chg.	2011
Light Vehicles			
Aston Martin	22	-12.0	25
Audi	1,679	-20.5	2,111
Bentley	93	55.0	60
BMW	*5,039*	*1.6*	*4,959*
Mini	*285*	*-21.5*	*363*
BMW Total	5,324	–	5,322
Brilliance (Jinbei)	1	–	–
Chery	5	-82.1	28
Chrysler	*13*	*-55.2*	*29*
Jeep	*25*	*47.1*	*17*
Chrysler Total	38	-17.4	46
Daihatsu	37	-7.5	40
Daimler (Mercedes-Benz)	4,325	3.8	4,166
Ferrari	80	-13.0	92
Alfa Romeo	*49*	*2.1*	*48*
Fiat	*24*	*-81.4*	*129*
Lancia	*1*	–	–
Fiat Total	74	-58.2	177
Ford	354	48.7	238
Foton	9	–	–
Cadillac	*1*	–	–
Chevrolet	*437*	*-32.2*	*645*
Opel	*99*	*59.7*	*62*
GM Total	537	-24.0	707
Hafei	2	–	–
Honda	515	-45.3	942
Hyundai	1,155	-10.4	1,289
Jaguar	*383*	*-1.5*	*389*
Land Rover	*384*	*64.8*	*233*
Jaguar Land Rover Total	767	23.3	622
Kia	600	-37.0	953
Lamborghini	56	14.3	49
Maserati	84	-7.7	91
Mazda	348	31.3	265
McLaren	33	–	–
MG Motors	1	–	–
Mitsubishi	61	-69.7	201
Morgan	2	–	–
Infiniti	*72*	*928.6*	*7*
Nissan	*687*	*-14.1*	*800*
Nissan Total	759	-5.9	807
Perodua	8	-46.7	15
Citroen	*357*	*3.2*	*346*
Peugeot	*560*	*21.7*	*460*
Peugeot Citroen Total	917	13.8	806
Porsche	507	-13.2	584
Lotus	*9*	*-35.7*	*14*
Proton	*53*	*-29.3*	*75*
Proton Total	62	-30.3	89
Renault	114	-56.7	263
Rolls-Royce	49	25.6	39
Saab	15	-64.3	42
Ssangyong	16	-15.8	19
Subaru	135	-37.5	216
Suzuki	144	-47.6	275
Tesla	1	–	–
Toyota	4,272	19.9	3,562
SEAT	*2*	–	–
Skoda	*62*	*-4.6*	*65*
Volkswagen	*3,567*	*11.3*	*3,204*
Volkswagen Total	3,631	11.1	3,269
Volvo	1,066	27.7	835
Other	5,393	-1.7	5,487
Total Light Vehicles	**33,291**	**-1.3**	**33,732**
Medium/Heavy Trucks			
Other	4,341	-28.6	6,077
Total Med./Hvy. Trks.	**4,341**	**-28.6**	**6,077**
Total Vehicles	**37,632**	**-5.5**	**39,809**

SOUTH KOREA

ECONOMIC SUMMARY

Common Language	Korean
Currency	Won
Steering	Left Hand Drive

2012 Vehicles in Operation

Passenger Cars	14,577,182
Commercial Vehicles	4,293,351
Total	18,870,533
Persons per Car	3.4
Persons per Vehicle	2.6

2012 GDP (billions of US $)

Nominal GDP	1,155.9
Real GDP	1,033.6
GDP % annual growth	2.0
GDP per capita	20,668

Current Account Balance (billions of US $)

2012:	43.1
2011:	26.1

Current Account Balance in % of GDP

2012:	3.7
2011:	2.3

Total Number of Fatal Road Accidents (most recent data)

2010:	5,505

Capital	Seoul
Population	48,955,203
Population Density	1,308.2
Surface Area sq km/miles	96,920 / 37,421

Road Infrastructure

Paved Road Surface %	78.3
Total Road Length in km/miles	103,029 / 64,019
Road Traffic (Vehicles per mile)	294.8

2012 Total Vehicle Percent Market Share by Company

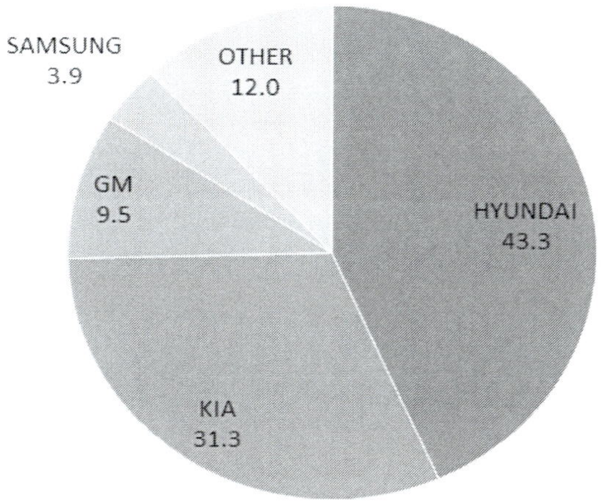

Total Vehicle Production and Sales 2002-2012

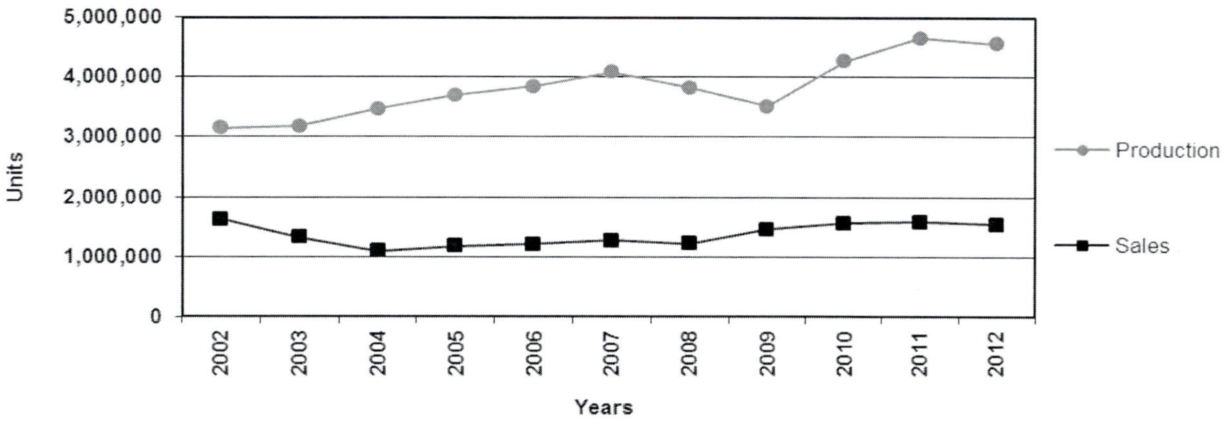

SOUTH KOREA

PRODUCTION

Year	Cars	Trucks	Buses	Total
2012	2,867,842	1,568,639	121,257	4,557,738
2011	2,901,707	1,610,087	145,300	4,657,094
2010	2,792,210	1,342,010	137,521	4,271,741
2009	2,374,869	1,029,200	108,857	3,512,926
2008	2,435,616	1,267,211	123,855	3,826,682
2007	2,509,440	1,467,498	109,370	4,086,308
2006	2,298,474	1,430,868	110,760	3,840,102
2005	2,194,613	1,389,722	115,015	3,699,350
2004	2,053,668	1,289,299	126,497	3,469,464
2003	1,825,178	1,185,186	167,506	3,177,870
2002	1,691,726	1,239,799	216,059	3,147,584
2001	1,715,475	998,241	232,613	2,946,329
2000	1,881,341	987,369	246,288	3,114,998
1999	2,361,735	253,097	228,282	2,843,114
1998	1,625,125	169,682	159,687	1,954,494
1997	2,308,476	266,928	242,871	2,818,275
1996	2,264,709	311,489	236,516	2,812,714
1995	2,003,146	312,705	210,549	2,526,400
1994	1,805,895	307,725	198,043	2,311,663
1993	1,592,669	277,200	180,189	2,050,058
1992	1,306,752	267,354	155,590	1,729,696
1991	1,158,245	233,860	105,713	1,497,818
1990	986,751	226,101	108,778	1,321,630
1989	871,898	177,391	80,181	1,129,470
1988	872,074	142,677	68,904	1,083,655
1987	793,125	128,183	58,431	979,739
1986	457,383	107,777	36,386	601,546
1985	264,458	84,614	29,090	378,162
1984	158,503	80,304	26,554	265,361
1983	121,987	73,438	25,594	221,019
1982	94,460	47,199	20,931	162,590

SOUTH KOREA

PRODUCTION BY MANUFACTURER

	2012	% Chg.	2011
Cars			
Chevrolet	*310,033*	*-6.4*	*331,380*
Daewoo	*258,743*	*-14.4*	*302,182*
GM Daewoo Total	568,776	-10.2	633,562
Hyundai	1,237,381	3.8	1,192,194
Kia	957,095	8.1	885,620
Samsung	100,013	-45.4	183,012
Ssangyong	4,577	-37.5	7,319
Total Cars	**2,867,842**	**-1.2**	**2,901,707**
Light Trucks			
Chevrolet	*199,673*	*33.7*	*149,333*
Daewoo	*6,620*	*-63.8*	*18,297*
GM Daewoo Total	206,293	23.1	167,630
Hyundai	552,995	-1.1	559,170
Kia	627,053	-10.0	696,684
Samsung	53,878	-12.0	61,248
Ssangyong	114,565	8.2	105,930
Tata Daewoo	3,086	-26.8	4,216
Total Light Trucks	**1,557,870**	**-2.3**	**1,594,878**
Medium/Heavy Trucks			
Hyundai	8,574	-14.9	10,079
Tata Daewoo	2,195	-57.2	5,130
Total Med./Hvy. Trks.	**10,769**	**-29.2**	**15,209**
Buses			
Daewoo Bus	2,721	-15.2	3,210
GM Daewoo	10,688	10.6	9,662
Hyundai	106,311	-18.7	130,811
Kia	1,537	-4.9	1,617
Total Buses	**121,257**	**-16.5**	**145,300**
Total Comm. Vehs.	**1,689,896**	**-3.7**	**1,755,387**
Total Vehicles	**4,557,738**	**-2.1**	**4,657,094**

SOUTH KOREA

SALES

Year	Cars	Trucks	Buses	Total
2012	995,339	481,830	64,546	1,541,715
2011	1,030,630	476,855	72,189	1,579,674
2010	1,031,949	453,778	70,261	1,555,988
2009	957,366	432,126	65,501	1,454,993
2008	802,324	351,398	62,409	1,216,131
2007	756,284	446,249	70,192	1,272,725
2006	708,588	422,119	74,077	1,204,784
2005	652,720	444,397	76,346	1,173,463
2004	514,485	488,143	92,024	1,094,652
2003	582,545	622,250	113,517	1,318,312
2002	704,537	763,302	154,429	1,622,268
2001	657,483	629,469	164,498	1,451,450
2000	623,955	647,684	158,821	1,430,460
1999	910,725	208,415	153,889	1,273,029
1998	568,063	136,657	75,185	779,905
1997	1,151,287	228,033	133,615	1,512,935
1996	1,238,940	263,924	141,268	1,644,132
1995	1,149,409	278,429	128,064	1,555,902
1994	1,140,399	278,101	137,102	1,555,602
1993	1,037,488	258,188	140,291	1,435,967
1992	876,262	256,410	135,702	1,268,374
1991	772,548	228,717	102,919	1,104,184
1990	965,798	228,048	107,531	1,301,377
1989	861,757	177,072	80,170	1,118,999
1988	888,072	142,767	68,771	1,099,610
1987	784,679	125,267	56,412	966,358
1986	455,342	103,793	35,485	594,620
1985	255,369	85,126	28,897	369,392
1984	156,039	79,947	26,458	262,444
1983	121,193	73,307	25,925	220,425
1982	93,612	47,551	21,001	162,164

SOUTH KOREA

SALES BY COMPANY

	2012	% Chg.	2011
Cars			
Audi	15,126	46.2	10,345
Bentley	135	32.4	102
BMW	*28,152*	*20.9*	*23,293*
Mini	*5,927*	*38.4*	*4,282*
BMW Total	34,079	23.6	27,575
Chrysler	4,123	24.3	3,316
Daimler (Mercedes-Benz)	20,389	4.3	19,542
Ford	5,126	22.5	4,184
Cadillac	*475*	*-36.8*	*752*
Chevrolet	*99,513*	*6.0*	*93,870*
Daewoo	*9,711*	*-32.8*	*14,451*
GM Total	109,699	0.6	109,073
Honda	3,944	25.1	3,153
Hyundai	393,707	-6.7	422,200
Jaguar	1,197	17.8	1,016
Kia	304,029	5.6	287,822
Mitsubishi	81	138.2	34
Infiniti	*1,103*	*-48.7*	*2,152*
Nissan	*2,398*	*-36.9*	*3,802*
Nissan Total	3,501	-41.2	5,954
Citroen	*255*	*–*	*6*
Peugeot	*2,407*	*-8.7*	*2,636*
Peugeot Citroen Total	2,662	0.8	2,642
Porsche	1,516	16.5	1,301
Rolls-Royce	27	–	27
Samsung	54,990	-45.9	101,603
Ssangyong	4,447	-36.7	7,030
Subaru	627	-5.6	664
Lexus	*4,976*	*21.0*	*4,111*
Toyota	*10,795*	*115.0*	*5,020*
Toyota Total	15,771	72.7	9,131
Volkswagen	18,395	47.9	12,436
Volvo	1,768	19.5	1,480
Total Cars	**995,339**	**-3.4**	**1,030,630**
Light Trucks			
Chevrolet	*22,570*	*16.5*	*19,376*
Daewoo	*5,874*	*8.0*	*5,439*
GM Total	28,444	14.6	24,815
Hyundai	216,019	11.1	194,382
Kia	176,474	-13.3	203,586
Land Rover	1,916	38.5	1,383
Samsung	4,936	-35.2	7,618
Ssangyong	43,253	36.8	31,621
Tata Daewoo	3,105	-24.4	4,106
Total Light Trucks	**474,147**	**1.4**	**467,511**
Medium/Heavy Trucks			
Hyundai	5,366	-14.6	6,280
Tata Daewoo	2,317	-24.4	3,064
Total Med./Hvy. Trks.	**7,683**	**-17.8**	**9,344**
Buses			
Daewoo Bus	2,270	31.2	1,730
GM Daewoo	8,034	6.1	7,569
Hyundai	52,685	-14.0	61,295
Kia	1,557	-2.4	1,595
Total Buses	**64,546**	**-10.6**	**72,189**
Total Comm. Vehs.	**546,376**	**-0.5**	**549,044**
Total Vehicles	**1,541,715**	**-2.4**	**1,579,674**

TAIWAN

ECONOMIC SUMMARY

Common Language	Mandarin Chinese
Currency	New Taiwan Dollar
Steering	Left Hand Drive

Capital	Taipei
Population	23,299,716
Population Density	1,870.7
Surface Area sq km/miles	32,260 / 12,455

2012 Vehicles in Operation

Passenger Cars	6,091,324
Commercial Vehicles	1,054,584
Total	7,145,908
Persons per Car	3.8
Persons per Vehicle	3.3

Road Infrastructure

Paved Road Surface %	98.9
Total Road Length in km/miles	41,475 / 25,771
Road Traffic (Vehicles per mile)	277.3

2012 GDP (billions of US $)

Nominal GDP	474.0
Real GDP	514.4
GDP % annual growth	1.3
GDP per capita	22,062

Current Account Balance (billions of US $)

2012:	49.6
2011:	41.2

Current Account Balance in % of GDP

2012:	10.5
2011:	8.9

Total Number of Fatal Road Accidents (most recent data)

2012:	2,040

2012 Total Vehicle Percent Market Share by Company

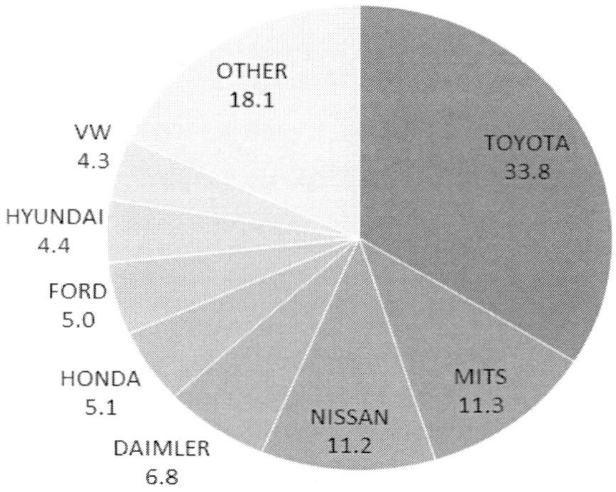

Total Vehicle Production and Sales 2002-2012

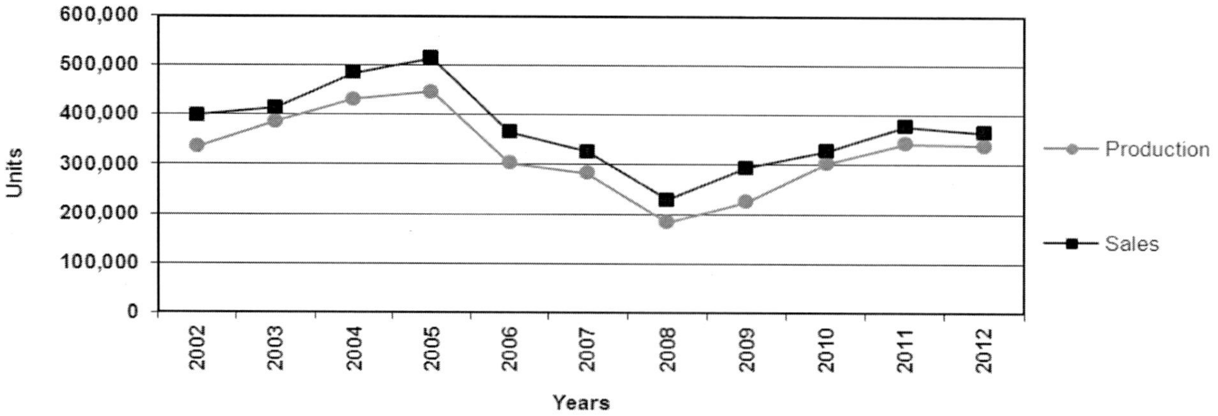

TAIWAN

PRODUCTION

Year	Cars	Light Trucks	Medium/Heavy Trucks	Buses	Total
2012	273,958	58,028	6,034	1,018	339,038
2011	280,641	57,058	4,769	828	343,296
2010	243,519	55,168	3,891	878	303,456
2009	183,820	38,323	3,507	706	226,356
2008	138,601	40,424	3,478	466	182,969
2007	212,685	67,622	2,663	469	283,439
2006	211,306	86,230	4,959	734	303,229
2005	323,819	117,322	4,608	596	446,345
2004	299,639	125,635	4,464	1,076	430,814
2003	264,837	117,771	3,248	830	386,686
2002	233,000	98,844	2,499	850	335,193
2001	195,109	73,986	2,028	581	271,704
2000	263,013	106,093	3,108	399	372,613
1999	245,969	99,864	4,440	*	350,273
1998	293,009	106,194	5,342	*	404,545
1997	269,801	104,833	6,152	*	380,786
1996	264,943	97,098	3,168	*	365,209
1995	282,006	118,274	6,200	*	406,480

*Included with Medium/Heavy Trucks.

TAIWAN

PRODUCTION BY MANUFACTURER

	2012	% Chg.	2011
Cars			
China Motor (Mitsubishi)	21,289	-24.3	28,125
Ford	*10,687*	*-17.8*	*13,000*
Mazda	*12,240*	*-15.8*	*14,529*
Ford Total	22,927	-16.7	27,529
Honda	18,532	-3.8	19,267
Hyundai	12,191	28.9	9,455
Prince (Chery)	–	–	244
Toyota	157,084	5.6	148,698
Luxgen	*4,457*	*4.0*	*4,286*
Nissan	*3,877*	*202.4*	*1,282*
Tobe	*1,095*	*30.4*	*840*
Yulon Total	9,429	47.1	6,408
Yulon Nissan	32,506	-20.6	40,915
Total Cars	**273,958**	**-2.4**	**280,641**
Light Trucks			
Mitsubishi	*21,625*	*-6.7*	*23,182*
Mitsubishi Fuso	*9,025*	*22.7*	*7,358*
China Motor Total	30,650	0.4	30,540
Ford	*8,269*	*-16.8*	*9,938*
Mazda	*3*	–	–
Ford Total	8,272	-16.8	9,938
Hino	2,328	70.7	1,364
Hyundai	3,165	8.1	2,929
Prince (Chery)	–	–	60
Toyota	9,529	194.7	3,233
Yulon (Luxgen)	4,084	-48.2	7,880
Yulon Nissan	–	–	1,114
Total Light Trucks	**58,028**	**1.7**	**57,058**
Medium/Heavy Trucks			
China Motor (Mitsubishi Fuso)	1,768	23.3	1,434
Formosa (DAF)	504	25.1	403
Hino	3,660	31.0	2,794
Isuzu	–	–	84
Yulon Volvo Truck (UD Trucks)	102	88.9	54
Total Med./Hvy. Trucks	**6,034**	**26.5**	**4,769**
Buses			
Hino	1,018	22.9	828
Total Buses	**1,018**	**22.9**	**828**
Total Comm. Vehs.	**65,080**	**3.9**	**62,655**
Total Vehicles	**339,038**	**-1.2**	**343,296**

TAIWAN

SALES

Year	Cars	Commercial Vehicles	Total
2012	281,700	84,171	365,871
2011	304,443	73,844	378,287
2010	265,751	61,864	327,615
2009	238,433	55,990	294,423
2008	176,705	52,792	229,497
2007	245,747	80,552	326,299
2006	261,514	104,383	365,897
2005	372,941	141,238	514,179
2004	268,288	216,004	484,292
2003	241,768	171,593	413,361
2002	246,050	152,832	398,882
2001	226,001	121,419	347,420
2000	287,163	134,113	421,276
1999	313,527	109,582	423,109
1998	359,449	115,694	475,143
1997	368,782	113,656	482,438
1996	368,336	103,231	471,567
1995	413,301	129,098	542,399

TAIWAN

SALES BY COMPANY

	2012	% Chg.	2011
Cars			
Audi	3,223	21.2	2,660
BMW	9,134	-3.5	9,463
Daihatsu	382	-25.7	514
Daimler (Mercedes-Benz)	10,786	2.8	10,494
Ford	11,031	-20.2	13,829
GM (Chevrolet)	667	-18.8	821
Honda	18,794	-3.6	19,497
Hyundai	12,725	11.5	11,411
Kia	287	-6.8	308
Mazda	13,222	-18.7	16,270
Mitsubishi	20,746	-21.9	26,550
Nissan	40,227	-8.0	43,722
Porsche	1,982	–	–
Subaru	1,611	-1.5	1,635
Suzuki	–	–	299
Toyota	112,076	-3.6	116,254
Skoda	*1,996*	*-2.6*	*2,049*
Volkswagen	*6,446*	*-15.6*	*7,638*
Volkswagen Total	8,442	-12.9	9,687
Volvo	4,181	5.7	3,954
Luxgen	*8,801*	*-32.2*	*12,979*
Tobe	*1,000*	*-10.9*	*1,122*
Yulon Total	9,801	-30.5	14,101
Other	2,383	-19.9	2,974
Total Cars	**281,700**	**-7.5**	**304,443**
Light Trucks			
Audi	806	7.5	750
BMW	3,508	11.2	3,154
Daihatsu	477	-22.4	615
Mercedes-Benz	*3,260*	*4.0*	*3,136*
Mitsubishi Fuso	*7,785*	*9.6*	*7,106*
Daimler Total	11,045	7.8	10,242
Ford	7,428	-10.7	8,317
GM (Chevrolet)	287	-6.8	308
Hyundai	3,289	7.4	3,063
Isuzu	–	–	9
Kia	762	-17.5	924
Mitsubishi	20,523	-4.1	21,405
Nissan	674	20.4	560
Suzuki	4,008	-2.5	4,109
Toyota	11,652	158.5	4,508
Volkswagen	7,151	59.4	4,486
Other	1,716	-20.4	2,156
Total Light Trucks	**73,326**	**13.5**	**64,606**
Medium/Heavy Trucks			
Daimler (Mitsubishi Fuso)	2,885	-9.3	3,181
Hino	6,467	66.4	3,886
Isuzu	–	–	107
PACCAR (DAF)	391	-2.7	402
Volvo Truck (UD Trucks)	141	605.0	20
Other	961	-41.5	1,642
Total Med./Hvy. Trucks	**10,845**	**17.4**	**9,238**
Total Comm. Vehs.	**84,171**	**14.0**	**73,844**
Total Vehicles	**365,871**	**-3.3**	**378,287**

THAILAND

ECONOMIC SUMMARY

Common Language	Thai
Currency	Baht
Steering	Right Hand Drive

2012 Vehicles in Operation

Passenger Cars	6,273,983
Commercial Vehicles	6,591,061
Total	12,865,044
Persons per Car	10.8
Persons per Vehicle	5.2

2012 GDP (billions of US $)

Nominal GDP	365.6
Real GDP	158.8
GDP % annual growth	6.4
GDP per capita	2,466

Current Account Balance (billions of US $)

2012:	2.7
2011:	5.9

Current Account Balance in % of GDP

2012:	0.7
2011:	1.7

Total Number of Fatal Road Accidents (most recent data)

2010:	13,766

Capital	Bangkok
Population	67,448,120
Population Density	341.9
Surface Area sq km/miles	510,890 / 197,255

Road Infrastructure

Paved Road Surface %	98.5
Total Road Length in km/miles	180,053 / 111,879
Road Traffic (Vehicles per mile)	115.0

2012 Total Vehicle Percent Market Share by Company

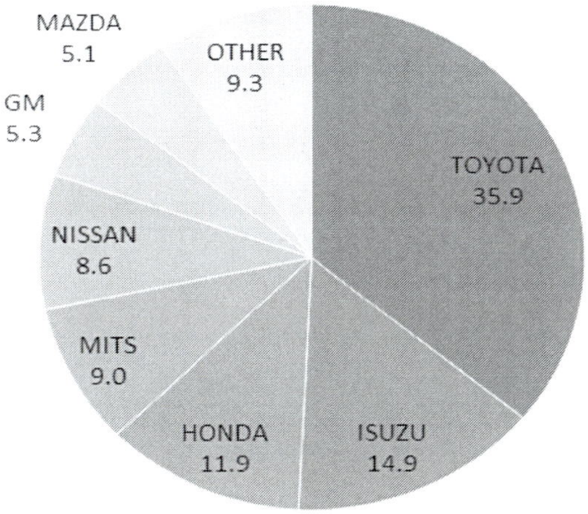

Total Vehicle Production and Sales 2002-2012

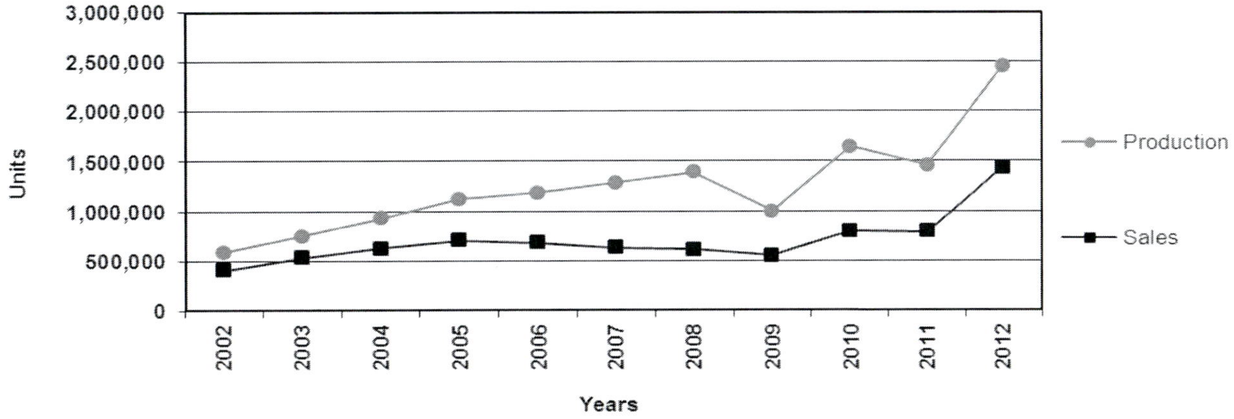

THAILAND

PRODUCTION

Year	Cars	Trucks	Buses	Total
2012	957,623	1,496,094	–	2,453,717
2011	537,987	919,808	–	1,457,795
2010	554,387	1,090,917	–	1,645,304
2009	313,665	685,713	–	999,378
2008	401,309	992,433	–	1,393,742
2007	315,444	971,902	–	1,287,346
2006	298,819	889,225	–	1,188,044
2005	277,603	847,713	–	1,125,316
2004	299,439	628,429	213	928,081
2003	251,691	498,731	90	750,512
2002	169,321	415,242	388	584,951
2001	156,066	303,081	271	459,418
2000	97,129	314,592	–	411,721
1999	78,538	248,614	81	327,233
1998	34,018	123,535	577	158,130
1997	114,018	245,731	554	360,303
1996	142,218	416,601	609	559,428
1995	129,265	394,689	1,726	525,680

THAILAND

PRODUCTION BY MANUFACTURER

	2012	% Chg.	2011
Cars			
BMW	6,065	39.7	4,342
Ford	*78,844*	*69.4*	*46,535*
Mazda	*55,560*	*19.4*	*46,535*
Ford Total	134,404	44.4	93,070
GM (Chevrolet)	41,712	42.6	29,254
Honda	222,263	96.8	112,961
Mercedes-Benz	5,223	9.3	4,779
Mitsubishi	84,134	–	7,471
Nissan	168,721	55.9	108,222
Suzuki	17,684	–	–
Toyota	277,417	56.0	177,888
Total Cars	**957,623**	**78.0**	**537,987**
Light Trucks			
Ford	*67,136*	*34.2*	*50,021*
Mazda	*67,136*	*34.2*	*50,021*
Ford Total	134,272	34.2	100,042
GM (Chevrolet)	82,897	127.4	36,447
Isuzu	272,776	85.2	147,293
Mercedes-Benz	4,307	-21.3	5,470
Mitsubishi	272,488	36.1	200,189
Nissan	82,756	7.5	76,982
Hino	*1,800*	*237.1*	*534*
Toyota	*610,011*	*80.5*	*337,925*
Toyota Total	611,811	80.8	338,459
Total Light Trucks	**1,461,307**	**61.5**	**904,882**
Medium/Heavy Trucks			
Isuzu	14,733	219.9	4,606
Mercedes-Benz	156	231.9	47
Mitsubishi Fuso	1,952	–	–
Scania	331	61.5	205
Toyota (Hino)	16,505	85.2	8,913
Volvo Truck	1,110	-3.9	1,155
Total Med./Hvy. Trks.	**34,787**	**133.1**	**14,926**
Total Comm. Vehs.	**1,496,094**	**62.7**	**919,808**
Total Vehicles	**2,453,717**	**68.3**	**1,457,795**

THAILAND

SALES

Year	Cars	Commercial Vehicles	Total
2012	672,460	763,875	1,436,335
2011	361,219	433,725	794,944
2010	346,644	453,713	800,357
2009	230,036	318,833	548,869
2008	226,805	388,465	615,270
2007	170,118	461,133	631,251
2006	191,763	490,398	682,161
2005	188,211	515,226	703,437
2004	209,110	416,916	626,026
2003	179,005	354,171	533,176
2002	126,253	282,989	409,242
2001	104,502	192,550	297,052
2000	83,106	179,083	262,189
1999	66,858	151,472	218,330
1998	46,309	98,646	144,955
1997	132,060	231,096	363,156
1996	172,730	416,396	589,126
1995	163,371	408,209	571,580

THAILAND

SALES BY COMPANY

	2012	% Chg.	2011
Cars			
Audi	23	-58.9	56
BMW	*5,614*	*94.8*	*2,882*
Mini	*142*	*-58.5*	*342*
BMW Total	5,756	78.5	3,224
Chery	35	-97.1	1,192
Daimler (Mercedes-Benz)	6,274	53.3	4,092
Ford	35,016	77.4	19,743
GM (Chevrolet)	21,486	28.6	16,712
Honda	163,442	109.7	77,930
Hyundai	–	–	137
Jaguar Land Rover	36	-18.2	44
Kia	684	150.5	273
Mazda	44,282	33.3	33,212
Mitsubishi	42,427	489.3	7,200
Nissan	103,504	116.4	47,831
Citroen	*121*	*-40.7*	*204*
Peugeot	*210*	*45.8*	*144*
Peugeot Citroen Total	331	-4.9	348
Porsche	–	–	122
Proton	1,789	-47.9	3,434
Ssangyong	1,775	-11.3	2,000
Subaru	118	-13.2	136
Suzuki	17,951	280.6	4,716
Toyota	224,816	63.8	137,265
Skoda	*64*	*73.0*	*37*
Volkswagen	*563*	*118.2*	*258*
Volkswagen Total	627	112.5	295
Volvo	1,654	114.0	773
Other	434	-10.3	484
Total Cars	**672,460**	**86.2**	**361,219**
Light Trucks			
BMW	500	-3.7	519
Mercedes-Benz	*404*	*92.4*	*210*
Mitsubishi Fuso	*131*	*-49.0*	*257*
Daimler Total	535	14.6	467
Dongfeng	2,158	38.2	1,562
Ford	20,885	120.8	9,457
GM (Chevrolet)	53,971	262.6	14,883
Hino	5,334	76.2	3,028
Honda	7,766	29.0	6,022
Hyundai	4,760	8.8	4,377
Isuzu	201,575	60.1	125,871
Jaguar Land Rover	47	20.5	39
Kia	836	31.9	634
Mazda	28,446	224.4	8,768
Mitsubishi	87,033	49.6	58,160
Nissan	20,346	-4.3	21,263
Suzuki	6,734	35.4	4,972
Tata	4,816	-7.5	5,205
Toyota	291,270	90.6	152,796
Volkswagen	249	-47.0	470
Volvo	200	-2.4	205
Volvo Truck (UD Trucks)	8	-82.2	45
Other	–	–	78
Total Light Trucks	**737,469**	**76.1**	**418,821**
Medium/Heavy Trucks			
Mercedes-Benz	*33*	*10.0*	*30*
Mitsubishi Fuso	*1,530*	*297.4*	*385*
Daimler Total	1,563	276.6	415
Hino	11,725	64.9	7,110
Isuzu	11,805	86.7	6,323
Scania	287	9.1	263
UD Trucks	*41*	*-36.9*	*65*
Volvo Truck	*910*	*39.1*	*654*
Volvo Truck Total	951	32.3	719
Other	75	1.4	74
Total Med./Hvy. Trks.	**26,406**	**77.2**	**14,904**
Total Comm. Vehs.	**763,875**	**76.1**	**433,725**
Total Vehicles	**1,436,335**	**80.7**	**794,944**

VIETNAM

ECONOMIC SUMMARY

Common Language	Vietnamese
Currency	Dong
Steering	Left Hand Drive

2012 Vehicles in Operation

Passenger Cars	178,000
Commercial Vehicles	192,000
Total	370,000
Persons per Car	519.5
Persons per Vehicle	249.9

2012 GDP (billions of US $)

Current GDP	138.1
Real GDP	29.5
GDP % annual growth	5.0
GDP per capita	326

Current Account Balance (billions of US $)

2012:	10.3
2011:	0.2

Current Account Balance in % of GDP

2012:	7.4
2011:	0.2

Total Number of Fatal Road Accidents (most recent data)

2010:	11,029

Capital	Hanoi
Population	92,477,857
Population Density	772.5
Surface Area sq km/miles	310,070 / 119,719

Road Infrastructure

Paved Road Surface %	74.2
Total Road Length in km/miles	180,549 / 112,188
Road Traffic (Vehicles per mile)	3.3

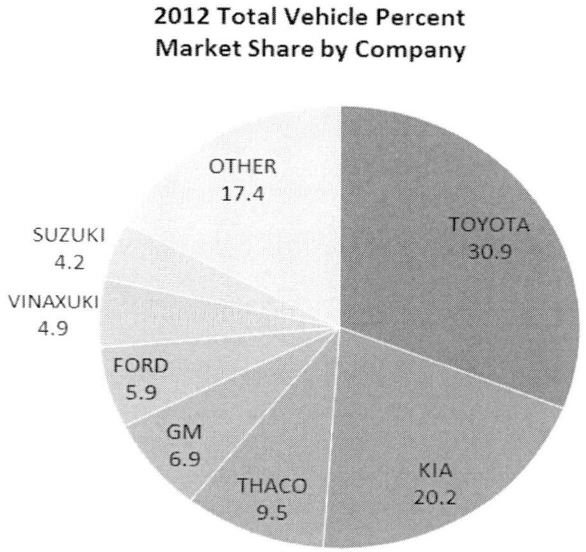

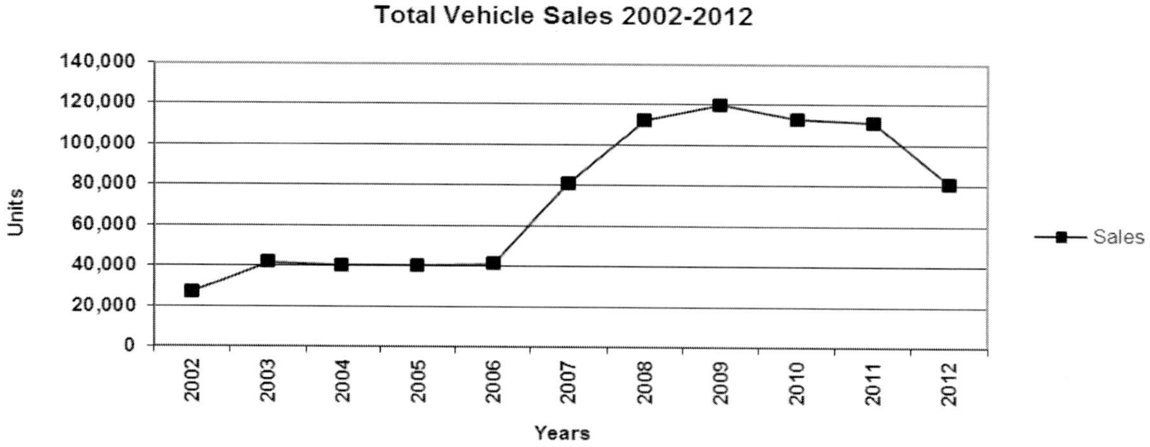

VIETNAM

SALES

Year	Cars	Commercial Vehicles	Total
2012	27,292	53,480	80,772
2011	41,429	69,687	111,116
2010	33,469	79,327	112,796
2009	34,656	85,212	119,868
2008	23,549	88,836	112,385
2007	17,027	63,972	80,999
2006	7,480	33,703	41,183
2005	11,619	28,530	40,149
2004	15,520	24,621	40,141
2003	20,485	21,072	41,557
2002	15,523	11,349	26,872
2001	9,428	10,128	19,556
2000	6,772	7,162	13,934

VIETNAM

SALES BY COMPANY

	2012	% Chg.	2011
Cars			
Chery	157	9.0	144
Daimler (Mercedes-Benz)	1,218	-33.4	1,828
Ford	1,908	-48.2	3,686
GM (Chevrolet)	3,852	-50.2	7,734
Honda	794	-50.0	1,589
Kia	6,669	-38.6	10,853
Lifan	–	–	4
Suzuki	15	–	–
Toyota	12,679	-18.7	15,591
Total Cars	**27,292**	**-34.1**	**41,429**
Light Trucks			
Brilliance (Jinbei)	460	21.1	380
Mercedes-Benz	*711*	*-33.4*	*1,068*
Mitsubishi Fuso	*454*	*-18.1*	*554*
Daimler Total	1,165	-28.2	1,622
Dongfeng	–	–	517
Ford	2,882	-42.5	5,011
GM (Chevrolet)	1,761	-32.7	2,616
Hino	286	-33.6	431
Honda	1,010	4.6	966
Hyundai	345	-71.2	1,197
Isuzu	911	-19.7	1,135
Kia	9,638	-22.9	12,508
Mekong	275	-35.3	425
Mitsubishi	1,141	-30.0	1,630
Nissan	313	-68.3	986
Pyeonghwa Motors (PMC)	189	-21.9	242
Sanyang	344	-50.1	690
Suzuki	3,394	-21.9	4,344

	2012	% Chg.	2011
THACO	2,172	-55.4	4,874
Toyota	12,248	-13.8	14,201
VEAM	1,817	40.5	1,293
Vinaxuki	3,993	-44.7	7,227
Total Light Trucks	**44,344**	**-28.8**	**62,295**
Medium/Heavy Trucks			
Dongfeng	–	–	349
Hino	346	-15.2	408
Isuzu	306	-35.3	473
Kamaz	30	-68.1	94
Mekong	–	–	3
SAMCO	27	-3.6	28
Scania	32	-60.0	80
THACO	4,454	54.8	2,878
VEAM	64	-40.7	108
Volvo Truck	–	–	71
Total Med./Hvy Trucks	**5,259**	**17.1**	**4,492**
Buses			
Hino	–	–	8
Hyundai	2,213	33.6	1,656
Isuzu	276	91.7	144
Kamaz	9	-52.6	19
SAMCO	311	-17.7	378
Scania	–	–	7
THACO	1,068	55.2	688
Total Buses	**3,877**	**33.7**	**2,900**
Total Comm. Vehicles	**53,480**	**-23.3**	**69,687**
Total Vehicles	**80,772**	**-27.3**	**111,116**

Ward's World Motor Vehicle Data 2013

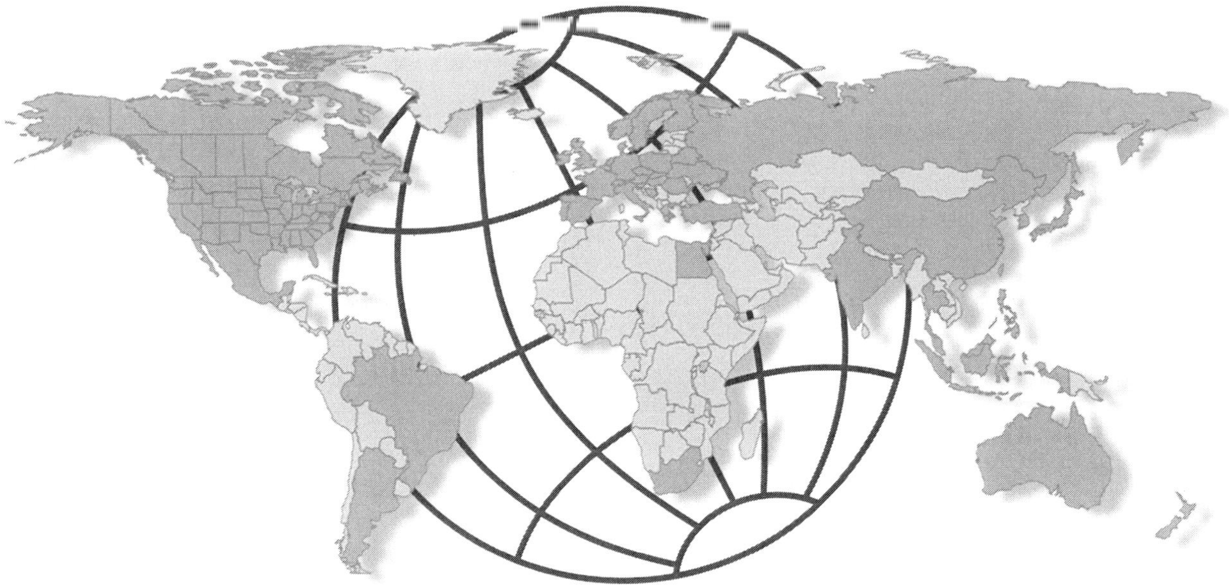

Europe Data

AUSTRIA

ECONOMIC SUMMARY

Common Language	German
Currency	Euro
Steering	Left Hand Drive
2012 Vehicles in Operation	
Passenger Cars	4,584,202
Commercial Vehicles	426,081
Total	5,010,283
Persons per Car	1.8
Persons per Vehicle	1.6
2012 GDP (billions of US $)	
Nominal GDP	398.6
Real GDP	359.2
GDP % annual growth	0.8
GDP per capita	42,432
Current Account Balance (billions of US $)	
2012:	7.9
2011:	2.4
Current Account Balance in % of GDP	
2012:	2.0
2011:	0.6
Total Number of Fatal Road Accidents (most recent data)	
2010:	552

Capital	Vienna
Population	8,221,646
Population Density	258.3
Surface Area sq km/miles	82,445 / 31,832
Road Infrastructure	
Paved Road Surface %	100.0
Total Road Length in km/miles	124,508 / 77,366
Road Traffic (Vehicles per mile)	64.8

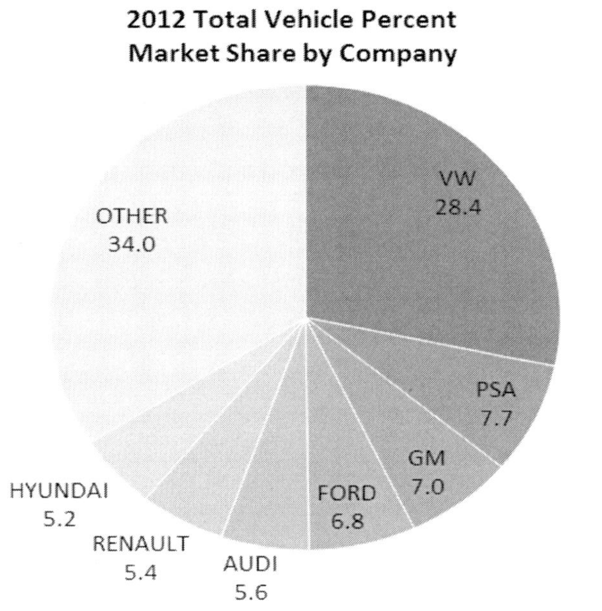

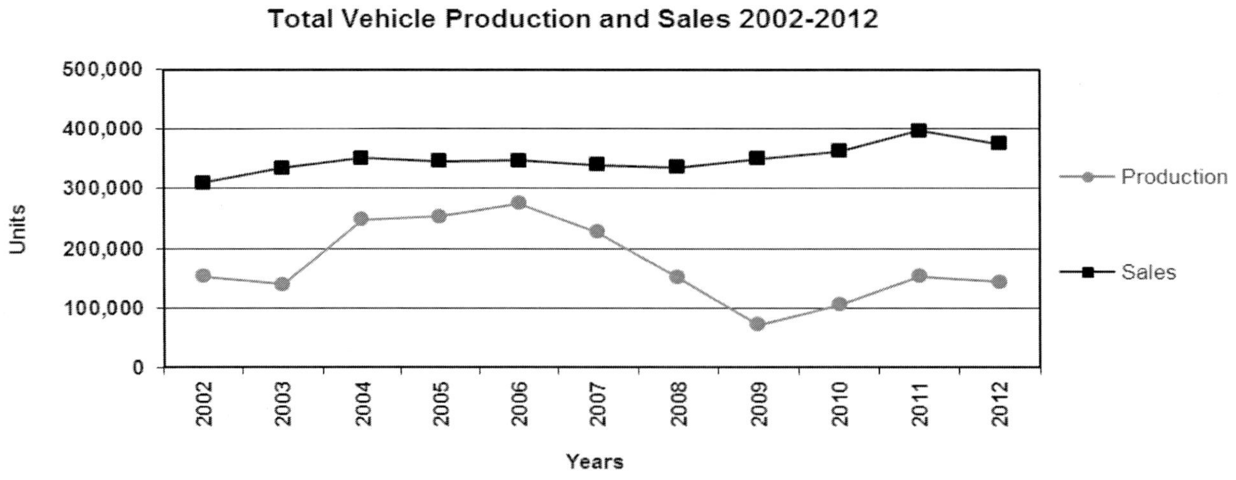

AUSTRIA

PRODUCTION

Year	Cars	Trucks	Buses	Total
2012	123,602	19,487	--	143,089
2011	130,343	22,162	--	152,505
2010	86,183	18,814	--	104,997
2009	56,620	15,714	--	72,334
2008	125,436	25,441	--	150,877
2007	199,969	28,097	--	228,066
2006	248,059	26,646	227	274,932
2005	230,508	22,567	122	253,197
2004	227,244	21,416	60	248,720
2003	118,650	20,868	138	139,656
2002	132,768	19,734	117	152,619
2001	131,098	24,257	48	155,403
2000	115,979	25,016	31	141,026
1999	123,586	15,674	71	139,331
1998	91,264	11,871	71	103,206
1997	97,774	10,099	116	107,989
1996	97,386	8,548	155	106,089
1995	59,196	8,983	191	68,370
1994	45,785	6,573	222	52,580
1993	40,777	7,656	230	48,663
1992	23,479	18,218	275	41,972
1991	13,682	22,293	252	36,227
1990	14,741	16,770	234	31,745
1989	6,638	11,505	251	18,394
1988	7,261	10,306	287	17,854
1987	6,958	9,982	276	17,216
1986	6,801	11,580	290	18,671
1985	7,118	10,932	274	18,324
1984	5,629	5,206	343	11,178
1983	6,148	5,769	291	12,208
1982	7,051	7,810	294	15,155

AUSTRIA

PRODUCTION BY MANUFACTURER

	2012	% Chg.	2011		2012	% Chg.	2011
Cars				**Medium/Heavy Trucks**			
Aston Martin	379	-55.6	853	MAN	19,487	-12.1	22,162
BMW (Mini)	103,545	0.9	102,643	**Total Med./Hvy. Trks.**	**19,487**	**-12.1**	**22,162**
Mercedes-Benz	9,850	38.3	7,122				
Peugeot	9,828	-50.2	19,725	**Total Comm. Vehs.**	**19,487**	**-12.1**	**22,162**
Total Cars	**123,602**	**-5.2**	**130,343**	**Total Vehicles**	**143,089**	**-6.2**	**152,505**

AUSTRIA

SALES

Year	Cars	Trucks	Buses	Total
2012	336,010	38,097	722	374,829
2011	356,145	39,908	602	396,655
2010	328,563	33,241	760	362,564
2009	319,403	30,372	654	350,429
2008	293,697	41,343	950	335,990
2007	298,182	40,724	771	339,677
2006	308,594	37,937	827	347,358
2005	307,915	37,001	802	345,718
2004	311,292	38,577	930	350,799
2003	300,121	33,457	754	334,332
2002	279,493	29,173	710	309,376
2001	293,528	31,944	606	326,078
2000	309,427	35,696	792	345,915
1999	314,182	33,407	799	348,388
1998	295,865	32,830	510	329,205
1997	275,001	29,825	543	305,369
1996	307,671	28,045	553	336,269
1995	279,610	27,304	570	307,484
1994	273,663	26,779	563	301,005
1993	285,162	26,430	611	312,203
1992	320,094	28,348	595	349,037
1991	303,723	29,134	479	333,336
1990	288,636	28,637	574	317,847
1989	276,065	25,109	569	301,743
1988	253,072	27,716	574	281,362
1987	243,221	23,792	567	267,580
1986	262,175	22,033	515	284,723
1985	242,670	20,906	577	264,153
1984	215,640	19,644	655	235,939
1983	256,676	18,928	544	276,148
1982	201,155	17,572	602	219,329

AUSTRIA

SALES BY COMPANY

	2012	% Chg.	2011
Cars			
Aston Martin	17	13.3	15
Audi	20,910	4.4	20,034
AutoVaz (Lada)	41	-54.4	90
Bentley	51	-7.3	55
BMW	*15,851*	*1.4*	*15,636*
Mini	*2,110*	*-17.1*	*2,544*
BMW Total	17,961	-1.2	18,180
Caterham	2	-33.3	3
Chrysler	*1*	*-99.5*	*208*
Dodge	*6*	*-80.6*	*31*
Jeep	*874*	*26.5*	*691*
Chrysler Total	881	-5.3	930
Dacia	6,222	760.6	723
Daihatsu	132	-23.3	172
Mercedes-Benz	*10,615*	*-2.9*	*10,935*
Smart	*738*	*-46.2*	*1,372*
Daimler Total	11,353	-7.8	12,307
Ferrari	46	24.3	37
Alfa Romeo	*2,330*	*-26.2*	*3,157*
Fiat	*12,422*	*-16.8*	*14,933*
Lancia	*1,350*	*82.7*	*739*
Fiat Total	16,102	-14.5	18,829
Fisker	12	–	–
Ford	21,624	-8.7	23,678
Cadillac	*15*	*-78.3*	*69*
Chevrolet	*4,713*	*1.1*	*4,661*
Hummer	*–*	*–*	*4*
Opel	*19,269*	*-25.7*	*25,919*
GM Total	23,997	-21.7	30,653
Honda	2,200	-3.8	2,287
Hyundai	19,286	6.4	18,120
Jaguar	*300*	*11.9*	*268*
Land Rover	*1,913*	*95.4*	*979*
Jaguar Land Rover Total	2,213	77.5	1,247
Kia	12,641	13.4	11,147
Lamborghini	16	-5.9	17
Maserati	13	18.2	11
Mazda	8,424	-11.5	9,524
Mitsubishi	3,772	-14.9	4,431
Morgan	13	-13.3	15

	2012	% Chg.	2011
Infiniti	*28*	*-36.4*	*44*
Nissan	*8,607*	*18.2*	*7,282*
Nissan Total	8,635	17.9	7,326
Citroen	*11,216*	*-16.2*	*13,381*
Peugeot	*14,264*	*-10.9*	*16,005*
Peugeot Citroen Total	25,480	-13.3	29,386
Piaggio	–	–	2
Porsche	960	22.0	787
Proton (Lotus)	8	-50.0	16
Renault	16,928	-27.4	23,324
Rolls-Royce	–	–	2
Saab	19	-88.7	168
Ssangyong	15	-64.3	42
Subaru	1,638	15.8	1,415
Suzuki	5,880	-18.1	7,179
Tata	20	122.2	9
Tesla	8	-11.1	9
Lexus	*285*	*-25.2*	*381*
Toyota	*7,393*	*-7.6*	*8,000*
Toyota Total	7,678	-8.4	8,381
SEAT	*14,316*	*-12.7*	*16,405*
Skoda	*21,698*	*6.6*	*20,364*
Volkswagen	*61,627*	*-5.4*	*65,150*
Volkswagen Total	97,641	-4.2	101,919
Volvo	3,126	-12.8	3,585
Other	45	-50.0	90
Total Cars	**336,010**	**-5.7**	**356,145**
Light Trucks			
AutoVaz (Lada)	95	48.4	64
Chrysler	*–*	*–*	*2*
Dodge	*111*	*-5.9*	*118*
Jeep	*15*	*–*	*–*
Chrysler Total	126	5.0	120
Dacia	359	–	–
Daimler (Mercedes-Benz)	2,506	-9.9	2,782
Fiat	2,596	-16.6	3,114
Ford	3,899	-5.9	4,143
Chevrolet	*3*	*-57.1*	*7*
Hummer	*–*	*–*	*1*
Opel	*2,052*	*-9.0*	*2,254*
GM Total	2,055	-9.2	2,262

AUSTRIA

SALES BY COMPANY

	2012	% Chg.	2011
Hyundai	82	7.9	76
Isuzu	215	-9.3	237
Iveco	1,007	5.4	955
Kia	69	392.9	14
Land Rover	295	13.0	261
Mahindra & Mahindra	2	-75.0	8
Mazda	50	-46.8	94
Mitsubishi	471	-16.9	567
Multicar	1	–	–
Nissan	842	2.7	820
Citroen	*1,542*	*-13.6*	*1,784*
Peugeot	*1,825*	*-7.3*	*1,969*
Peugeot Citroen Total	3,367	-10.3	3,753
Piaggio	22	-42.1	38
Renault	3,402	3.1	3,300
Ssangyong	–	–	9
Suzuki	199	40.1	142
Tata	13	-35.0	20
Toyota	1,144	-15.9	1,361
SEAT	*25*	*19.0*	*21*
Skoda	*204*	*-16.7*	*245*
Volkswagen	*8,463*	*4.1*	*8,131*
Volkswagen Total	8,692	3.5	8,397
Volvo	2	100.0	1
Other	9	–	9
Total Light Trucks	**31,520**	**-3.2**	**32,547**
Medium/Heavy Trucks			
Mercedes-Benz	*1,015*	*4.1*	*975*
Mitsubishi Fuso	*10*	*-28.6*	*14*
Daimler Total	1,025	3.6	989
Fiat	1	–	–
Ford	4	300.0	1
GM (Opel)	2	-66.7	6
Isuzu	1	–	1
Iveco	592	-6.6	634
MAN	*2,297*	*-17.1*	*2,772*
Volkswagen Truck & Bus	*31*	*210.0*	*10*
MAN Total	2,328	-16.3	2,782
Multicar	7	250.0	2

	2012	% Chg.	2011
PACCAR (DAF)	1,001	-4.7	1,050
Peugeot	–	–	1
Scania	740	-5.4	782
Toyota	–	–	1
Renault Truck	*239*	*-19.8*	*298*
UD Trucks	*5*	*400.0*	*1*
Volvo Truck	*629*	*-21.6*	*802*
Volvo Truck Total	873	-20.7	1,101
Other	3	-72.7	11
Total Med./Hvy. Trks.	**6,577**	**-10.7**	**7,361**
Buses			
Mercedes-Benz	–	–	*198*
Setra	–	–	*152*
Daimler Total	–	–	350
Fiat	–	–	2
Ford	–	–	1
Isuzu	–	–	1
Irisbus	–	–	*17*
Iveco	–	–	*12*
Iveco Total	–	–	29
King Long	–	–	5
MAN	–	–	*78*
Neoplan	–	–	*11*
Volkswagen Truck & Bus	–	–	*1*
MAN Total	–	–	90
Scania	–	–	11
Temsa	–	–	5
Van Hool	–	–	13
Bova	–	–	*3*
VDL	–	–	*6*
VDL Total	–	–	9
Renault Truck	–	–	*5*
Volvo Truck	–	–	*53*
Volvo Truck Total	–	–	58
Other	722	–	28
Total Buses	**722**	**19.9**	**602**
Total Comm. Vehs.	**38,819**	**-4.2**	**40,510**
Total Vehicles	**374,829**	**-5.5**	**396,655**

BELARUS

ECONOMIC SUMMARY

Common Language	Belarusian/Russian
Currency	Ruble
Steering	Left Hand Drive

2012 Vehicles in Operation

Passenger Cars	2,797,000
Commercial Vehicles	175,000
Total	2,972,000
Persons per Car	3.4
Persons per Vehicle	3.2

2012 GDP (billions of US $)

Nominal GDP	63.3
Real GDP	18.7
GDP % annual growth	1.5
GDP per capita	1,976

Total Number of Fatal Road Accidents (most recent data)

2010: 1,190

Capital	Minsk
Population	9,625,888
Population Density	122.9
Surface Area sq km/miles	202,900 / 78,340

Road Infrastructure

Paved Road Surface %	88.6
Total Road Length in km/miles	94,797 / 58,904
Road Traffic (Vehicles per mile)	50.5

Current Account Balance (billions of US $)

2012: -1.8
2011: -5.8

Current Account Balance in % of GDP

2012: -2.9
2011: -9.7

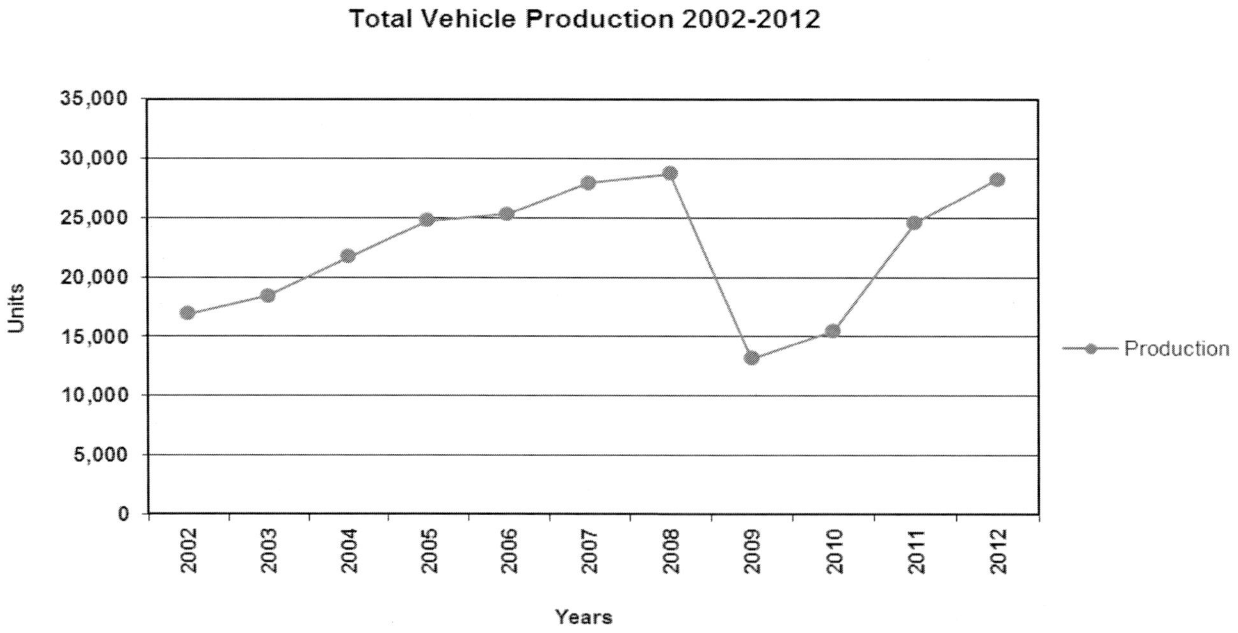

BELARUS

PRODUCTION

Year	Cars	Trucks	Buses	Total
2012	270	25,858	2,124	28,252
2011	230	22,261	2,082	24,573
2010	227	13,203	2,046	15,476
2009	161	11,457	1,514	13,132
2008	240	26,291	2,198	28,729
2007	232	25,548	2,160	27,940
2006	36	23,175	2,104	25,315
2005	–	22,817	1,950	24,767
2004	–	21,105	580	21,685
2003	–	17,933	480	18,413
2002	–	16,478	446	16,924
2001	–	15,996	429	16,425
2000	–	14,134	500	14,634
1999	195	12,846	–	13,041
1998	1,258	12,185	–	13,443
1997	881	12,627	–	13,508
1996	–	10,389	–	10,389
1995	–	12,958	–	12,958

PRODUCTION/ASSEMBLY BY MANUFACTURER

	2012	% Chg.	2011
Cars			
ZAO (Iran Khodro)	190	-17.4	230
Other	80	–	–
Total Cars	**270**	**17.4**	**230**
Light Trucks			
ZAO	10	–	–
Total Light Trucks	**10**	**–**	**–**
Medium/Heavy Trucks			
BelAZ	1,649	14.6	1,439
MAZ	22,854	13.4	20,152
MKZT	372	27.4	292
MOAZ	41	-26.8	56
Other	932	189.4	322
Total Med./Hvy. Trks.	**25,848**	**16.1**	**22,261**
Buses			
MAZ	2,022	2.3	1,976
Other	102	-3.8	106
Total Buses	**2,124**	**2.0**	**2,082**
Total Comm. Vehs.	**27,982**	**14.9**	**24,343**
Total Vehicles	**28,252**	**15.0**	**24,573**

BELGIUM

ECONOMIC SUMMARY

Common Language	Dutch/French/German
Currency	Euro
Steering	Left Hand Drive

2012 Vehicles in Operation

Passenger Cars	5,392,908
Commercial Vehicles	846,083
Total	6,238,991
Persons per Car	1.9
Persons per Vehicle	1.7

2012 GDP (billions of US $)

Nominal GDP	484.7
Real GDP	478.1
GDP % annual growth	-0.2
GDP per capita	43,088

Current Account Balance (billions of US $)

2012:	-2.4
2011:	-7.3

Current Account Balance in % of GDP

2012:	-0.5
2011:	-1.4

Total Number of Fatal Road Accidents (most recent data)

2010:	812

Capital	Brussels
Population	10,444,268
Population Density	893.4
Surface Area sq. km/miles	30,278 / 11,690

Road Infrastructure

Paved Road Surface %	78.2
Total Road Length in km/miles	154,575 / 96,048
Road Traffic (Vehicles per mile)	65.0

2012 Total Vehicle Percent Market Share by Company

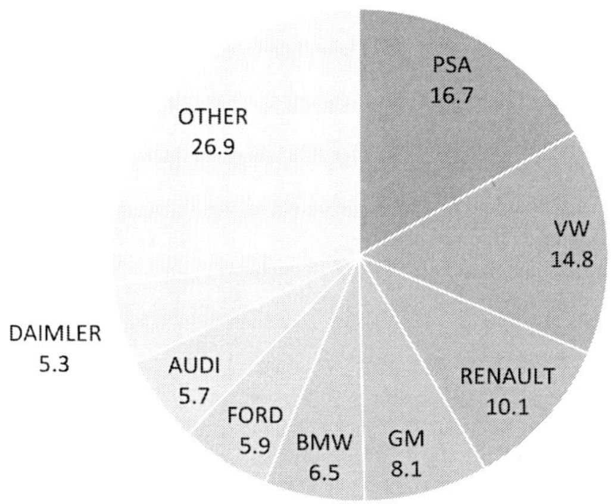

Total Vehicle Production and Sales 2002-2012

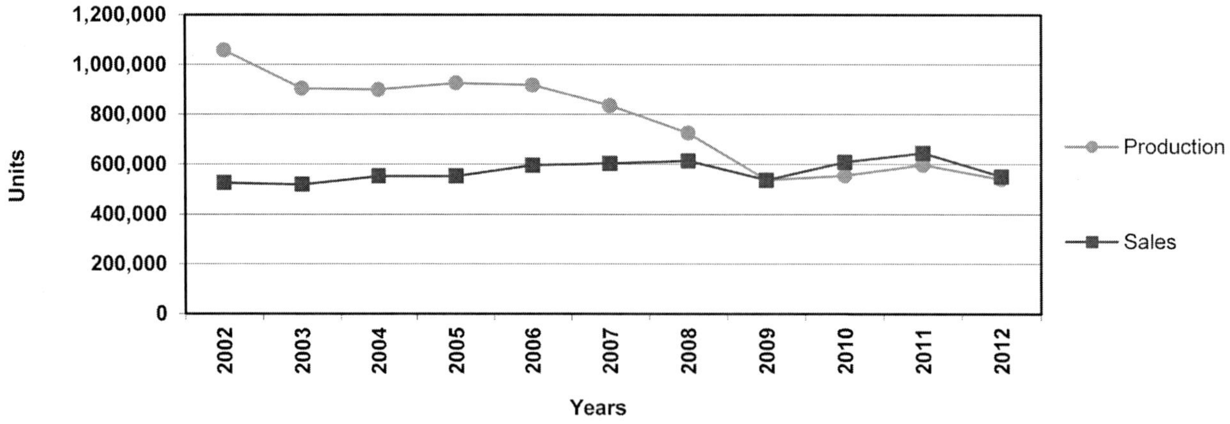

BELGIUM

PRODUCTION

Year	Cars	Trucks	Buses	Total
2012	507,204	31,500	1,000	539,704
2011	562,191	33,220	1,250	596,661
2010	528,224	25,326	1,100	554,650
2009	524,595	12,039	720	537,354
2008	680,149	43,081	1,286	724,516
2007	789,674	43,516	1,213	834,403
2006	881,929	34,984	1,143	918,056
2005	895,109	30,379	1,040	926,528
2004	857,119	41,875	1,279	900,273
2003	791,703	111,578	1,103	904,384
2002	936,903	115,940	4,358	1,057,201
2001	1,058,656	124,834	3,767	1,187,257
2000	912,233	119,562	1,499	1,033,294
1999	917,514	96,643	2,275	1,016,432
1998	938,881	90,860	2,823	1,032,564
1997	1,005,000	94,000	2,000	1,101,000
1996	1,144,106	88,160	1,518	1,233,784
1995	1,168,354	102,787	1,221	1,272,362
1994	1,202,031	68,001	1,404	1,271,436
1993	1,086,298	58,212	1,502	1,146,012
1992	1,075,547	74,914	1,756	1,152,217
1991	1,050,647	92,276	1,471	1,144,394
1990	1,160,412	86,375	1,503	1,248,290
1989	1,143,711	98,793	1,172	1,243,676
1988	1,137,612	88,580	998	1,227,190
1987	1,123,409	69,170	950	1,193,529
1986	1,021,643	64,312	883	1,086,838
1985	986,182	46,037	811	1,033,030
1984	865,297	52,112	*	917,409
1983	972,255	37,433	*	1,009,688
1982	950,410	33,177	*	983,587

* Included with Trucks.

BELGIUM

PRODUCTION BY MANUFACTURER

	2012	% Chg.	2011		2012	% Chg.	2011
Cars				**Total Med./Hvy. Trks.**	**31,500**	**-5.2**	**33,220**
Ford	126,791	-28.9	178,380				
Volkswagen (Audi)	123,111	4.7	117,566	Buses			
Volvo	257,302	-3.4	266,245	Van Hool	1,000	-20.0	1,250
Total Cars	**507,204**	**-9.8**	**562,191**	**Total Buses**	**1,000**	**-20.0**	**1,250**
Medium/Heavy Trucks				**Total Comm. Vehs.**	**32,500**	**-5.7**	**34,470**
Volvo Truck	31,500	-5.2	33,220	**Total Vehicles**	**539,704**	**-9.5**	**596,661**

BELGIUM

SALES

Year	Cars	Light Trucks	Medium/Heavy Trucks	Buses	Total
2012	486,737	54,608	8,475	701	550,521
2011	572,211	61,428	9,883	716	644,238
2010	547,347	52,509	7,649	1,019	608,524
2009	476,194	51,250	8,358	979	536,781
2008	535,947	64,639	12,106	1,206	613,898
2007	524,795	65,392	12,178	1,095	603,460
2006	526,141	57,917	10,396	1,144	595,598
2005	480,088	59,593	11,820	1,183	552,684
2004	484,757	56,601	10,070	1,176	552,604
2003	458,796	49,351	9,497	1,524	519,168
2002	467,567	46,067	9,751	2,509	525,894
2001	488,683	55,225	11,974	3,771	559,653
2000	515,204	50,620	11,486	1,129	578,439
1999	489,621	54,151	11,894	NA	555,666
1998	452,129	46,711	10,425	NA	509,265
1997	396,000	51,000	*	NA	447,000
1996	397,359	32,663	4,496	NA	434,518
1995	358,868	29,622	4,717	NA	393,207
1994	387,348	25,850	3,880	NA	417,078
1993	375,409	25,813	4,400	NA	405,622
1992	466,195	29,962	5,599	NA	501,756
1991	462,125	34,412	6,441	NA	502,978
1990	473,506	30,979	7,055	NA	511,540
1989	439,756	31,501	7,549	NA	478,806
1988	427,172	26,046	6,509	NA	459,727
1987	406,239	26,454	6,120	NA	438,813
1986	395,039	24,126	4,898	NA	424,063
1985	359,831	22,519	4,930	NA	387,280
1984	352,227	18,673	4,227	NA	375,127
1983	338,933	28,245	*	NA	367,178
1982	344,690	27,907	*	NA	372,597

* Included with Light Trucks. NA is not available.

BELGIUM

SALES BY COMPANY

	2012	% Chg.	2011
Cars			
Aston Martin	37	-59.8	92
Audi	31,330	-1.1	31,688
AutoVaz (Lada)	26	-23.5	34
Bentley	34	-8.1	37
BMW	*30,442*	*3.2*	*29,499*
Mini	*5,443*	*-19.5*	*6,758*
BMW Total	35,885	-1.0	36,257
Caterham	1	-50.0	2
Chrysler	*2*	*-99.0*	*196*
Dodge	*–*	*–*	*64*
Jeep	*641*	*31.4*	*488*
Chrysler Total	643	-14.0	748
Dacia	12,337	-11.6	13,960
Daihatsu	45	-69.4	147
Mercedes-Benz	*21,297*	*5.1*	*20,268*
Smart	*746*	*-42.6*	*1,299*
Daimler Total	22,043	2.2	21,567
Ferrari	88	-30.2	126
Alfa Romeo	*3,778*	*-39.1*	*6,205*
Fiat	*13,383*	*2.1*	*13,111*
Lancia	*1,911*	*12.4*	*1,700*
Fiat Total	19,072	-9.3	21,016
Fisker	54	–	–
Ford	28,754	-26.2	38,936
Cadillac	*8*	*-57.9*	*19*
Chevrolet	*5,044*	*16.9*	*4,316*
Opel	*35,769*	*-16.5*	*42,852*
GM Total	40,821	-13.5	47,187
Honda	2,439	-26.9	3,335
Hyundai	14,606	0.7	14,502
Iveco	–	–	1
Jaguar	*793*	*30.6*	*607*
Land Rover	*3,942*	*76.4*	*2,235*
Jaguar Land Rover Total	4,735	66.6	2,842
Kia	11,869	-4.1	12,376
Lamborghini	8	-33.3	12
Maserati	45	-44.4	81
Mazda	3,174	-0.4	3,187
Mitsubishi	3,132	-34.0	4,747
Morgan	19	-5.0	20
Infiniti	*46*	*-62.6*	*123*
Nissan	*17,100*	*-7.2*	*18,435*
Nissan Total	17,146	-7.6	18,558
Citroen	*37,238*	*-18.3*	*45,593*
Peugeot	*39,932*	*-18.0*	*48,716*
Peugeot Citroen Total	77,170	-18.2	94,309
Porsche	1,591	1.7	1,565
Proton (Lotus)	37	5.7	35
Renault	45,996	-25.2	61,470
Rolls-Royce	6	-25.0	8
Saab	49	-94.2	846
Ssangyong	402	22.6	328
Subaru	516	11.2	464
Suzuki	5,023	-29.9	7,161
Tesla	33	73.7	19
Lexus	*1,292*	*15.2*	*1,122*
Toyota	*16,178*	*-26.1*	*21,892*
Toyota Total	17,470	-24.1	23,014
SEAT	*5,966*	*-42.6*	*10,389*
Skoda	*16,636*	*-15.3*	*19,652*
Volkswagen	*52,104*	*-15.8*	*61,904*
Volkswagen Total	74,706	-18.7	91,945
Volvo	15,345	-21.5	19,559
Other	50	66.7	30
Total Cars	**486,737**	**-14.9**	**572,211**
Light Trucks			
Audi	–	–	2
AutoVaz (Lada)	14	-46.2	26
BMW	*12*	*500.0*	*2*
Mini	*2*	*-33.3*	*3*
BMW Total	14	180.0	5
Chrysler	*–*	*–*	*16*
Dodge	*–*	*–*	*1*
Jeep	*111*	*79.0*	*62*
Chrysler Total	111	40.5	79
Dacia	354	-37.3	565
Mercedes-Benz	*5,313*	*-4.8*	*5,580*
Mitsubishi Fuso	*53*	*–*	*–*
Daimler Total	5,366	-3.8	5,580
Fiat	*3,325*	*-18.0*	*4,054*
Lancia	*24*	*–*	*–*
Fiat Total	3,349	-17.4	4,054
Ford	3,792	-17.3	4,587
Cadillac	*1*	*–*	*–*

BELGIUM

SALES BY COMPANY

	2012	% Chg.	2011
Chevrolet	*15*	*150.0*	*6*
Opel	*3,559*	*-24.9*	*4,740*
GM Total	3,575	-24.7	4,746
Hyundai	305	-40.2	510
Isuzu	475	5.1	452
Iveco	1,198	-15.2	1,413
Kia	51	2.0	50
Land Rover	556	-1.4	564
Mazda	12	-86.5	89
Micro-Vett	–	–	1
Mitsubishi	650	-37.3	1,037
Infiniti	*1*	–	–
Nissan	*1,860*	*-7.1*	*2,003*
Nissan Total	1,861	-7.1	2,003
Citroen	*7,715*	*-8.1*	*8,397*
Peugeot	*6,980*	*-14.5*	*8,159*
Peugeot Citroen Total	14,695	-11.2	16,556
Piaggio	4	-69.2	13
Porsche	–	–	1
Renault	9,627	-10.6	10,768
Ssangyong	133	250.0	38
Subaru	12	9.1	11
Suzuki	77	6.9	72
Toyota	1,354	6.1	1,276
SEAT	*6*	*20.0*	*5*
Skoda	*38*	*216.7*	*12*
Volkswagen	*6,804*	*-0.4*	*6,831*
Volkswagen Total	6,848	–	6,848
Volvo	57	-8.1	62
Other	118	490.0	20
Total Light Trucks	**54,608**	**-11.1**	**61,428**
Medium/Heavy Trucks			
Ashok Leyland (Avia)	5	–	–
Mercedes-Benz	*1,418*	*-18.9*	*1,749*
Mitsubishi Fuso	*21*	*-43.2*	*37*
Daimler Total	1,439	-19.4	1,786
Ford	26	13.0	23
GM (Opel)	11	83.3	6
Isuzu	10	233.3	3
Iveco	818	-3.5	848
MAN	*1,416*	*-18.2*	*1,732*
Volkswagen Truck & Bus	*112*	*119.6*	*51*
MAN Total	1,528	-14.3	1,783
Multicar	1	-50.0	2
PACCAR (DAF)	1,670	-9.9	1,854
Citroen	*3*	–	*3*
Peugeot	*3*	–	*3*
Peugeot Citroen Total	6	–	6
Scania	803	-22.9	1,041
Renault Truck	*514*	*-27.8*	*712*
UD Trucks	*28*	*-24.3*	*37*
Volvo Truck	*1,579*	*-9.6*	*1,747*
Volvo Truck Total	2,121	-15.0	2,496
Other	37	5.7	35
Total Med./Hvy. Trks.	**8,475**	**-14.2**	**9,883**
Buses			
Evobus	–	–	*1*
Mercedes-Benz	*258*	*44.9*	*178*
Setra	*18*	*125.0*	*8*
Daimler Total	276	47.6	187
Ford	6	–	–
GM (Opel)	1	–	–
Irisbus	*46*	*-6.1*	*49*
Iveco	*10*	*400.0*	*2*
Iveco Total	56	9.8	51
King Long	1	–	–
MAN	*34*	*9.7*	*31*
Neoplan	*4*	*-60.0*	*10*
Volkswagen Truck & Bus	*4*	*100.0*	*2*
MAN Total	42	-2.3	43
Peugeot Citroen	3	–	3
Scania	3	–	3
Temsa	45	-4.3	47
Van Hool	78	-60.6	198
VDL	110	-23.6	144
Renault Truck	*13*	*333.3*	*3*
Volvo Truck	*5*	*-16.7*	*6*
Volvo Truck Total	18	100.0	9
Other	62	100.0	31
Total Buses	**701**	**-2.1**	**716**
Total Comm. Vehs.	**63,784**	**-11.4**	**72,027**
Total Vehicles	**550,521**	**-14.5**	**644,238**

CZECH REPUBLIC

ECONOMIC SUMMARY

Common Language	Czech
Currency	Koruna
Steering	Left Hand Drive

2012 Vehicles in Operation

Passenger Cars	4,638,372
Commercial Vehicles	707,784
Total	5,346,156
Persons per Car	2.2
Persons per Vehicle	1.9

2012 GDP (billions of US $)

Nominal GDP	196.1
Real GDP	188.4
GDP % annual growth	-1.2
GDP per capita	17,854

Current Account Balance (billions of US $)

2012:	-5.3
2011:	-6.3

Current Account Balance in % of GDP

2012:	-2.7
2011:	-2.9

Total Number of Fatal Road Accidents (most recent data)

2010:	802

Capital	Prague
Population	10,162,921
Population Density	340.8
Surface Area sq km/miles	77,247 / 29,825

Road Infrastructure

Paved Road Surface %	100.0
Total Road Length in km/miles	128,582 / 79,897
Road Traffic (Vehicles per mile)	66.9

2012 Total Vehicle Percent Market Share by Company

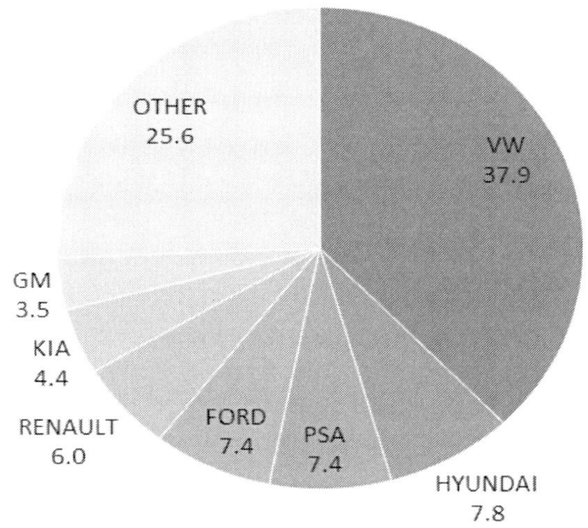

Total Vehicle Production and Sales 2002-2012

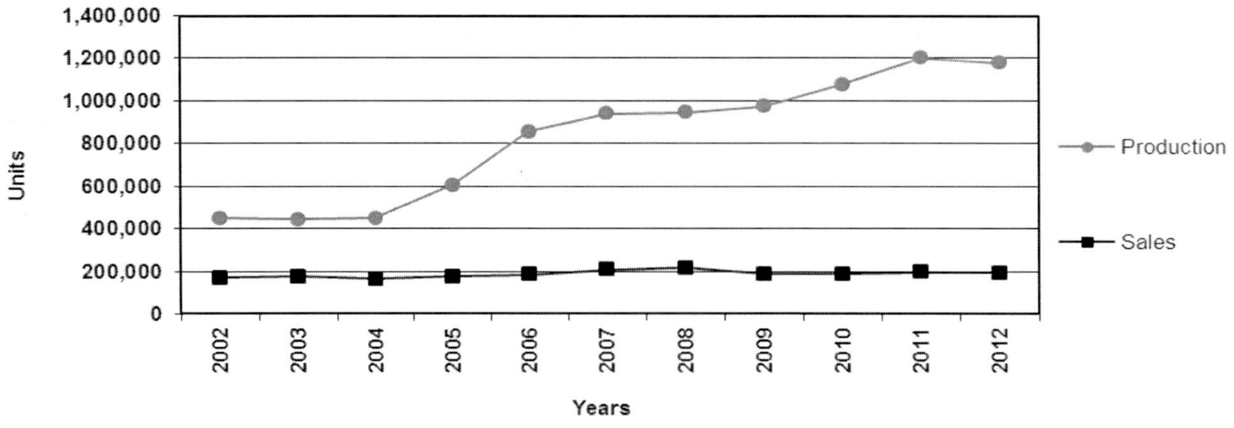

CZECH REPUBLIC

PRODUCTION

Year	Cars	Trucks	Buses	Total
2012	1,171,774	3,947	3,217	1,178,938
2011	1,191,968	4,306	3,560	1,199,834
2010	1,069,518	4,156	2,711	1,076,385
2009	967,760	3,741	3,068	974,569
2008	933,312	9,014	3,496	945,822
2007	925,778	9,392	3,357	938,527
2006	848,922	3,056	2,929	854,907
2005	599,472	3,264	2,194	604,930
2004	442,812	3,322	1,972	448,106
2003	436,297	3,635	1,785	441,717
2002	441,312	3,964	1,811	447,087
2001	456,927	6,788	1,566	465,281
2000	428,224	25,844	1,424	455,492
1999	348,751	26,064	1,269	376,084
1998	368,527	41,200	1,240	410,967
1997	321,077	46,312	1,092	368,481
1996	240,053	30,697	1,276	272,026
1995	193,297	22,173	944	216,414
1994	173,659	5,423	1,005	180,087
1993	219,712	8,518	917	229,147
1992	200,220	15,428	1,030	216,678
1991	172,726	23,139	1,671	197,536
1990	187,773	46,738	3,197	237,708
1989	183,609	55,961	3,201	242,771
1988	159,160	55,771	3,329	218,260
1987	172,355	54,782	3,350	230,487
1986	185,030	56,318	3,410	244,758
1985	180,701	54,270	3,386	238,357
1984	180,471	46,872	3,240	230,583
1983	177,505	43,194	3,150	223,849
1982	173,517	47,270	3,662	224,449

CZECH REPUBLIC

PRODUCTION BY MANUFACTURER

	2012	% Chg.	2011
Cars			
Hyundai	303,035	20.7	251,146
Citroen	*71,630*	*-20.6*	*90,227*
Peugeot	*71,655*	*-20.6*	*90,251*
Toyota	*71,630*	*-20.6*	*90,227*
TPCA Total	214,915	-20.6	270,705
Volkswagen (Skoda)	653,813	-2.4	670,114
Other	11	266.7	3
Total Cars	**1,171,774**	**-1.7**	**1,191,968**
Light Trucks			
Volkswagen (Skoda)	2,493	-17.3	3,013
Total Light Trucks	**2,493**	**-17.3**	**3,013**

	2012	% Chg.	2011
Medium/Heavy Trucks			
Avia	958	62.1	591
Tatra	496	-29.3	702
Total Med./Hvy. Trks.	**1,454**	**12.5**	**1,293**
Buses			
Iveco (Irisbus)	2,760	-7.1	2,970
Other	457	-22.5	590
Total Buses	**3,217**	**-9.6**	**3,560**
Total Comm. Vehs.	**7,164**	**-8.9**	**7,866**
Total Vehicles	**1,178,938**	**-1.7**	**1,199,834**

SALES

Year	Cars	Trucks	Buses	Total
2012	174,009	19,055	731	193,795
2011	173,282	21,231	837	195,350
2010	169,236	17,021	751	187,008
2009	161,659	24,187	775	186,621
2008	143,661	70,567	1,191	215,419
2007	132,542	73,625	949	207,116
2006	123,987	59,418	944	184,349
2005	127,376	47,676	816	175,868
2004	125,768	36,750	958	163,476
2003	149,636	24,763	986	175,385
2002	147,818	20,896	987	169,701
2001	152,171	21,601	979	174,751
2000	148,730	21,295	760	170,785
1999	146,170	20,901	707	167,778
1998	141,155	19,525	629	161,309
1997	171,548	25,146	NA	196,694
1996	153,962	24,534	NA	178,496

CZECH REPUBLIC

SALES BY COMPANY

	2012	% Chg.	2011
Cars			
Audi	3,593	5.2	3,416
AutoVaz (Lada)	93	-9.7	103
BMW	*3,761*	*20.6*	*3,119*
Mini	*297*	*91.6*	*155*
BMW Total	4,058	23.9	3,274
Chrysler	*–*	*–*	*34*
Jeep	*378*	*93.8*	*195*
Chrysler Total	378	65.1	229
Dacia	3,803	-5.6	4,030
Mercedes-Benz	*3,156*	*15.0*	*2,745*
Smart	*71*	*-10.1*	*79*
Daimler Total	3,227	14.3	2,824
Alfa Romeo	*228*	*-8.8*	*250*
Fiat	*2,530*	*15.8*	*2,185*
Lancia	*140*	*59.1*	*88*
Fiat Total	2,898	14.9	2,523
Fisker	1	–	–
Ford	12,523	-11.6	14,170
Chevrolet	*2,811*	*-5.0*	*2,959*
Opel	*3,800*	*-14.1*	*4,423*
GM Total	6,611	-10.4	7,382
Honda	1,659	7.9	1,538
Hyundai	15,091	25.7	12,010
Iveco	–	–	3
Jaguar	*74*	*-5.1*	*78*
Land Rover	*569*	*58.9*	*358*
Jaguar Land Rover Total	643	47.5	436
Kia	8,548	0.4	8,514
Lifan	–	–	3
Mazda	1,170	15.5	1,013
Mitsubishi	1,056	-29.9	1,506
Infiniti	*56*	*-18.8*	*69*
Nissan	*3,115*	*67.4*	*1,861*
Nissan Total	3,171	64.3	1,930
Citroen	*5,644*	*10.2*	*5,122*
Peugeot	*6,629*	*-8.9*	*7,280*
Peugeot Citroen Total	12,273	-1.0	12,402
Porsche	239	17.7	203
Renault	10,388	-15.3	12,269
Saab	2	-91.7	24
Ssangyong	15	-31.8	22
Subaru	815	8.1	754
Suzuki	2,309	-19.7	2,876
Lexus	*160*	*14.3*	*140*
Toyota	*3,894*	*2.5*	*3,800*
Toyota Total	4,054	2.9	3,940
SEAT	*2,665*	*-18.6*	*3,272*
Skoda	*53,353*	*1.8*	*52,408*
Volkswagen	*15,008*	*2.1*	*14,693*
Volkswagen Total	71,026	0.9	70,373
Volvo	1,458	-9.7	1,615
Other	2,907	-25.5	3,900
Total Cars	**174,009**	**0.4**	**173,282**
Light Trucks			
AutoVaz (Lada)	–	–	10
Chrysler (Jeep)	1	-75.0	4
Dacia	305	-17.1	368
Mercedes-Benz	*798*	*-10.0*	*887*
Mitsubishi Fuso	*28*	*12.0*	*25*
Daimler Total	826	-9.4	912
Fiat	1,624	13.6	1,429
Ford	1,758	-28.7	2,467
GM (Opel)	155	-25.5	208
GoNow	13	–	–
Hyundai	61	-46.5	114
Isuzu	35	6.1	33
Iveco	510	1.6	502
Jaguar Land Rover	23	-14.8	27
Kia	–	–	3
Mazda	1	-80.0	5
Mitsubishi	79	-36.8	125
Nissan	115	-29.0	162
Citroen	*996*	*-39.9*	*1,658*
Peugeot	*1,152*	*-18.4*	*1,411*
Peugeot Citroen Total	2,148	-30.0	3,069
Renault	1,192	-0.7	1,200
Suzuki	1	–	–
Toyota	232	11.0	209
Skoda	*275*	*-12.1*	*313*
Volkswagen	*2,062*	*31.4*	*1,569*
Volkswagen Total	2,337	24.2	1,882
Volvo Truck (Renault Truck)	74	-27.5	102

CZECH REPUBLIC

SALES BY COMPANY

	2012	% Chg.	2011
Other	331	-24.4	438
Total Light Trucks	**11,821**	**-10.9**	**13,269**
Medium/Heavy Trucks			
Ashok Leyland (Avia)	159	5.3	151
Chrysler (Dodge)	1	-50.0	2
Mercedes-Benz	*1,512*	*-8.9*	*1,660*
Mitsubishi Fuso	*59*	*25.5*	*47*
Daimler Total	1,571	-8.0	1,707
Fiat	2	–	–
Ford	3	–	3
GM (Opel)	–	–	2
Hyundai	–	–	1
Isuzu	6	-33.3	9
Iveco	695	-10.3	775
MAN	*1,131*	*-19.8*	*1,410*
Volkswagen Truck & Bus	*9*	*–*	*9*
MAN Total	1,140	-19.7	1,419
MAZ	1	–	–
Multicar	3	-40.0	5
Nissan	2	-71.4	7
PACCAR (DAF)	1,348	0.8	1,337
Citroen	*1*	*–*	*–*
Peugeot	*–*	*–*	*1*
Peugeot Citroen Total	1	–	1
Renault	4	100.0	2
Scania	755	-3.3	781
Tatra	85	-7.6	92
Renault Truck	*405*	*-32.3*	*598*
Volvo Truck	*1,037*	*-1.7*	*1,055*
Volvo Truck Total	1,442	-12.8	1,653
Other	16	6.7	15
Total Med./Hvy. Trks.	**7,234**	**-9.1**	**7,962**

	2012	% Chg.	2011
Buses			
Mercedes-Benz	*88*	*18.9*	*74*
Setra	*23*	*-14.8*	*27*
Daimler Total	111	9.9	101
Fiat	1	–	1
Ford	6	200.0	2
Isuzu	3	-40.0	5
Irisbus	*178*	*-2.7*	*183*
Iveco	*27*	*92.9*	*14*
Iveco Total	205	4.1	197
MAN	*32*	*23.1*	*26*
Neoplan	*4*	*-50.0*	*8*
MAN Total	36	5.9	34
Citroen	*1*	*–*	*–*
Peugeot	*1*	*–*	*–*
Peugeot Citroen Total	2	–	–
Scania	–	–	8
Bova	*–*	*–*	*4*
VDL	*9*	*800.0*	*1*
VDL Total	9	80.0	5
Skoda	*1*	*–*	*–*
Volkswagen	*10*	*–*	*–*
Volkswagen Total	11	–	–
Renault Truck	*8*	*33.3*	*6*
Volvo Truck	*22*	*-29.0*	*31*
Volvo Truck Total	30	-18.9	37
Other	317	-29.1	447
Total Buses	**731**	**-12.7**	**837**
Total Comm. Vehs.	**19,786**	**-10.3**	**22,068**
Total Vehicles	**193,795**	**-0.8**	**195,350**

DENMARK

ECONOMIC SUMMARY

Common Language	Danish
Currency	Krone
Steering	Left Hand Drive

2012 Vehicles in Operation

Passenger Cars	2,237,122
Commercial Vehicles	473,042
Total	2,710,164
Persons per Car	2.5
Persons per Vehicle	2.1

2012 GDP (billions of US $)

Nominal GDP	313.6
Real GDP	2,049.8
GDP % annual growth	-0.6
GDP per capita	367,315

Current Account Balance (billions of US $)

2012:	16.5
2011:	18.8

Current Account Balance in % of GDP

2012:	5.3
2011:	5.6

Total Number of Fatal Road Accidents (most recent data)

2010:	255

Capital	Copenhagen
Population	5,556,452
Population Density	339.2
Surface Area sq km/miles	42,434 / 16,383

Road Infrastructure

Paved Road Surface %	100.0
Total Road Length in km/miles	73,197 / 45,482
Road Traffic (Vehicles per mile)	59.6

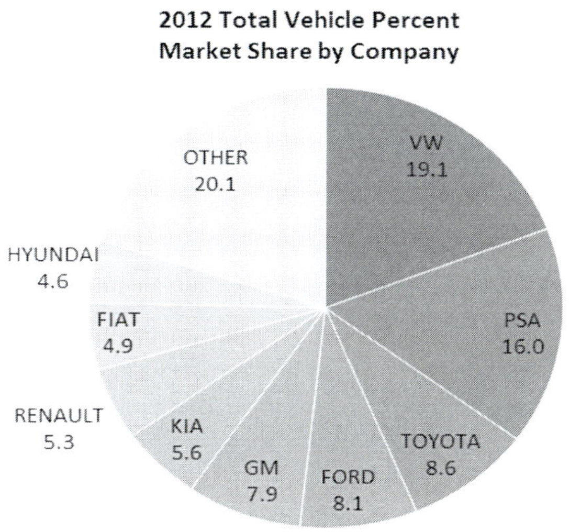

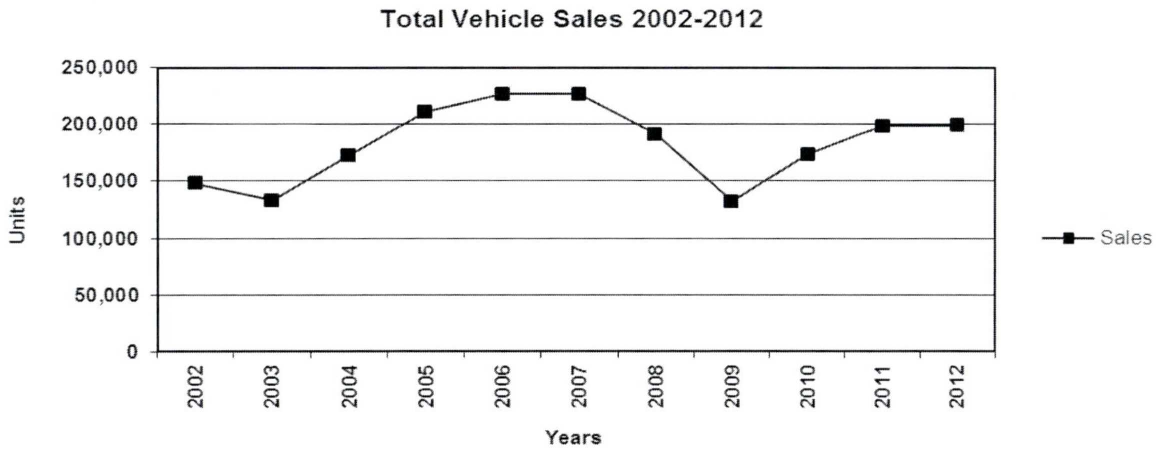

DENMARK

SALES

Year	Cars	Trucks	Buses	Total
2012	170,763	27,854	529	199,146
2011	169,974	27,956	515	198,445
2010	153,562	18,948	727	173,237
2009	112,436	18,452	864	131,752
2008	150,185	40,180	779	191,144
2007	162,668	62,931	561	226,160
2006	156,930	68,618	650	226,198
2005	148,819	61,639	571	211,029
2004	121,472	50,615	566	172,653
2003	96,067	36,109	597	132,773
2002	111,598	35,589	698	147,885
2001	96,184	35,643	466	132,293
2000	112,692	36,267	646	149,605
1999	143,511	37,162	653	181,326
1998	162,406	34,372	853	197,631
1997	152,643	35,336	568	188,547
1996	142,299	31,393	755	174,447
1995	135,692	30,252	642	166,586
1994	139,621	23,888	642	164,151
1993	83,758	22,124	464	106,346
1992	85,385	22,388	453	108,226
1991	84,146	23,075	370	107,591
1990	80,913	22,590	441	103,944
1989	78,435	21,133	471	100,039
1988	88,603	22,277	489	111,369
1987	124,097	37,240	574	161,911
1986	169,386	44,082	658	214,126
1985	157,474	37,891	567	195,932
1984	134,345	31,634	533	166,512
1983	116,238	21,716	507	138,461
1982	85,405	17,695	643	103,743

DENMARK

SALES BY COMPANY

	2012	% Chg.	2011
Cars			
Aston Martin	1	–	1
Audi	5,211	3.1	5,056
AutoVaz (Lada)	–	–	1
Bentley	1	-80.0	5
BMW	*2,909*	*-8.8*	*3,190*
Mini	*341*	*-30.7*	*492*
BMW Total	3,250	-11.7	3,682
Chrysler	*3*	*-70.0*	*10*
Dodge	–	–	*5*
Jeep	*3*	*200.0*	*1*
Chrysler Total	6	-62.5	16
Mercedes-Benz	*3,638*	*9.8*	*3,312*
Smart	*191*	*-45.9*	*353*
Daimler Total	3,829	4.5	3,665
Ferrari	13	44.4	9
Alfa Romeo	*472*	*-51.5*	*974*
Fiat	*7,351*	*14.0*	*6,451*
Lancia	–	–	*1*
Fiat Total	7,823	5.3	7,426
Ford	13,123	-16.7	15,759
Cadillac	*3*	*50.0*	*2*
Chevrolet	*6,108*	*-29.4*	*8,657*
GMC	*1*	–	*1*
Hummer	*1*	–	–
Opel	*7,363*	*-25.3*	*9,856*
GM Total	13,476	-27.2	18,516
Honda	2,112	21.4	1,739
Hyundai	8,919	-6.6	9,552
Iveco	1	–	1
Jaguar	*36*	*5.9*	*34*
Land Rover	*46*	*-35.2*	*71*
Jaguar Land Rover Total	82	-21.9	105
Kia	10,979	76.3	6,228
Maserati	2	-33.3	3
Mazda	2,763	-33.4	4,147
MG	–	–	*1*
Rover	–	–	*1*
MG Motors Total	–	–	2
Mitsubishi	438	-74.7	1,729
Morgan	2	–	2
Infiniti	*1*	–	–
Nissan	*4,469*	*18.1*	*3,783*
Nissan Total	4,470	18.2	3,783
Citroen	*13,055*	*-1.6*	*13,268*
Peugeot	*15,237*	*-8.5*	*16,658*
Peugeot Citroen Total	28,292	-5.5	29,926
Porsche	35	-7.9	38
Proton (Lotus)	1	–	–
Renault	8,695	0.3	8,667
Saab	19	-90.4	198
Ssangyong	–	–	3
Subaru	51	-37.8	82
Suzuki	6,420	-23.6	8,400
Tesla	12	50.0	8
Lexus	*39*	*-40.9*	*66*
Toyota	*15,915*	*23.5*	*12,883*
Toyota Total	15,954	23.2	12,949
SEAT	*3,648*	*77.8*	*2,052*
Skoda	*9,041*	*30.8*	*6,913*
Volkswagen	*20,273*	*19.2*	*17,003*
Volkswagen Total	32,962	26.9	25,968
Volvo	1,646	-23.0	2,138
Wiesmann	1	–	–
Other	174	2.4	170
Total Cars	**170,763**	**0.5**	**169,974**
Light Trucks			
Audi	218	69.0	129
BMW	*58*	*-4.9*	*61*
Mini	*21*	*5.0*	*20*
BMW Total	79	-2.5	81
Chrysler	–	–	*1*
Dodge	*1*	*-50.0*	*2*
Jeep	–	–	*2*
Chrysler Total	1	-80.0	5
Mercedes-Benz	*2,467*	*10.3*	*2,236*
Mitsubishi Fuso	*9*	*200.0*	*3*
Daimler Total	2,476	10.6	2,239
Alfa Romeo	*15*	*-16.7*	*18*
Fiat	*1,915*	*4.7*	*1,829*
Fiat Total	1,930	4.5	1,847
Ford	2,912	-0.7	2,933
GAZ (LDV)	–	–	4
Chevrolet	*22*	–	–
Opel	*2,148*	*-11.9*	*2,437*

DENMARK

SALES BY COMPANY

	2012	% Chg.	2011
GM Total	2,170	-11.0	2,437
Honda	17	30.8	13
Hyundai	185	-17.0	223
Iveco	823	8.9	756
Jaguar Land Rover	70	40.0	50
Kia	89	14.1	78
Mazda	45	-44.4	81
Mitsubishi	54	-47.1	102
Nissan	1,108	33.0	833
Citroen	*1,521*	*10.5*	*1,376*
Peugeot	*1,995*	*20.0*	*1,663*
Peugeot Citroen Total	3,516	15.7	3,039
Porsche	1	-66.7	3
Renault	1,829	16.3	1,572
Subaru	13	-27.8	18
Suzuki	140	-42.1	242
Lexus	*1*	*-80.0*	*5*
Toyota	*1,166*	*-45.5*	*2,138*
Toyota Total	1,167	-45.5	2,143
SEAT	*247*	*-27.8*	*342*
Skoda	*52*	*-33.3*	*78*
Volkswagen	*4,814*	*-1.5*	*4,889*
Volkswagen Total	5,113	-3.7	5,309
Volvo	112	-12.5	128
Other	16	–	16
Total Light Trucks	**24,084**	**-0.8**	**24,281**
Medium/Heavy Trucks			
Mercedes-Benz	*583*	*22.2*	*477*
Mitsubishi Fuso	*18*	*50.0*	*12*
Daimler Total	601	22.9	489
Dennis Eagle	–	–	9
Fiat	5	150.0	2
Ford	2	-50.0	4
GM (Opel)	1	–	–
Isuzu	1	–	–
Iveco	299	0.7	297
MAN	*662*	*-20.5*	*833*
Volkswagen Truck & Bus	*14*	*7.7*	*13*
MAN Total	676	-20.1	846
Modec	–	–	1
PACCAR (DAF)	253	0.4	252
Peugeot Citroen	2	–	–
Scania	779	3.0	756
Toyota	4	–	–
Renault Truck	*144*	*41.2*	*102*
Volvo Truck	*968*	*11.0*	*872*
Volvo Truck Total	1,112	14.2	974
Other	35	-22.2	45
Total Med./Hvy. Trks.	**3,770**	**2.6**	**3,675**
Buses			
Mercedes-Benz	*155*	*-9.9*	*172*
Setra	*14*	*-22.2*	*18*
Daimler Total	169	-11.1	190
Fiat	43	95.5	22
Ford	7	–	–
GM (Opel)	1	–	1
Irisbus	*70*	*-2.8*	*72*
Iveco	*6*	*100.0*	*3*
Iveco Total	76	1.3	75
MAN	*7*	*-80.6*	*36*
Neoplan	*8*	*-11.1*	*9*
Volkswagen Truck & Bus	*1*	–	*1*
MAN Total	16	-65.2	46
Optare	–	–	1
Scania	89	39.1	64
Temsa	2	100.0	1
Bova	–	–	*1*
VDL	*39*	*56.0*	*25*
VDL Total	39	50.0	26
Renault Truck	*4*	–	–
Volvo Truck	*63*	*6.8*	*59*
Volvo Truck Total	67	13.6	59
Other	20	-33.3	30
Total Buses	**529**	**2.7**	**515**
Total Comm. Vehs.	**28,383**	**-0.3**	**28,471**
Total Vehicles	**199,146**	**0.4**	**198,445**

FINLAND

ECONOMIC SUMMARY

Common Language	Finnish/Swedish
Currency	Euro
Steering	Left Hand Drive

2012 Vehicles in Operation

Passenger Cars	3,036,618
Commercial Vehicles	518,024
Total	3,554,642
Persons per Car	1.7
Persons per Vehicle	1.5

2012 GDP (billions of US $)

Nominal GDP	250.1
Real GDP	212.7
GDP % annual growth	-0.2
GDP per capita	39,194

Current Account Balance (billions of US $)

2012:	-4.3
2011:	-4.2

Current Account Balance in % of GDP

2012:	-1.7
2011:	-1.6

Total Number of Fatal Road Accidents (most recent data)

2011:	292

Capital	Helsinki
Population	5,266,114
Population Density	44.9
Surface Area sq km/miles	303,815 / 117,303

Road Infrastructure

Paved Road Surface %	65.2
Total Road Length in km/miles	78,162 / 48,568
Road Traffic (Vehicles per mile)	73.2

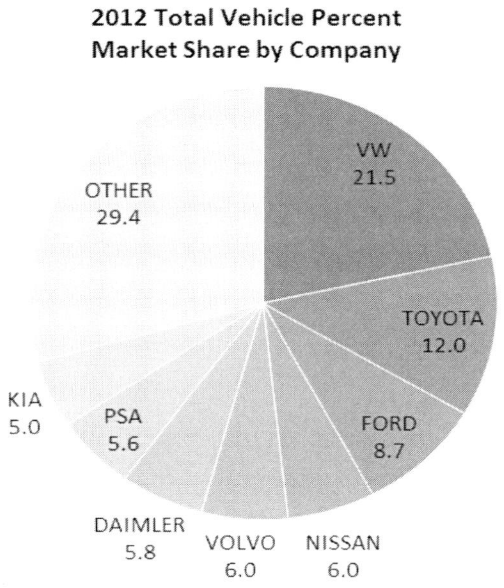

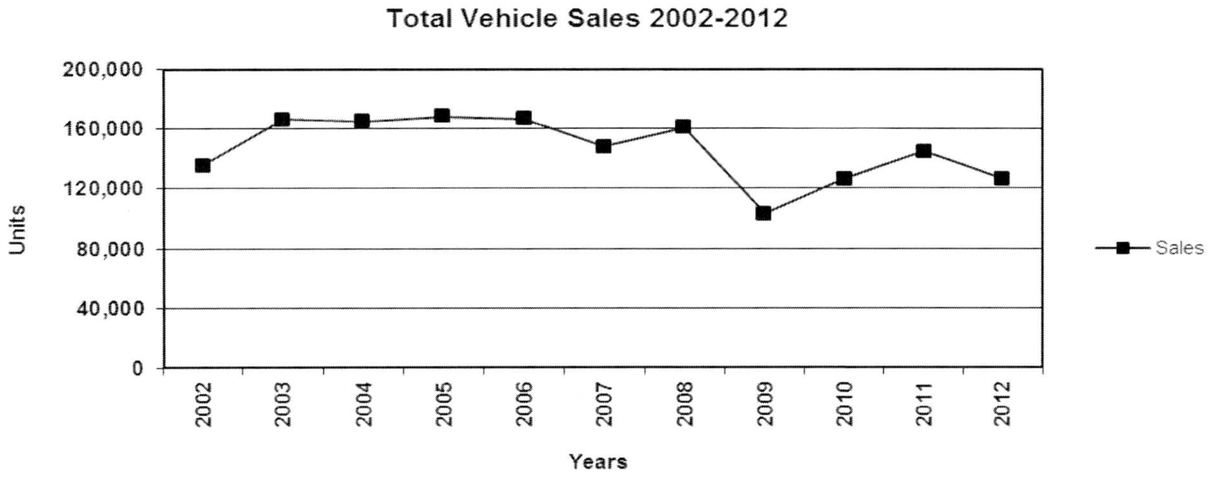

FINLAND

SALES

Year	Cars	Trucks	Buses	Total
2012	111,249	14,721	533	126,503
2011	126,116	17,904	398	144,418
2010	111,956	13,915	513	126,384
2009	90,574	11,850	592	103,016
2008	139,647	20,733	596	160,976
2007	125,608	21,738	496	147,842
2006	145,700	20,525	448	166,673
2005	148,161	19,549	411	168,121
2004	142,642	21,589	510	164,741
2003	147,405	18,340	474	166,219
2002	117,034	17,856	477	135,367
2001	109,428	17,692	389	127,509
2000	134,603	18,120	542	153,265
1999	136,324	19,368	541	156,233
1998	125,751	18,992	517	145,260
1997	104,507	16,000	466	120,973
1996	95,830	12,615	454	108,899
1995	79,890	10,256	315	90,461
1994	67,201	5,479	203	72,883
1993	55,836	5,031	191	61,058
1992	68,547	9,306	200	78,053
1991	92,485	14,824	343	107,652
1990	139,041	29,172	452	168,665
1989	177,610	35,384	557	213,551
1988	174,479	25,854	570	200,903
1987	152,327	21,538	550	174,415
1986	144,021	19,038	490	163,549
1985	138,976	17,607	471	157,054
1984	127,655	17,682	482	145,819
1983	120,174	17,556	596	138,326
1982	128,761	17,674	535	146,970

FINLAND

SALES BY COMPANY

	2012	% Chg	2011
Cars			
Aston Martin	–	–	1
Audi	6,178	-0.7	6,219
BMW	*4,198*	*4.9*	*4,001*
Mini	*269*	*3.9*	*259*
BMW Total	4,467	4.9	4,260
Chrysler	–	–	*14*
Dodge	–	–	*35*
Jeep	*251*	*182.0*	*89*
Chrysler Total	251	81.9	138
Dacia	613	-26.7	836
Mercedes-Benz	*4,913*	*-1.2*	*4,975*
Smart	*88*	*-12.0*	*100*
Daimler Total	5,001	-1.5	5,075
Alfa Romeo	*311*	*-36.7*	*491*
Fiat	*684*	*-37.0*	*1,086*
Fiat Total	995	-36.9	1,577
Fisker	3	–	–
Ford	*8,503*	*-20.4*	*10,688*
Lincoln	*1*	–	–
Ford Total	8,504	-20.4	10,688
Chevrolet	*1,127*	*-20.9*	*1,425*
Opel	*3,824*	*-27.1*	*5,247*
GM Total	4,951	-25.8	6,672
Honda	3,140	-4.6	3,292
Hyundai	4,390	-1.7	4,466
Jaguar	*97*	*-14.2*	*113*
Land Rover	*328*	*98.8*	*165*
Jaguar Land Rover Total	425	52.9	278
Kia	6,361	3.8	6,129
Mazda	1,345	-36.5	2,119
Mitsubishi	1,291	-13.8	1,498
Nissan	7,196	-3.8	7,477
Citroen	*2,982*	*-16.4*	*3,567*
Peugeot	*3,326*	*-19.8*	*4,149*
Peugeot Citroen Total	6,308	-18.2	7,716
Porsche	84	-16.0	100
Renault	1,267	-32.3	1,872
Saab	39	-86.6	290

	2012	% Chg.	2011
Subaru	1,025	-19.7	1,276
Suzuki	814	-15.9	968
Lexus	*203*	*-6.5*	*217*
Toyota	*13,388*	*0.5*	*13,325*
Toyota Total	13,591	0.4	13,542
SEAT	*1,449*	*-23.9*	*1,903*
Skoda	*8,686*	*-15.0*	*10,217*
Volkswagen	*13,914*	*-18.9*	*17,156*
Volkswagen Total	24,049	-17.9	29,276
Volvo	7,553	-13.1	8,691
Other	1,408	-15.2	1,660
Total Cars	**111,249**	**-11.8**	**126,116**
Light Trucks			
Audi	14	133.3	6
BMW	8	14.3	7
Chrysler	*1*	–	–
Dodge	*5*	–	*5*
Jeep	–	–	*1*
Chrysler Total	6	–	6
Dacia	121	-30.1	173
Mercedes-Benz	*1,553*	*-20.5*	*1,953*
Mitsubishi Fuso	*4*	*33.3*	*3*
Daimler Total	1,557	-20.4	1,956
Alfa Romeo	–	–	*1*
Fiat	*434*	*-25.7*	*584*
Fiat Total	434	-25.8	585
Ford	2,480	12.3	2,208
Chevrolet	*17*	*13.3*	*15*
GMC	*1*	–	–
Opel	*559*	*-16.1*	*666*
GM Total	577	-15.3	681
Honda	4	-55.6	9
Hyundai	9	-71.9	32
Isuzu	138	-1.4	140
Iveco	86	21.1	71
Jaguar Land Rover	52	100.0	26
Kia	17	21.4	14
Mazda	1	-88.9	9

FINLAND

SALES BY COMPANY

	2012	% Chg	2011
Mitsubishi	86	24.6	69
Nissan	435	-41.2	740
Citroen	*408*	*-11.9*	*463*
Peugeot	*339*	*-0.6*	*341*
Peugeot Citroen Total	747	-7.1	804
Renault	476	-27.5	657
Subaru	6	-25.0	8
Suzuki	–	–	1
Toyota	1,539	-53.2	3,290
SEAT	*1*	*-50.0*	*2*
Skoda	*36*	*-10.0*	*40*
Volkswagen	*3,143*	*-12.5*	*3,592*
Volkswagen Total	3,180	-12.5	3,634
Volvo	20	-39.4	33
Other	5	–	5
Total Light Trucks	**11,998**	**-20.9**	**15,164**
Medium/Heavy Trucks			
Mercedes-Benz	*582*	*3.9*	*560*
Mitsubishi Fuso	*37*	*105.6*	*18*
Daimler Total	619	7.1	578
Dennis Eagle	1	–	–
Fiat	1	–	–
Iveco	77	-26.7	105
MAN	150	-5.7	159
Mitsubishi	–	–	1
PACCAR (DAF)	102	39.7	73
Scania	738	-5.0	777
Sisu	92	24.3	74

	2012	% Chg.	2011
Renault Truck	*73*	*21.7*	*60*
UD Trucks	*–*	*–*	*1*
Volvo Truck	*865*	*-3.8*	*899*
Volvo Truck Total	938	-2.3	960
Other	5	-61.5	13
Total Med./Hvy. Trks.	**2,723**	**-0.6**	**2,740**
Buses			
Mercedes-Benz	*165*	*1.9*	*162*
Setra	*–*	*–*	*3*
Daimler Total	165	–	165
Fiat	8	-33.3	12
Ford	18	260.0	5
Irisbus	*3*	*-93.3*	*45*
Iveco	*1*	*-88.9*	*9*
Iveco Total	4	-92.6	54
King Long (Golden Dragon)	–	–	4
MAN (VW Truck & Bus)	17	88.9	9
Peugeot	–	–	3
Scania	61	-11.6	69
Solaris	2	–	–
VDL	113	–	5
Volvo Truck	125	145.1	51
Other	20	-4.8	21
Total Buses	**533**	**33.9**	**398**
Total Comm. Vehs.	**15,254**	**-16.7**	**18,302**
Total Vehicles	**126,503**	**-12.4**	**144,418**

FRANCE

ECONOMIC SUMMARY

Common Language	French
Currency	Euro
Steering	Left Hand Drive

2012 Vehicles in Operation

Passenger Cars	31,600,000
Commercial Vehicles	6,538,000
Total	38,138,000
Persons per Car	2.1
Persons per Vehicle	1.7

2012 GDP (billions of US $)

Nominal GDP	2,608.7
Real GDP	2,281.5
GDP % annual growth	0.0
GDP per capita	37,558

Current Account Balance (billions of US $)

2012:	-62.9
2011:	-54.2

Current Account Balance in % of GDP

2012:	-2.4
2011:	-2.0

Total Number of Fatal Road Accidents (most recent data)

2010:	3,992

Capital	Paris
Population	65,951,611
Population Density	266.7
Surface Area sq km/miles	640,427 / 247,269

Road Infrastructure

Paved Road Surface %	100.0
Total Road Length in km/miles	1,027,183 / 638,261
Road Traffic (Vehicles per mile)	59.8

2012 Total Vehicle Percent Market Share by Company

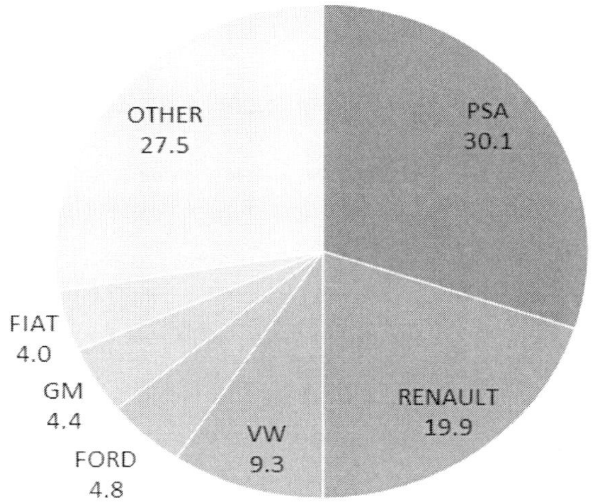

Total Vehicle Production and Sales 2002-2012

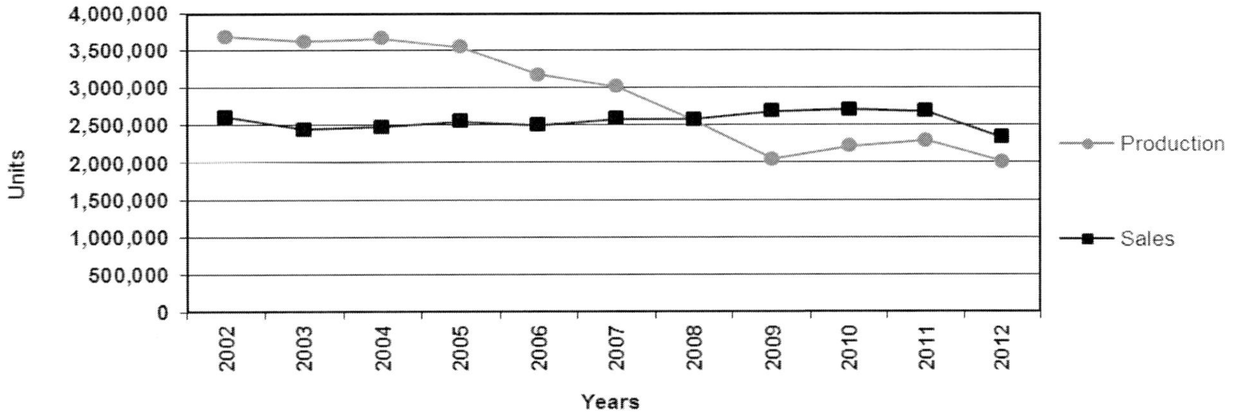

FRANCE

PRODUCTION

Year	Cars	Trucks	Buses	Total
2012	1,682,814	324,701	2,988	2,010,503
2011	1,931,030	360,457	3,402	2,294,889
2010	1,914,029	301,299	3,430	2,218,758
2009	1,815,011	224,223	4,007	2,043,241
2008	2,144,744	418,667	4,149	2,567,560
2007	2,550,869	460,649	4,336	3,015,854
2006	2,728,196	442,010	4,054	3,174,260
2005	3,112,956	432,360	3,687	3,549,003
2004	3,227,416	435,582	2,992	3,665,990
2003	3,220,329	397,334	2,393	3,620,056
2002	3,282,858	406,483	2,589	3,691,930
2001	3,181,549	443,297	3,563	3,628,409
2000	2,879,810	465,078	3,473	3,348,361
1999	2,784,469	382,548	2,700	3,169,717
1998	2,603,453	338,437	2,856	2,944,746
1997	2,258,782	322,360	2,777	2,583,919
1996	2,087,505	440,355	2,610	2,530,470
1995	3,050,929	421,140	2,636	3,474,705
1994	3,175,213	381,028	2,197	3,558,438
1993	2,836,280	317,007	2,430	3,155,717
1992	3,329,490	435,732	2,578	3,767,800
1991	3,187,634	420,544	2,457	3,610,635
1990	3,294,815	471,641	2,537	3,768,993
1989	3,409,017	508,080	2,679	3,919,776
1988	3,223,987	472,044	2,434	3,698,465
1987	3,051,830	438,824	2,556	3,493,210
1986	2,773,094	418,699	2,822	3,194,615
1985	2,632,366	380,753	2,987	3,016,106
1984	2,713,289	346,230	2,633	3,062,152
1983	2,960,823	372,473	2,566	3,335,862
1982	2,777,125	369,072	2,610	3,148,807

FRANCE

PRODUCTION BY MANUFACTURER

	2012	% Chg.	2011
Cars			
Citroen	*443,475*	*-14.2*	*516,994*
Peugeot	*573,961*	*-19.9*	*716,461*
Peugeot Citroen Total	1,017,436	-17.5	1,233,455
Renault	359,532	-19.2	444,862
Smart	104,154	0.6	103,560
Toyota	201,692	35.2	149,153
Total Cars	**1,682,814**	**-12.9**	**1,931,030**
Light Trucks			
Fiat	15,144	-23.5	19,786
Iveco	250	-16.4	299
Citroen	*51,134*	*5.3*	*48,540*
Peugeot	*45,634*	*8.4*	*42,115*
Peugeot Citroen Total	96,768	6.7	90,655
Mercedes-Benz	*14,129*	–	–
Renault	*158,910*	*-21.1*	*201,457*
Renault Total	173,039	-14.1	201,457
Total Light Trucks	**285,201**	**-8.6**	**312,197**
Medium/Heavy Trucks			
Renault Truck	30,500	-16.8	36,641
Scania	9,000	-22.5	11,619
Total Med./Hvy. Trks.	**39,500**	**-18.2**	**48,260**
Buses			
Evobus	588	4.4	563
Heuliez	400	-14.9	470
Iveco (Irisbus)	2,000	-15.6	2,369
Total Buses	**2,988**	**-12.2**	**3,402**
Total Comm. Vehs.	**327,689**	**-9.9**	**363,859**
Total Vehicles	**2,010,503**	**-12.4**	**2,294,889**

FRANCE

SALES

Year	Cars	Trucks	Buses	Total
2012	1,898,760	427,427	5,544	2,331,731
2011	2,204,229	476,618	6,206	2,687,053
2010	2,251,669	451,833	5,382	2,708,884
2009	2,268,672	409,517	6,664	2,684,853
2008	2,050,341	517,691	5,655	2,573,687
2007	2,064,543	514,001	5,491	2,584,035
2006	2,000,549	493,154	5,243	2,498,946
2005	2,067,789	475,349	4,773	2,547,911
2004	2,013,709	455,638	4,213	2,473,560
2003	2,009,246	427,084	4,362	2,440,692
2002	2,145,071	456,336	4,601	2,606,008
2001	2,254,732	491,572	4,691	2,750,995
2000	2,133,884	472,884	4,320	2,611,088
1999	2,148,423	429,070	4,435	2,581,928
1998	1,943,553	394,498	4,085	2,342,136
1997	1,713,030	352,068	3,124	2,068,222
1996	2,132,091	374,045	3,991	2,510,127
1995	1,930,504	353,795	3,981	2,288,280
1994	1,972,918	322,415	3,579	2,298,912
1993	1,721,222	282,082	3,932	2,007,236
1992	2,105,700	356,685	3,954	2,466,339
1991	2,031,274	388,984	4,056	2,424,314
1990	2,309,130	442,773	4,210	2,756,113
1989	2,274,307	444,334	4,075	2,722,716
1988	2,217,149	425,400	4,077	2,646,626
1987	2,105,180	413,935	4,131	2,523,246
1986	1,911,521	382,961	7,612	2,302,094
1985	1,766,328	333,895	8,269	2,108,492
1984	1,757,673	307,990	8,149	2,073,812
1983	2,017,617	337,387	8,927	2,363,931
1982	2,056,490	352,954	9,714	2,419,158

FRANCE

SALES BY COMPANY

	2012	% Chg.	2011
Cars			
Aston Martin	104	-44.4	187
Audi	62,054	5.2	58,970
AutoVaz (Lada)	248	-38.8	405
Bentley	98	38.0	71
BMW	*48,045*	*3.8*	*46,305*
Mini	*21,483*	*-1.0*	*21,702*
BMW Total	69,528	2.2	68,007
Chrysler	*8*	*-95.7*	*184*
Dodge	*7*	*-95.2*	*147*
Jeep	*3,228*	*22.4*	*2,637*
Chrysler Total	3,243	9.3	2,968
Dacia	80,790	-9.2	88,980
Daihatsu	352	62.2	217
Mercedes-Benz	*47,567*	*9.2*	*43,545*
Smart	*5,441*	*-20.1*	*6,810*
Daimler Total	53,008	5.3	50,355
Ferrari	291	-4.6	305
Alfa Romeo	*10,323*	*-36.4*	*16,232*
Fiat	*43,554*	*-24.0*	*57,326*
Lancia	*5,248*	*31.2*	*4,000*
Fiat Total	59,125	-23.8	77,558
Ford	92,468	-19.8	115,357
Cadillac	*21*	*-43.2*	*37*
Chevrolet	*24,739*	*4.3*	*23,708*
Opel	*71,666*	*-23.8*	*94,102*
GM Total	96,426	-18.2	117,847
Honda	8,406	-4.4	8,793
Hyundai	28,733	42.2	20,204
Jaguar	*897*	*-10.4*	*1,001*
Land Rover	*7,770*	*80.0*	*4,317*
Jaguar Land Rover Total	8,667	63.0	5,318
Kia	33,018	18.1	27,961
Lamborghini	39	56.0	25
Maserati	98	-8.4	107
Mazda	5,107	-21.5	6,509
Mitsubishi	3,639	-17.0	4,386
Morgan	87	-28.1	121
Infiniti	*441*	*-0.9*	*445*
Nissan	*69,692*	*-2.9*	*71,767*
Nissan Total	70,133	-2.9	72,212
Citroen	*266,430*	*-17.5*	*323,076*
Peugeot	*305,440*	*-17.4*	*369,761*
Peugeot Citroen Total	571,870	-17.5	692,837
Porsche	3,336	22.0	2,734
Proton (Lotus)	74	-53.2	158
Renault	343,345	-24.7	455,705
Rolls-Royce	22	10.0	20
Saab	40	-89.4	377
Ssangyong	290	-48.2	560
Subaru	971	16.8	831
Suzuki	16,026	-16.7	19,233
Lexus	*2,456*	*-14.5*	*2,872*
Toyota	*68,007*	*1.0*	*67,320*
Toyota Total	70,463	0.4	70,192
SEAT	*24,180*	*-27.3*	*33,268*
Skoda	*22,464*	*6.0*	*21,185*
Volkswagen	*154,434*	*-5.6*	*163,584*
Volkswagen Total	201,078	-7.8	218,037
Volvo	13,396	-11.8	15,192
Other	2,187	46.8	1,490
Total Cars	**1,898,760**	**-13.9**	**2,204,229**
Light Trucks			
Audi	791	-23.6	1,035
AutoVaz (Lada)	166	-9.3	183
BMW	*78*	*-58.7*	*189*
Mini	*39*	–	–
BMW Total	117	-38.1	189
Chrysler	–	–	*12*
Dodge	*22*	*22.2*	*18*
Jeep	*674*	*35.6*	*497*
Chrysler Total	696	32.1	527
Dacia	3,732	-29.6	5,298
Daihatsu	2	–	–
Mercedes-Benz	*18,275*	*-9.0*	*20,073*
Mitsubishi Fuso	*268*	–	–
Daimler Total	18,543	-7.6	20,073
Alfa Romeo	*113*	*-33.1*	*169*
Fiat	*34,036*	*-8.4*	*37,152*
Lancia	*48*	*336.4*	*11*
Fiat Total	34,197	-8.4	37,332
Ford	18,478	-9.7	20,473
GAZ (LDV)	1	-50.0	2
Chevrolet	*30*	*275.0*	*8*
Hummer	*4*	*-85.2*	*27*
Opel	*7,257*	*-4.0*	*7,560*
GM Total	7,291	-4.0	7,595
GoNow	4	–	–
Honda	23	-30.3	33

FRANCE

SALES BY COMPANY

	2012	% Chg.	2011
Hyundai	276	51.6	182
Isuzu	1,788	-6.1	1,904
Iveco	11,385	-12.1	12,954
Jaguar Land Rover	1,478	-0.7	1,489
Kia	277	163.8	105
Mahindra & Mahindra	1	-88.9	9
Mazda	160	-62.3	424
Mitsubishi	1,716	-38.2	2,776
Nissan	9,076	-5.6	9,616
Citroen	*65,346*	*-13.9*	*75,876*
Peugeot	*63,671*	*-11.7*	*72,071*
Peugeot Citroen Total	129,017	-12.8	147,947
Piaggio	341	-14.3	398
Porsche	6	–	–
Renault	121,226	-10.2	134,967
Santana	–	–	3
Ssangyong	23	15.0	20
Subaru	14	–	–
Suzuki	324	-6.6	347
Toyota	4,505	9.5	4,115
SEAT	*384*	*-45.0*	*698*
Skoda	*251*	*-17.2*	*303*
Volkswagen	*14,815*	*-0.5*	*14,895*
Volkswagen Total	15,450	-2.8	15,896
Volvo	159	-31.8	233
Volvo Truck (Renault Truck)	2,221	-7.2	2,394
Other	565	-23.2	736
Total Light Trucks	**384,049**	**-10.5**	**429,255**
Medium/Heavy Trucks			
Mercedes-Benz	*7,100*	*0.2*	*7,087*
Mitsubishi Fuso	*26*	*-75.9*	*108*
Daimler Total	7,126	-1.0	7,195
Dennis Eagle	17	-22.7	22
Ford	1	–	–
Isuzu	83	-6.7	89
Iveco	4,502	-9.8	4,991
MAN	*4,540*	*-4.7*	*4,765*
Volkswagen Truck & Bus	–	–	*2*
MAN Total	4,540	-4.8	4,767
Mitsubishi	20	–	–
DAF	*5,545*	*-11.1*	*6,240*
Kenworth	*4*	*-87.5*	*32*
PACCAR Total	5,549	-11.5	6,272

	2012	% Chg	2011
Peugeot Citroen	2	–	–
Scania	2,823	-23.1	3,670
Renault Truck	*12,929*	*-9.9*	*14,343*
UD Trucks	*44*	*10.0*	*40*
Volvo Truck	*5,564*	*-4.5*	*5,825*
Volvo Truck Total	18,537	-8.3	20,208
Other	178	19.5	149
Total Med./Hvy. Trks.	**43,378**	**-8.4**	**47,363**
Buses			
BMC	16	-74.6	63
Breda	5	-37.5	8
Mercedes-Benz	*1,640*	*14.1*	*1,437*
Setra	*206*	*-15.6*	*244*
Daimler Total	1,846	9.8	1,681
Isuzu	5	400.0	1
Heuliez	*415*	*-7.4*	*448*
Irisbus	*1,889*	*-10.9*	*2,119*
Iveco	*299*	*8.3*	*276*
Iveco Total	2,603	-8.4	2,843
King Long	28	–	–
MAN	*156*	*-68.4*	*493*
Neoplan	*31*	*40.9*	*22*
Volkswagen Truck & Bus	*22*	–	*22*
MAN Total	209	-61.1	537
Otokar	88	–	–
PACCAR (DAF)	–	–	41
Scania	19	-76.8	82
Solaris	74	–	–
Temsa	174	-36.0	272
Van Hool	98	-44.0	175
Bova	*34*	*-60.5*	*86*
VDL	*50*	–	–
VDL Total	84	-2.3	86
Renault Truck	*24*	*-72.4*	*87*
Volvo Truck	*29*	*-31.0*	*42*
Volvo Truck Total	53	-58.9	129
Other	242	-16.0	288
Total Buses	**5,544**	**-10.7**	**6,206**
Total Comm. Vehs.	**432,971**	**-10.3**	**482,824**
Total Vehicles	**2,331,731**	**-13.2**	**2,687,053**

GERMANY

ECONOMIC SUMMARY

Common Language	German
Currency	Euro
Steering	Left Hand Drive

2012 Vehicles in Operation

Passenger Cars	43,431,124
Commercial Vehicles	3,107,000
Total	46,538,124
Persons per Car	1.9
Persons per Vehicle	1.7

2012 GDP (billions of US $)

Nominal GDP	3,400.6
Real GDP	3,263.4
GDP % annual growth	0.9
GDP per capita	39,838

Current Account Balance (billions of US $)

2012: 238.5
2011: 224.3

Current Account Balance in % of GDP

2012: 7.0
2011: 6.2

Total Number of Fatal Road Accidents (most recent data)

2012: 3,601

Capital	Berlin
Population	81,147,265
Population Density	602.8
Surface Area sq km/miles	348,672 / 134,623

Road Infrastructure

Paved Road Surface %	100.0
Total Road Length in km/miles	644,480 / 400,461
Road Traffic (Vehicles per mile)	116.2

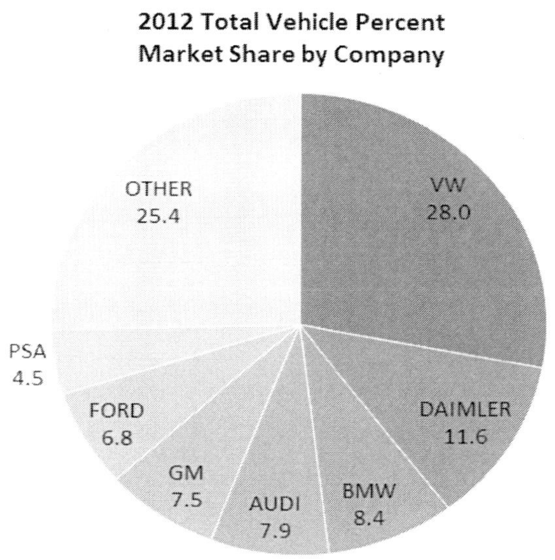

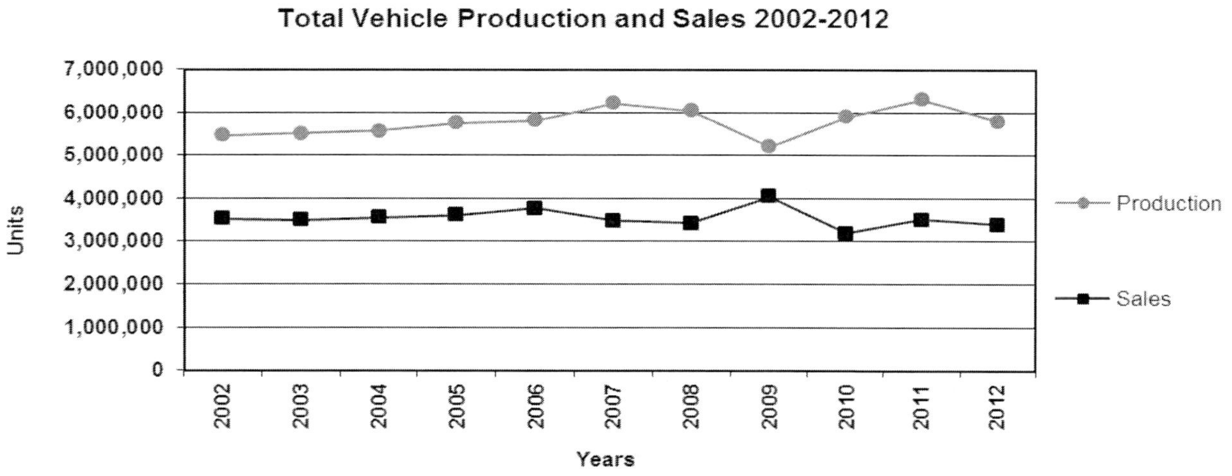

GERMANY

PRODUCTION

Year	Cars	Trucks	Buses	Total
2012	5,388,459	403,249	5,763	5,797,471
2011	5,871,918	422,821	6,840	6,301,579
2010	5,552,409	346,640	6,936	5,905,985
2009	4,964,523	237,548	7,786	5,209,857
2008	5,532,032	503,660	10,038	6,045,730
2007	5,709,139	495,236	9,085	6,213,460
2006	5,398,508	411,816	9,290	5,819,614
2005	5,350,187	398,733	8,790	5,757,710
2004	5,192,101	367,869	9,984	5,569,954
2003	5,145,403	350,803	10,423	5,506,629
2002	5,123,238	336,326	9,745	5,469,309
2001	5,301,189	378,548	11,940	5,691,677
2000	5,131,918	381,179	13,518	5,526,615
1999	5,309,524	366,343	11,825	5,687,692
1998	5,348,115	365,688	12,985	5,726,788
1997	4,678,022	333,337	11,569	5,022,928
1996	4,539,583	292,973	10,353	4,842,909
1995	4,360,235	297,241	9,888	4,667,364
1994	4,093,685	252,626	9,827	4,356,138
1993	3,794,491	225,729	11,580	4,031,800
1992	4,863,721	319,640	10,581	5,193,942
1991	4,676,666	346,741	11,043	5,034,450
1990	4,660,657	305,318	10,577	4,976,552
1989	4,563,673	278,101	9,873	4,851,647
1988	4,346,283	268,734	10,297	4,625,314
1987	4,373,629	250,971	9,474	4,634,074
1986	4,310,828	275,502	10,633	4,596,963
1985	4,166,686	268,367	10,867	4,445,920
1984	3,790,164	245,837	9,461	4,045,462
1983	3,877,641	263,527	13,271	4,154,439
1982	3,764,423	287,079	14,150	4,065,652

GERMANY

PRODUCTION BY MANUFACTURER

	2012	% Chg.	2011
Cars			
BMW	1,043,732	3.8	1,005,259
Ford	685,358	-4.9	720,411
GM (Opel)	387,525	-12.3	442,040
Mercedes-Benz	963,633	-3.4	997,628
Porsche	149,767	19.8	125,005
Audi	*814,854*	*-28.1*	*1,133,343*
Volkswagen	*1,343,590*	*-7.2*	*1,448,232*
Volkswagen Total	2,158,444	-16.4	2,581,575
Total Cars	**5,388,459**	**-8.2**	**5,871,918**
Light Trucks			
Ford	8,646	-9.3	9,535
Freightliner	*2,820*	*-5.9*	*2,996*
Mercedes-Benz	*140,905*	*-6.1*	*149,979*
Mercedes-Benz Total	143,725	-6.0	152,975
Multicar	1,005	-29.6	1,427
Volkswagen	107,232	-3.3	110,835
Total Light Trucks	**260,608**	**-5.2**	**274,772**
Medium/Heavy Trucks			
MAN	46,773	-3.1	48,275
Mercedes-Benz	95,868	-3.9	99,774
Total Med./Hvy. Trks.	**142,641**	**-3.7**	**148,049**
Buses			
Evobus	3,830	-15.4	4,528
MAN	1,678	-7.9	1,821
Neoplan	255	-48.1	491
Total Buses	**5,763**	**-15.7**	**6,840**
Total Comm. Vehs.	**409,012**	**-4.8**	**429,661**
Total Vehicles	**5,797,471**	**-8.0**	**6,301,579**

GERMANY

SALES

Year	Cars	Trucks	Buses	Total
2012	3,082,504	306,358	5,140	3,394,002
2011	3,173,634	329,775	5,047	3,508,456
2010	2,916,260	276,913	5,244	3,198,417
2009	3,807,175	236,544	5,640	4,049,359
2008	3,090,040	329,084	5,915	3,425,039
2007	3,148,163	328,642	5,474	3,482,279
2006	3,467,961	298,720	5,713	3,772,394
2005	3,342,122	267,344	5,432	3,614,898
2004	3,266,826	278,003	5,398	3,550,227
2003	3,236,558	259,076	5,669	3,501,303
2002	3,252,898	264,830	5,737	3,523,465
2001	3,341,718	290,480	6,121	3,638,319
2000	3,378,343	308,574	6,243	3,693,160
1999	3,802,176	318,580	6,321	4,127,077
1998	3,735,987	291,045	5,802	4,032,834
1997	3,528,179	258,409	5,514	3,792,102
1996	3,496,320	243,343	5,876	3,745,539
1995	3,314,061	255,160	5,352	3,574,573
1994	3,209,224	254,698	6,184	3,470,106
1993	3,194,204	253,277	7,679	3,455,160
1992	3,929,558	329,618	7,491	4,266,667
1991	4,158,674	326,256	6,048	4,490,978
1990	3,040,783	198,834	4,565	3,244,182
1989	2,831,740	168,652	4,448	3,004,840
1988	2,807,939	156,004	4,750	2,968,693
1987	2,915,654	148,109	4,850	3,068,613
1986	2,829,438	139,082	4,258	2,972,778
1985	2,379,261	129,788	4,046	2,513,095
1984	2,393,939	126,726	3,771	2,524,436
1983	2,426,774	138,796	5,336	2,570,906
1982	2,155,537	118,632	4,917	2,279,086

Note: Figures exclude East Germany prior to 1990.

GERMANY

SALES BY COMPANY

	2012	% Chg.	2011
Cars			
Aston Martin	322	-37.8	518
Audi	266,582	6.3	250,708
AutoVaz (Lada)	1,575	-22.5	2,031
Bentley	387	40.2	276
BMW	*248,192*	*-3.5*	*257,094*
Mini	*36,302*	*-10.0*	*40,345*
BMW Total	284,494	-4.4	297,439
Chrysler	*379*	*-53.0*	*806*
Dodge	*70*	*-81.8*	*384*
Jeep	*6,584*	*65.0*	*3,990*
Chrysler Total	7,033	35.8	5,180
Dacia	44,904	9.0	41,208
Daihatsu	1,263	-67.9	3,932
Mercedes-Benz	*283,006*	*-0.9*	*285,651*
Smart	*28,053*	*-4.8*	*29,470*
Daimler Total	311,059	-1.3	315,121
Ferrari	774	12.2	690
Alfa Romeo	*7,502*	*-28.4*	*10,480*
Fiat	*72,755*	*-9.2*	*80,125*
Lancia	*2,979*	*27.1*	*2,344*
Fiat Total	83,236	-10.4	92,949
Ford	206,128	-10.7	230,939
Cadillac	*149*	*39.3*	*107*
Chevrolet	*29,584*	*-1.3*	*29,986*
Opel	*213,627*	*-16.1*	*254,605*
Other GM	*168*	*-59.4*	*414*
GM Total	243,528	-14.6	285,112
Honda	26,306	-10.8	29,487
Hyundai	100,875	16.1	86,866
Iveco	554	10.8	500
Jaguar	*3,197*	*3.3*	*3,094*
Land Rover	*11,113*	*66.5*	*6,674*
Jaguar Land Rover Total	14,310	46.5	9,768
Kia	54,798	30.3	42,065
Lamborghini	106	23.3	86
Maserati	238	-20.4	299
Mazda	40,702	-1.5	41,330
Mitsubishi	23,565	-21.6	30,050
Morgan	111	-2.6	114
Infiniti	*874*	*-10.9*	*981*
Nissan	*62,210*	*-11.0*	*69,893*
Nissan Total	63,084	-11.0	70,874
Citroen	*61,038*	*-8.5*	*66,711*
Peugeot	*71,435*	*-13.7*	*82,743*
Peugeot Citroen Total	132,473	-11.4	149,454
Porsche	20,516	9.8	18,690
Proton (Lotus)	56	-33.3	84
Renault	105,836	-11.2	119,158
Rolls-Royce	57	-9.5	63
Saab	56	-88.8	501
Ssangyong	676	91.5	353
Subaru	9,457	52.3	6,210
Suzuki	30,843	-10.4	34,418
Lexus	*2,841*	*5.4*	*2,696*
Toyota	*80,993*	*0.6*	*80,508*
Toyota Total	83,834	0.8	83,204
SEAT	*68,109*	*15.4*	*59,030*
Skoda	*147,197*	*3.2*	*142,611*
Volkswagen	*672,921*	*-2.0*	*686,772*
Volkswagen Total	888,227	–	888,413
Volvo	32,750	-3.4	33,888
Wiesmann	107	-11.6	121
Other	1,682	9.6	1,535
Total Cars	**3,082,504**	**-2.9**	**3,173,634**
Light Trucks			
BMW	99	15.1	86
Chrysler	925	-0.1	926
Daimler (Mercedes-Benz)	70,441	-4.6	73,818
Fiat	17,020	-15.1	20,044
Ford	24,444	-11.6	27,656

SALES BY COMPANY

	2012	% Chg.	2011
Chevrolet	*76*	*-44.1*	*136*
Opel	*10,809*	*-20.9*	*13,667*
GM Total	10,885	-21.1	13,803
Hyundai	540	-34.5	824
Land Rover	1,755	5.9	1,658
Mazda	28	-92.0	350
Mitsubishi	2,266	-24.8	3,013
Nissan	4,933	4.5	4,722
Citroen	*11,126*	*-3.4*	*11,513*
Peugeot	*10,218*	*-11.1*	*11,495*
Peugeot Citroen Total	21,344	-7.2	23,008
Piaggio	441	5.0	420
Renault	18,708	-6.3	19,965
Saab	133	-4.3	139
Santana	–	–	22
Ssangyong	72	–	–
Tata	–	–	112
Toyota	2,129	-13.1	2,450
Skoda	*990*	*-2.2*	*1,012*
Volkswagen	*60,128*	*5.0*	*57,253*
Volkswagen Total	61,118	4.9	58,265
Other	2,417	-2.2	2,471
Total Light Trucks	**239,698**	**-5.5**	**253,752**
<u>Medium/Heavy Trucks</u>			
Daimler (Mercedes-Benz)	9,248	-12.2	10,539
Isuzu	768	-17.1	926
Iveco	12,972	-13.7	15,027
MAN	20,598	-14.1	23,976
Multicar	793	-34.2	1,206
PACCAR (DAF)	6,635	-15.2	7,822
Scania	4,695	-9.6	5,195
Renault Truck	*2,874*	*2.4*	*2,807*
Volvo Truck	*5,075*	*-15.3*	*5,990*
Volvo Truck Total	7,949	-9.6	8,797
Other	3,002	18.4	2,535
Total Med./Hvy. Trks.	**66,660**	**-12.3**	**76,023**
<u>Buses</u>			
Evobus	*2,050*	*-4.7*	*2,151*
Mercedes-Benz	*357*	*0.8*	*354*
Daimler Total	2,407	-3.9	2,505
Fiat	24	-72.4	87
Ford	310	496.2	52
Isuzu	–	–	1
Iveco	293	36.9	214
MAN	*1,258*	*-0.9*	*1,269*
Neoplan	–	–	*50*
Volkswagen Truck & Bus	*56*	*64.7*	*34*
MAN Total	1,314	-2.9	1,353
Citroen	*8*	*33.3*	*6*
Peugeot	*10*	*-78.3*	*46*
Peugeot Citroen Total	18	-65.4	52
Renault	23	91.7	12
Scania	37	-37.3	59
Temsa	66	-32.7	98
Van Hool	–	–	51
Bova	*50*	*-35.9*	*78*
VDL	*128*	*70.7*	*75*
VDL Total	178	16.3	153
Renault Truck	*3*	–	*3*
Volvo Truck	*88*	*8.6*	*81*
Volvo Truck Total	91	8.3	84
Other	379	16.3	326
Total Buses	**5,140**	**1.8**	**5,047**
Total Comm. Vehs.	**311,498**	**-7.0**	**334,822**
Total Vehicles	**3,394,002**	**-3.3**	**3,508,456**

GREECE

ECONOMIC SUMMARY

Common Language	Greek
Currency	Euro
Steering	Left Hand Drive

2012 Vehicles in Operation

Passenger Cars	5,167,557
Commercial Vehicles	1,345,880
Total	6,513,437
Persons per Car	2.1
Persons per Vehicle	1.7

2012 GDP (billions of US $)

Nominal GDP	249.2
Real GDP	222.7
GDP % annual growth	-6.4
GDP per capita	19,707

Current Account Balance (billions of US $)

2012:	-7.2
2011:	-28.7

Current Account Balance in % of GDP

2012:	-2.9
2011:	-9.9

Total Number of Fatal Road Accidents (most recent data)

2010:	1,258

Capital	Athens
Population	10,772,967
Population Density	213.6
Surface Area sq km/miles	130,647 / 50,443

Road Infrastructure

Paved Road Surface %	91.8
Total Road Length in km/miles	117,533 / 73,031
Road Traffic (Vehicles per mile)	89.2

2012 Total Vehicle Percent Market Share by Company

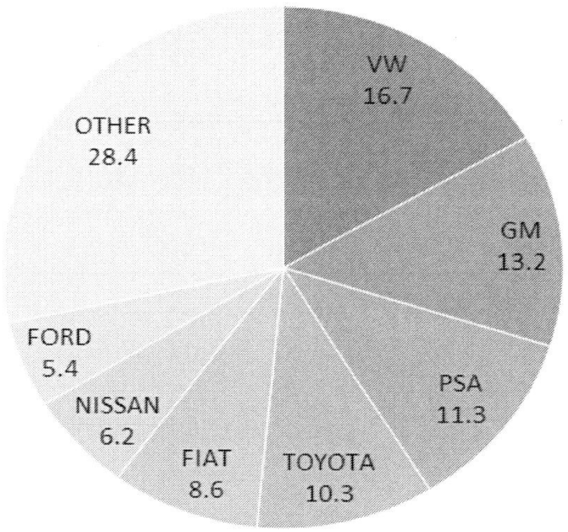

Total Vehicle Sales 2002-2012

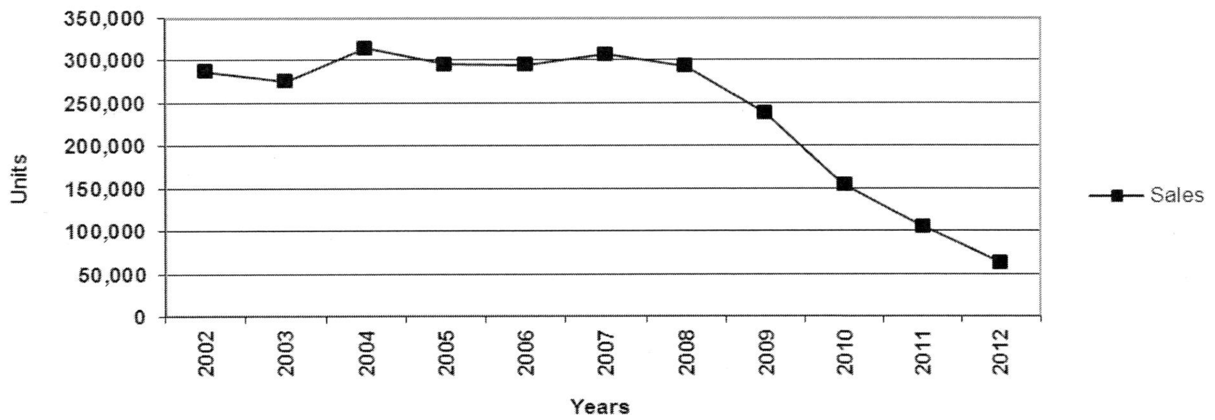

GREECE

SALES

Year	Cars	Trucks	Buses	Total
2012	58,482	3,918	118	62,518
2011	97,682	6,893	83	104,658
2010	141,499	11,938	403	153,840
2009	220,548	16,426	1,012	237,986
2008	267,242	25,002	575	292,819
2007	279,794	26,504	627	306,925
2006	267,706	25,922	474	294,102
2005	269,733	24,897	660	295,290
2004	289,753	22,494	1,508	313,755
2003	257,293	17,681	912	275,886
2002	268,489	18,162	343	286,994
2001	280,295	19,883	862	301,040
2000	290,225	22,327	624	313,176
1999	261,716	21,134	582	283,432
1998	179,220	19,065	359	198,644
1997	159,900	16,700	*	176,600
1996	141,589	35,185	758	177,532
1995	133,757	36,934	1,130	171,821
1994	118,088	25,813	1,246	145,147
1993	155,646	36,348	1,232	193,226
1992	206,652	45,578	1,340	253,570
1991	176,862	39,605	1,052	217,519
1990	132,487	43,964	936	177,387
1989	102,314	42,123	771	145,208
1988	76,043	34,229	465	110,737
1987	74,180	30,398	346	104,924
1986	96,181	27,674	368	124,223
1985	109,435	31,456	619	141,510
1984	85,945	35,091	393	121,429
1983	75,222	43,650	922	119,794
1982	91,927	57,403	722	150,052

* Included with Trucks.

GREECE

SALES BY COMPANY

	2012	% Chg.	2011
Cars			
Aston Martin	2	100.0	1
Audi	1,811	-27.9	2,513
AutoVaz (Lada)	–	–	16
BMW	*1,329*	*-34.2*	*2,021*
Mini	*667*	*-29.5*	*946*
BMW Total	1,996	-32.7	2,967
Caterham	–	–	1
Chrysler	*76*	*-65.5*	*220*
Dodge	*–*	*–*	*4*
Chrysler Total	76	-66.1	224
Dacia	347	-55.4	778
Daihatsu	188	-80.1	943
Mercedes-Benz	*1,262*	*-39.3*	*2,078*
Smart	*541*	*-63.7*	*1,492*
Daimler Total	1,803	-49.5	3,570
Dongfeng	–	–	4
Ferrari	1	–	–
Alfa Romeo	*855*	*-49.1*	*1,679*
Fiat	*3,311*	*-43.1*	*5,816*
Lancia	*787*	*271.2*	*212*
Fiat Total	4,953	-35.7	7,707
Ford	3,006	-53.2	6,427
Chevrolet	*930*	*-50.0*	*1,860*
Opel	*7,066*	*-38.2*	*11,440*
GM Total	7,996	-39.9	13,300
Honda	657	-44.7	1,189
Hyundai	2,618	-46.4	4,885
Jaguar	*3*	*-40.0*	*5*
Land Rover	*12*	*-60.0*	*30*
Jaguar Land Rover Total	15	-57.1	35
Kia	1,515	-34.9	2,328
Lifan	8	–	–
Maserati	–	–	1
Mazda	222	-50.7	450
Mitsubishi	608	-57.8	1,440
Infiniti	*1*	*-50.0*	*2*
Nissan	*3,143*	*-44.7*	*5,687*
Nissan Total	3,144	-44.7	5,689
Citroen	*4,206*	*16.5*	*3,609*
Peugeot	*2,543*	*-25.6*	*3,419*
Peugeot Citroen Total	6,749	-4.0	7,028
Piaggio	–	–	1
Porsche	25	-32.4	37
Renault	1,046	-31.3	1,523
Saab	27	-82.2	152
Ssangyong	23	475.0	4
Subaru	41	-87.9	339
Suzuki	2,278	-26.6	3,103
Lexus	*61*	*-58.2*	*146*
Toyota	*5,744*	*-52.6*	*12,128*
Toyota Total	5,805	-52.7	12,274
SEAT	*1,243*	*-59.0*	*3,029*
Skoda	*2,837*	*-39.5*	*4,688*
Volkswagen	*6,009*	*-36.5*	*9,466*
Volkswagen Total	10,089	-41.3	17,183
Volvo	1,380	-4.5	1,445
Other	53	-57.6	125
Total Cars	**58,482**	**-40.1**	**97,682**
Light Trucks			
Dacia	43	-71.3	150
Daimler (Mercedes-Benz)	43	-66.4	128
Fiat	439	-27.3	604
Ford	342	-46.6	641
GM (Opel)	239	-37.9	385
Hyundai	1	–	–
Isuzu	235	-39.1	386
Iveco	6	-14.3	7
Mahindra & Mahindra	13	-77.2	57
Mazda	37	-58.9	90
Mitsubishi	160	-41.0	271
Nissan	702	-28.1	976
Citroen	*171*	*-26.0*	*231*
Peugeot	*156*	*-44.7*	*282*
Peugeot Citroen Total	327	-36.3	513
Piaggio	56	9.8	51
Renault	53	6.0	50

GREECE

SALES BY COMPANY

	2012	% Chg.	2011
Ssangyong	4	-33.3	6
Toyota	648	-56.2	1,479
UAZ	–	–	1
SEAT	*2*	*100.0*	*1*
Skoda	*–*	*–*	*2*
Volkswagen	*351*	*-35.1*	*541*
Volkswagen Total	353	-35.1	544
Other	6	-66.7	18
Total Light Trucks	**3,707**	**-41.7**	**6,357**
Medium/Heavy Trucks			
BMC	–	–	5
Mercedes-Benz	*–*	*–*	*131*
Mitsubishi Fuso	*–*	*–*	*24*
Daimler Total	–	–	155
Fiat	–	–	11
Ford	–	–	4
GM (Opel)	–	–	2
Isuzu	–	–	9
Iveco	–	–	63
MAN	*–*	*–*	*89*
Volkswagen Truck & Bus	*–*	*–*	*16*
MAN Total	–	–	105
PACCAR (DAF)	–	–	62
Peugeot Citroen	–	–	1
Scania	–	–	61
Renault Truck	*–*	*–*	*6*
UD Trucks	*–*	*–*	*13*
Volvo Truck	*–*	*–*	*39*
Volvo Truck Total	–	–	58
Other	211	–	–
Total Med./Hvy. Trks.	**211**	**-60.6**	**536**
Buses			
Evobus	*–*	*–*	*2*
Mercedes-Benz	*–*	*–*	*37*
Setra	*–*	*–*	*1*
Daimler Total	–	–	40
Fiat	–	–	6
GM (Opel)	–	–	1
Isuzu	–	–	1
Irisbus	*–*	*–*	*3*
Iveco	*–*	*–*	*10*
Iveco Total	–	–	13
MAN	*–*	*–*	*5*
Neoplan	*–*	*–*	*3*
MAN Total	–	–	8
Temsa	–	–	1
VDL (Bova)	–	–	1
Renault Truck	*–*	*–*	*1*
Volvo Truck	*–*	*–*	*5*
Volvo Truck Total	–	–	6
Other	118	–	6
Total Buses	**118**	**42.2**	**83**
Total Comm. Vehs.	**4,036**	**-42.1**	**6,976**
Total Vehicles	**62,518**	**-40.3**	**104,658**

HUNGARY

ECONOMIC SUMMARY

Common Language	Hungarian
Currency	Forint
Steering	Left Hand Drive

2012 Vehicles in Operation

Passenger Cars	2,986,028
Commercial Vehicles	431,706
Total	3,417,734
Persons per Car	3.3
Persons per Vehicle	2.9

2012 GDP (billions of US $)

Nominal GDP	126.9
Real GDP	99.1
GDP % annual growth	-1.7
GDP per capita	9,945

Current Account Balance (billions of US $)

2012: 2.2
2011: 1.3

Current Account Balance in % of GDP

2012: 1.7
2011: 0.9

Total Number of Fatal Road Accidents (most recent data)

2012: 605

Capital	Budapest
Population	9,939,470
Population Density	287.3
Surface Area sq km/miles	89,608 / 34,597

Road Infrastructure

Paved Road Surface %	38.0
Total Road Length in km/miles	197,519 / 122,734
Road Traffic (Vehicles per mile)	27.8

2012 Total Vehicle Percent Market Share by Company

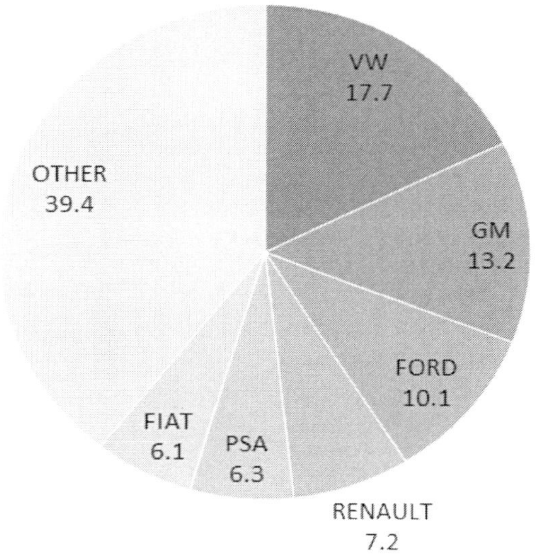

Total Vehicle Production and Sales 2002-2012

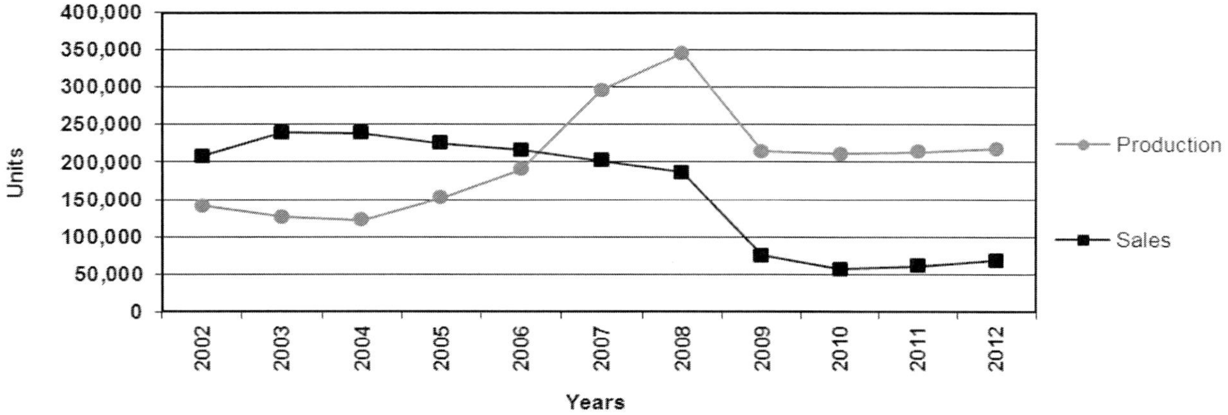

HUNGARY

PRODUCTION

Year	Cars	Trucks	Buses	Total
2012	215,440	2,295	105	217,840
2011	211,218	2,210	103	213,531
2010	208,572	2,760	130	211,462
2009	212,773	1,607	433	214,813
2008	342,049	3,092	638	345,779
2007	292,027	3,551	494	296,072
2006	187,633	1,996	614	190,243
2005	148,533	2,455	1,027	152,015
2004	118,590	3,175	901	122,666
2003	122,338	2,655	1,123	126,116
2002	138,239	2,157	1,117	141,513
2001	140,401	2,108	1,804	144,313
2000	134,023	1,621	1,748	137,392
1999	125,889	1,307	990	128,186
1998	89,733	1,561	1,636	92,930
1997	76,300	1,177	2,315	79,792
1996	63,033	35	800	63,868
1995	51,034	85	2,000	53,119
1994	19,408	81	1,576	21,065
1993	13,021	60	3,181	16,262

PRODUCTION BY MANUFACTURER

	2012	% Chg.	2011		2012	% Chg.	2011
Cars				**Buses**			
Daimler (Mercedes)	41,035	–	–	NABI	100	5.3	95
Suzuki	140,852	-18.0	171,700	Other	5	-37.5	8
Volkswagen (Audi)	33,553	-15.1	39,518	**Total Buses**	**105**	**1.9**	**103**
Total Cars	**215,440**	**2.0**	**211,218**				
Trucks							
RABA	105	5.0	100	**Total Comm. Vehs.**	**2,400**	**3.8**	**2,313**
Schwarzmuller	2,190	3.8	2,110				
Total Trucks	**2,295**	**3.8**	**2,210**	**Total Vehicles**	**217,840**	**2.0**	**213,531**

HUNGARY

SALES

Year	Cars	Commercial Vehicles	Total
2012	53,060	15,334	68,394
2011	45,094	15,932	61,026
2010	46,069	10,424	56,493
2009	60,189	14,561	74,750
2008	155,403	30,950	186,353
2007	171,661	30,619	202,280
2006	187,842	28,154	215,996
2005	199,068	26,528	225,596
2004	208,044	30,145	238,189
2003	208,441	31,080	239,521
2002	172,338	35,117	207,455
2001	148,293	29,693	177,986
2000	133,236	50,915	184,151
1999	129,318	53,120	182,438
1998	112,713	55,848	168,561
1997	85,435	36,994	122,429
1996	103,469	38,160	141,629
1995	118,895	41,339	160,234

HUNGARY

SALES BY COMPANY

	2012	% Chg.	2011
Cars			
Audi	1,636	24.6	1,313
AutoVaz (Lada)	11	10.0	10
Bentley	2	-33.3	3
BMW	*1,389*	*32.4*	*1,049*
Mini	*126*	*12.5*	*112*
BMW Total	1,515	30.5	1,161
Chrysler	*12*	*-87.5*	*96*
Dodge	*6*	*-64.7*	*17*
Jeep	*51*	*-12.1*	*58*
Chrysler Total	69	-59.6	171
Dacia	2,159	141.5	894
Daihatsu	–	–	5
Mercedes-Benz	*1,087*	*-2.0*	*1,109*
Smart	*8*	*-93.9*	*132*
Daimler Total	1,095	-11.8	1,241
Ferrari	3	-25.0	4
Alfa Romeo	*87*	*-23.7*	*114*
Fiat	*2,512*	*82.6*	*1,376*
Lancia	*112*	*-35.6*	*174*
Fiat Total	2,711	62.9	1,664
Ford	5,196	7.4	4,840
Cadillac	*2*	*-50.0*	*4*
Chevrolet	*896*	*37.8*	*650*
Opel	*7,160*	*37.8*	*5,196*
Saturn	–	–	–
GM Total	8,058	37.7	5,850
Honda	1,065	-2.6	1,093
Hyundai	1,234	37.0	901
Jaguar	*17*	*-37.0*	*27*
Land Rover	*84*	*47.4*	*57*
Jaguar Land Rover Total	101	20.2	84
Kia	1,798	117.9	825
Mazda	486	-14.9	571
Mitsubishi	169	-15.9	201
Nissan	2,346	40.7	1,667
Citroen	*1,200*	*-13.1*	*1,381*
Peugeot	*1,524*	*-20.1*	*1,907*
Peugeot Citroen Total	2,724	-17.2	3,288
Porsche	45	2.3	44
Renault	3,431	-5.6	3,636
Saab	–	–	105
Ssangyong	8	–	–
Subaru	100	-52.4	210
Suzuki	3,300	106.5	1,598
Lexus	*109*	*-4.4*	*114*
Toyota	*2,213*	*-13.0*	*2,544*
Toyota Total	2,322	-12.6	2,658
SEAT	*446*	*-39.8*	*741*
Skoda	*5,340*	*13.8*	*4,691*
Volkswagen	*4,546*	*0.1*	*4,540*
Volkswagen Total	10,332	3.6	9,972
Volvo	1,137	8.8	1,045
Other	7	-82.5	40
Total Cars	**53,060**	**17.7**	**45,094**
Commercial Vehicles			
Daimler (Mercedes-Benz)	1,101	-4.3	1,150
Fiat	1,460	-3.3	1,510
Ford	1,745	-3.2	1,803
Daewoo	*30*	*-9.1*	*33*
Opel	*940*	*-4.1*	*980*
GM Total	970	-4.2	1,013
Hyundai	544	-3.7	565
Iveco	620	-4.6	650
Kia	395	2.3	386
Land Rover	9	-10.0	10
MAN	661	-1.8	673
Mazda	135	-10.6	151
Mitsubishi	314	-4.0	327
Nissan	400	-2.7	411
PACCAR (DAF)	414	-3.7	430
Citroen	*901*	*-3.8*	*937*
Peugeot	*665*	*-3.8*	*691*
Peugeot Citroen Total	1,566	-3.8	1,628
Renault	1,465	-5.1	1,543
Scania	206	-0.5	207
Ssangyong	7	-41.7	12
Tatra	34	-2.9	35
Toyota	596	-1.2	603
Skoda	*148*	*-9.8*	*164*
Volkswagen	*1,648*	*-2.9*	*1,697*
Volkswagen Total	1,796	-3.5	1,861
Renault Truck	*371*	*-3.9*	*386*
Volvo Truck	*508*	*-3.8*	*528*
Volvo Truck Total	879	-3.8	914
Other	17	-66.0	50
Total Comm. Vehs.	**15,334**	**-3.8**	**15,932**
Total Vehicles	**68,394**	**12.1**	**61,026**

Note: Some individual company figures estimated for Cars in 2011 and for Commercial Vehicles in 2012.

IRELAND

ECONOMIC SUMMARY

Common Language	English/Gaelic	**Capital**	Dublin
Currency	Euro	**Population**	4,775,982
Steering	Right Hand Drive	**Population Density**	179.6
		Surface Area sq km/miles	68,883 / 26,595

2012 Vehicles in Operation

Passenger Cars	1,895,000
Commercial Vehicles	409,000
Total	2,304,000
Persons per Car	2.5
Persons per Vehicle	2.1

Road Infrastructure

Paved Road Surface %	100.0
Total Road Length in km/miles	96,036 / 59,674
Road Traffic (Vehicles per mile)	38.6

2012 GDP (billions of US $)

Nominal GDP	210.4
Real GDP	211.7
GDP % annual growth	0.9
GDP per capita	46,173

Current Account Balance (billions of US $)

2012:	10.4
2011:	2.5

Current Account Balance in % of GDP

2012:	4.9
2011:	1.1

Total Number of Fatal Road Accidents (most recent data)

2010:	212

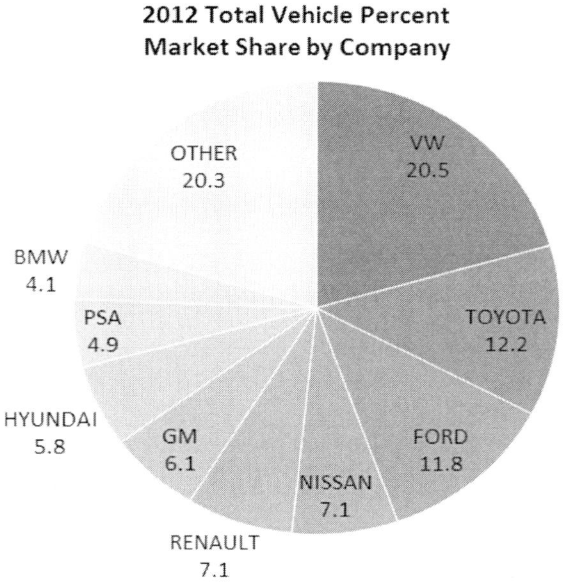

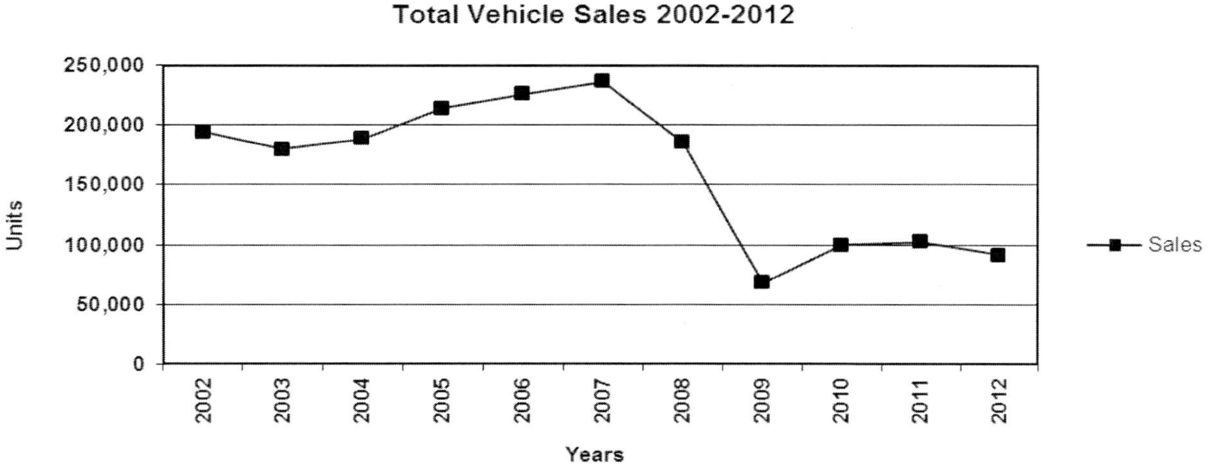

IRELAND

SALES

Year	Cars	Commercial Vehicles	Total
2012	79,498	12,230	91,728
2011	89,896	12,552	102,448
2010	88,373	11,613	99,986
2009	57,460	10,533	67,993
2008	151,607	34,053	185,660
2007	186,540	49,834	236,374
2006	178,826	46,853	225,679
2005	171,732	41,974	213,706
2004	154,136	34,416	188,552
2003	145,331	34,511	179,842
2002	156,115	38,325	194,440
2001	164,772	42,817	207,589
2000	230,486	45,614	276,100
1999	174,232	38,623	212,855
1998	145,715	31,383	177,098
1997	136,662	23,146	159,808
1996	115,104	19,258	134,362
1995	86,924	16,373	103,297
1994	80,402	12,845	93,247
1993	64,149	9,887	74,036
1992	68,359	11,883	80,242
1991	67,749	15,893	83,642
1990	83,420	23,979	107,399
1989	78,383	20,122	98,505
1988	61,888	14,557	76,445
1987	54,341	12,533	66,874
1986	58,760	13,891	72,651
1985	59,592	15,009	74,601
1984	55,893	14,928	70,821
1983	61,094	13,502	74,596
1982	73,330	12,428	85,758

IRELAND

SALES BY COMPANY

	2012	% Chg.	2011
Cars			
Audi	3,647	7.1	3,406
BMW	*3,329*	*2.0*	*3,263*
Mini	*440*	*6.5*	*413*
BMW Total	3,769	2.5	3,676
Chrysler	*115*	*666.7*	*15*
Jeep	–	–	*3*
Chrysler Total	115	538.9	18
Daihatsu	1	-80.0	5
Daimler (Mercedes-Benz)	1,834	-3.7	1,905
Alfa Romeo	*120*	*-29.4*	*170*
Fiat	*664*	*-22.3*	*855*
Fiat Total	784	-23.5	1,025
Ford	8,563	-17.9	10,428
Chevrolet	*28*	*-93.4*	*422*
Opel	*5,102*	*-20.0*	*6,377*
GM Total	5,130	-24.5	6,799
Honda	1,102	11.0	993
Hyundai	5,206	33.7	3,895
Jaguar	*200*	*60.0*	*125*
Land Rover	*498*	*13.7*	*438*
Jaguar Land Rover Total	698	24.0	563
Kia	3,117	17.8	2,646
Mazda	1,433	-20.2	1,796
Mitsubishi	361	-40.6	608
Nissan	6,023	-10.1	6,699
Perodua	8	-66.7	24
Citroen	*1,161*	*-6.1*	*1,236*
Peugeot	*2,226*	*-18.1*	*2,719*
Peugeot Citroen Total	3,387	-14.4	3,955
Porsche	15	–	–
Renault	5,255	-38.6	8,564
Saab	6	-95.7	138
Subaru	143	-31.3	208

	2012	% Chg.	2011
Suzuki	630	4.7	602
Lexus	*264*	*-35.9*	*412*
Toyota	*9,956*	*-14.5*	*11,646*
Toyota Total	10,220	-15.2	12,058
SEAT	*1,360*	*-22.7*	*1,759*
Skoda	*5,078*	*9.4*	*4,641*
Volkswagen	*10,051*	*-10.2*	*11,196*
Volkswagen Total	16,489	-6.3	17,596
Volvo	1,154	-8.7	1,264
Other	408	-60.2	1,025
Total Cars	**79,498**	**-11.6**	**89,896**
Light Trucks			
Chrysler	*2*	*-80.0*	*10*
Dodge	–	–	*7*
Chrysler Total	2	-88.2	17
Daimler (Mercedes-Benz)	289	-13.5	334
Fiat	405	2.0	397
Ford	2,278	-7.1	2,452
Chevrolet	–	–	*1*
Opel	*491*	*-5.0*	*517*
GM Total	491	-5.2	518
Hyundai	157	60.2	98
Isuzu	34	-92.6	458
Iveco	17	30.8	13
Jaguar Land Rover	442	187.0	154
Kia	38	280.0	10
Mazda	–	–	1
Mitsubishi	138	-19.8	172
Nissan	521	-21.8	666
Citroen	*554*	*22.8*	*451*
Peugeot	*508*	*4.7*	*485*
Peugeot Citroen Total	1,062	13.5	936
Piaggio	–	–	4

IRELAND

SALES BY COMPANY

	2012	% Chg.	2011		2012	% Chg.	2011
Renault	1,266	-23.1	1,647	PACCAR (DAF)	117	-20.9	148
Subaru	14	-44.0	25	Scania	245	34.6	182
Suzuki	–	–	3	*Renault Truck*	*117*	*-2.5*	*120*
Toyota	967	1.4	954	*Volvo Truck*	*229*	*30.9*	*175*
Volkswagen	2,325	30.3	1,784	Volvo Truck Total	346	17.3	295
Other	428	-43.3	755	Other	28	-69.2	91
Total Light Trucks	**10,874**	**-4.6**	**11,398**	**Total Med./Hvy. Trks.**	**1,120**	**2.4**	**1,094**
Medium/Heavy Trucks				**Buses**			
Ashok Leyland (Avia)	8	300.0	2	Daimler (Mercedes-Benz)	13	–	1
Mercedes-Benz	*107*	*7.0*	*100*	PACCAR (DAF)	–	–	1
Mitsubishi Fuso	*26*	*-46.9*	*49*	Scania	49	–	3
Daimler Total	133	-10.7	149	VDL	12	200.0	4
Dennis Eagle	31	–	–	Other	162	217.6	51
Hino	48	33.3	36	**Total Buses**	**236**	**293.3**	**60**
Isuzu	47	-38.2	76				
Iveco	31	-43.6	55	**Total Comm. Vehs.**	**12,230**	**-2.6**	**12,552**
MAN	86	43.3	60	**Total Vehicles**	**91,728**	**-10.5**	**102,448**

ITALY

ECONOMIC SUMMARY

Common Language	Italian
Currency	Euro
Steering	Left Hand Drive

2012 Vehicles in Operation

Passenger Cars	37,078,274
Commercial Vehicles	4,921,712
Total	41,999,986
Persons per Car	1.7
Persons per Vehicle	1.5

2012 GDP (billions of US $)

Nominal GDP	2,014.1
Real GDP	1,836.8
GDP % annual growth	-2.4
GDP per capita	30,200

Current Account Balance (billions of US $)

2012: -10.7
2011: -67.4

Current Account Balance in % of GDP

2012: -0.5
2011: -3.1

Total Number of Fatal Road Accidents (most recent data)

2010: 4,090

Capital	Rome
Population	61,482,297
Population Density	541.3
Surface Area sq km/miles	294,140 / 113,586

Road Infrastructure

Paved Road Surface %	100.0
Total Road Length in km/miles	487,700 / 303,042
Road Traffic (Vehicles per mile)	138.6

2012 Total Vehicle Percent Market Share by Company

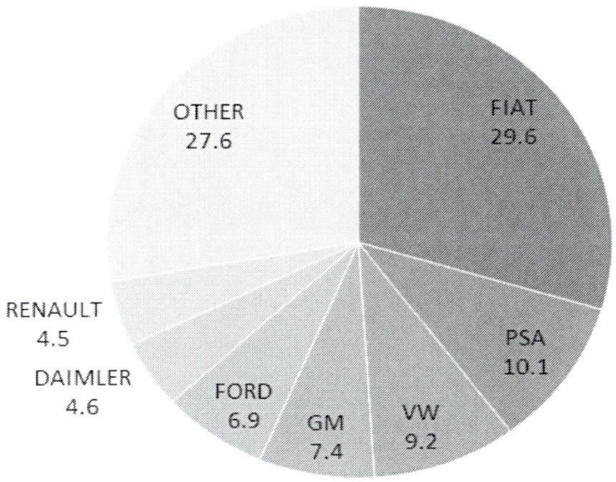

Total Vehicle Production and Sales 2002-2012

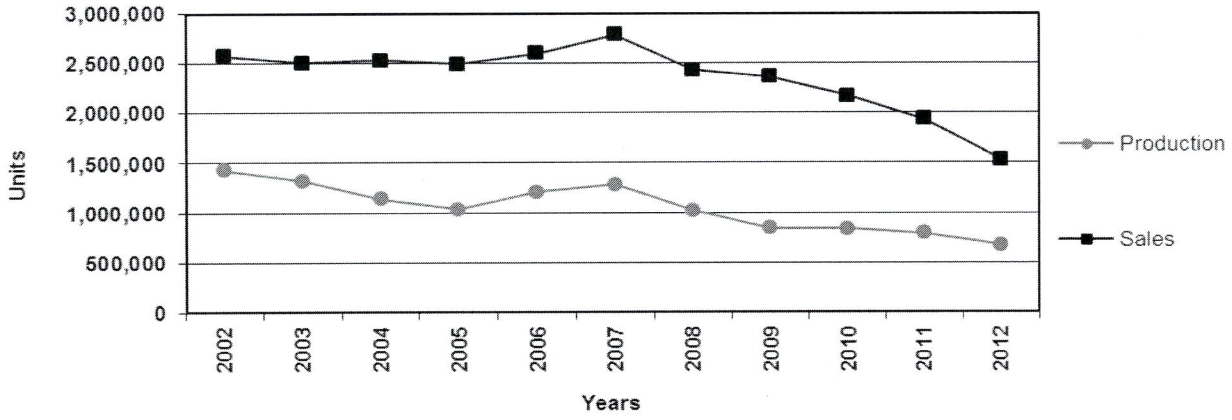

ITALY

PRODUCTION

Year	Cars	Trucks	Buses	Total
2012	396,817	274,462	489	671,768
2011	485,606	303,919	823	790,348
2010	573,169	262,028	1,068	836,265
2009	661,100	181,135	1,004	843,239
2008	659,221	363,209	1,344	1,023,774
2007	910,860	372,003	1,449	1,284,312
2006	892,502	316,225	2,867	1,211,594
2005	725,528	309,365	3,459	1,038,352
2004	833,578	305,451	2,915	1,141,944
2003	1,026,454	292,327	2,850	1,321,631
2002	1,125,769	298,536	2,776	1,427,081
2001	1,271,780	305,704	2,212	1,579,696
2000	1,422,284	312,868	3,163	1,738,315
1999	1,410,459	287,727	3,070	1,701,256
1998	1,402,382	286,373	3,982	1,692,737
1997	1,562,865	250,206	3,439	1,816,510
1996	1,317,995	223,391	3,979	1,545,365
1995	1,422,359	233,857	11,054	1,667,270
1994	1,340,878	191,238	2,353	1,534,469
1993	1,117,053	156,421	3,968	1,277,442
1992	1,476,627	200,345	9,515	1,686,487
1991	1,632,904	235,733	9,652	1,878,289
1990	1,874,672	239,718	6,460	2,120,850
1989	1,971,969	243,683	5,122	2,220,774
1988	1,884,313	221,924	4,782	2,111,019
1987	1,713,300	195,587	3,725	1,912,612
1986	1,652,452	175,528	3,720	1,831,700
1985	1,389,156	176,868	6,883	1,572,907
1984	1,439,283	155,505	6,389	1,601,177
1983	1,395,531	172,220	7,400	1,575,151
1982	1,297,351	149,005	6,687	1,453,043

PRODUCTION BY MANUFACTURER

	2012	% Chg.	2011
Cars			
Ferrari	7,663	4.8	7,314
Alfa Romeo	*92,053*	*-32.4*	*136,127*
Fiat	*262,041*	*-0.6*	*263,652*
Lancia	*26,659*	*-62.2*	*70,614*
Fiat Total	380,753	-19.1	470,393
Lamborghini	2,197	26.4	1,738
Maserati	6,204	0.7	6,161
Total Cars	**396,817**	**-18.3**	**485,606**
Light Trucks			
Faam	42	-8.7	46
Fiat	123,683	-9.2	136,184
Iveco	23,859	-22.2	30,665
Citroen	*40,827*	*-6.6*	*43,735*
Peugeot	*49,899*	*-9.2*	*54,983*
Peugeot Citroen Total	90,726	-8.1	98,718
Piaggio	2,876	-39.2	4,729
Total Light Trucks	**241,186**	**-10.8**	**270,342**
Medium/Heavy Trucks			
Iveco	33,276	-0.9	33,577
Total Med./Hvy. Trks.	**33,276**	**-0.9**	**33,577**
Buses			
Breda	31	-89.8	303
Iveco	458	1.6	451
Other	–	–	69
Total Buses	**489**	**-40.6**	**823**
Total Comm. Vehs.	**274,951**	**-9.8**	**304,742**
Total Vehicles	**671,768**	**-15.0**	**790,348**

ITALY

SALES

Year	Cars	Trucks	Buses	Total
2012	1,402,089	129,116	2,131	1,533,336
2011	1,748,143	190,015	3,063	1,941,221
2010	1,960,282	204,868	3,769	2,168,919
2009	2,158,010	200,329	3,086	2,361,425
2008	2,160,131	265,352	4,252	2,429,735
2007	2,490,570	288,714	4,414	2,783,698
2006	2,321,099	269,839	4,742	2,595,680
2005	2,235,882	251,953	5,073	2,492,908
2004	2,272,375	250,876	5,097	2,528,348
2003	2,249,648	248,541	4,279	2,502,468
2002	2,269,535	299,624	4,989	2,574,148
2001	2,413,378	259,148	5,575	2,678,101
2000	2,427,197	272,545	5,109	2,704,851
1999	2,349,200	191,864	4,746	2,545,810
1998	2,378,592	175,875	4,180	2,558,647
1997	2,403,697	153,226	3,451	2,560,374
1996	1,732,187	170,859	3,255	1,906,301
1995	1,732,202	159,644	1,985	1,893,831
1994	1,672,409	132,470	1,519	1,806,398
1993	1,693,323	133,684	1,665	1,828,672
1992	2,389,395	180,446	2,675	2,572,516
1991	2,220,321	185,472	3,098	2,408,891
1990	2,283,362	195,470	4,153	2,482,985
1989	2,296,784	197,401	3,322	2,497,507
1988	2,119,193	176,018	3,800	2,299,011
1987	1,929,568	132,471	3,545	2,065,584
1986	1,769,191	102,667	3,412	1,875,270
1985	1,653,217	107,047	3,873	1,764,137
1984	1,572,402	104,102	3,978	1,680,482
1983	1,451,512	107,317	4,286	1,563,115
1982	1,851,174	139,087	5,721	1,995,982

ITALY

SALES BY COMPANY

	2012	% Chg.	2011
Cars			
Aston Martin	57	–	–
Audi	51,062	-15.7	60,578
AutoVaz (Lada)	69	6.2	65
BMW	*42,462*	*-16.2*	*50,666*
Mini	*17,827*	*-17.4*	*21,594*
BMW Total	60,289	-16.6	72,260
Chrysler	–	–	*696*
Dodge	–	–	*42*
Jeep	*6,555*	*-16.8*	*7,878*
Chrysler Total	6,555	-23.9	8,616
Dacia	26,034	0.6	25,875
Daihatsu	1,415	-55.4	3,172
Mercedes-Benz	*42,316*	*-17.0*	*50,983*
Smart	*21,473*	*-11.0*	*24,121*
Daimler Total	63,789	-15.1	75,104
Ferrari	248	-56.5	570
Alfa Romeo	*42,174*	*-27.5*	*58,148*
Fiat	*294,778*	*-18.8*	*363,017*
Lancia	*71,418*	*-15.8*	*84,856*
Fiat Total	408,370	-19.3	506,021
Ford	98,997	-32.5	146,764
Chevrolet	*31,155*	*-4.5*	*32,611*
Opel	*78,889*	*-32.4*	*116,651*
GM Total	110,044	-26.3	149,262
Great Wall	10	-99.4	1,671
Honda	5,314	-40.5	8,928
Hyundai	43,435	–	43,448
Jaguar	*1,194*	*-5.1*	*1,258*
Land Rover	*12,540*	*38.5*	*9,053*
Jaguar Land Rover Total	13,734	33.2	10,311
Kia	27,043	39.1	19,441
Lamborghini	60	-16.7	72
Mahindra & Mahindra	60	500.0	10
Maserati	115	-72.2	414
Mazda	4,303	-34.7	6,593
Mitsubishi	3,553	-52.7	7,517
Infiniti	*240*	*-56.1*	*547*
Nissan	*49,040*	*-22.4*	*63,189*
Nissan Total	49,280	-22.7	63,736
Citroen	*69,435*	*-14.9*	*81,600*
Peugeot	*69,075*	*-11.7*	*78,235*
Peugeot Citroen Total	138,510	-13.3	159,835
Porsche	3,366	-21.7	4,299
Renault	59,818	-28.1	83,222
Saab	42	-95.2	881
Ssangyong	1,169	-39.0	1,915
Subaru	3,691	-23.2	4,809
Suzuki	14,731	-31.9	21,631
Tata	256	-76.8	1,105
Lexus	*1,146*	*-23.2*	*1,492*
Toyota	*55,178*	*-18.1*	*67,380*
Toyota Total	56,324	-18.2	68,872
SEAT	*10,552*	*-34.4*	*16,087*
Skoda	*10,421*	*-27.3*	*14,332*
Volkswagen	*113,514*	*-18.2*	*138,770*
Volkswagen Total	134,487	-20.5	169,189
Volvo	14,818	-19.7	18,448
Other	1,041	-70.3	3,509
Total Cars	**1,402,089**	**-19.8**	**1,748,143**
Light Trucks			
Audi	10	66.7	6
BMW	13	62.5	8
Chrysler (Jeep)	6	200.0	2
Dacia	808	-17.3	977
Daihatsu	5	-68.8	16
Daimler (Mercedes-Benz)	4,774	-24.6	6,332
Alfa Romeo	*31*	*-32.6*	*46*
Fiat	*45,525*	*-37.2*	*72,537*
Fiat Total	45,556	-37.2	72,583
Ford	6,037	-27.8	8,357
Chevrolet	*14*	*-60.0*	*35*
Opel	*3,962*	*-14.9*	*4,656*
GM Total	3,976	-15.2	4,691
GoNow	114	-34.5	174
Great Wall	970	10.2	880
Hyundai	60	5.3	57

ITALY

SALES BY COMPANY

	2012	% Chg.	2011
Isuzu	1,257	-33.4	1,888
Iveco	7,345	-38.8	12,008
Jaguar	*35*	*105.9*	*17*
Land Rover	*1,223*	*-23.9*	*1,607*
Jaguar Land Rover Total	1,258	-22.5	1,624
Kia	35	12.9	31
Mahindra & Mahindra	219	-61.8	573
Mazda	93	-70.7	317
Mitsubishi	756	-52.7	1,598
Nissan	4,843	-24.8	6,444
Citroen	*7,332*	*-38.8*	*11,977*
Peugeot	*9,684*	*-28.4*	*13,529*
Peugeot Citroen Total	17,016	-33.3	25,506
Piaggio	1,886	-38.2	3,051
Renault	9,689	-26.4	13,170
Ssangyong	52	36.8	38
Subaru	28	-64.1	78
Suzuki	62	-51.6	128
Tata	541	-7.8	587
Toyota	653	-24.5	865
SEAT	*260*	*4.4*	*249*
Skoda	*429*	*-28.4*	*599*
Volkswagen	*5,941*	*-16.9*	*7,151*
Volkswagen Total	6,630	-17.1	7,999
Volvo	100	-37.1	159
Other	583	48.0	394
Total Light Trucks	**115,375**	**-32.3**	**170,541**
Medium/Heavy Trucks			
Mercedes-Benz	*1,559*	*-32.5*	*2,309*
Mitsubishi Fuso	*106*	*-45.9*	*196*
Daimler Total	1,665	-33.5	2,505
Fiat	94	-8.7	103
Ford	7	–	7
GM (Opel)	–	–	1
Isuzu	306	-24.6	406
Iveco	5,905	-24.2	7,793
MAN	*1,047*	*-17.6*	*1,270*
Volkswagen Truck & Bus	*10*	*-85.1*	*67*

	2012	% Chg	2011
MAN Total	1,057	-20.9	1,337
PACCAR (DAF)	1,228	-36.5	1,935
Peugeot	–	–	8
Scania	1,029	-41.3	1,753
Renault Truck	*1,274*	*-26.1*	*1,724*
UD Trucks	*79*	*-36.3*	*124*
Volvo Truck	*1,026*	*-39.3*	*1,690*
Volvo Truck Total	2,379	-32.8	3,538
Other	71	-19.3	88
Total Med./Hvy. Trks.	**13,741**	**-29.4**	**19,474**
Buses			
BMC	22	-43.6	39
Breda	7	-97.5	275
Daimler (Evobus)	526	-36.7	831
Fiat	76	-41.5	130
Ford	5	66.7	3
Isuzu	24	20.0	20
Iveco (Irisbus)	929	-8.8	1,019
King Long	84	-10.6	94
Neoplan	*104*	*10.6*	*94*
Volkswagen Truck & Bus	*20*	*-55.6*	*45*
MAN Total	124	-10.8	139
Citroen	*9*	*-25.0*	*12*
Peugeot	*1*	*-85.7*	*7*
Peugeot Citroen Total	10	-47.4	19
Scania	61	-50.0	122
Temsa	76	90.0	40
Toyota	–	–	1
Van Hool	4	-84.6	26
VDL	49	-15.5	58
Renault Truck	*69*	*23.2*	*56*
Volvo Truck	*21*	*-52.3*	*44*
Volvo Truck Total	90	-10.0	100
Other	44	-70.1	147
Total Buses	**2,131**	**-30.4**	**3,063**
Total Comm. Vehs.	**131,247**	**-32.0**	**193,078**
Total Vehicles	**1,533,336**	**-21.0**	**1,941,221**

LITHUANIA

ECONOMIC SUMMARY

Common Language	Lithuanian
Currency	Litas
Steering	Left Hand Drive

2012 Vehicles in Operation

Passenger Cars	1,745,000
Commercial Vehicles	142,000
Total	1,887,000
Persons per Car	2.0
Persons per Vehicle	1.9

2012 GDP (billions of US $)

Nominal GDP	42.2
Real GDP	32.0
GDP % annual growth	3.6
GDP per capita	10,646

Current Account Balance (billions of US $)

2012:	-0.4
2011:	-1.6

Current Account Balance in % of GDP

2012:	-0.9
2011:	-3.7

Total Number of Fatal Road Accidents (most recent data)

2012:	301

Capital	Vilnius
Population	3,515,858
Population Density	145.3
Surface Area sq km/miles	62,680 / 24,201

Road Infrastructure

Paved Road Surface %	87.7
Total Road Length in km/miles	82,131 / 51,034
Road Traffic (Vehicles per mile)	37.0

2012 Total Vehicle Percent Market Share by Company

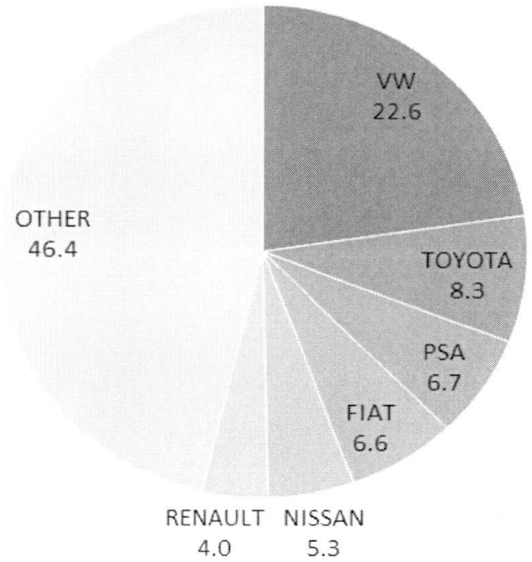

Total Vehicle Sales 2005-2012

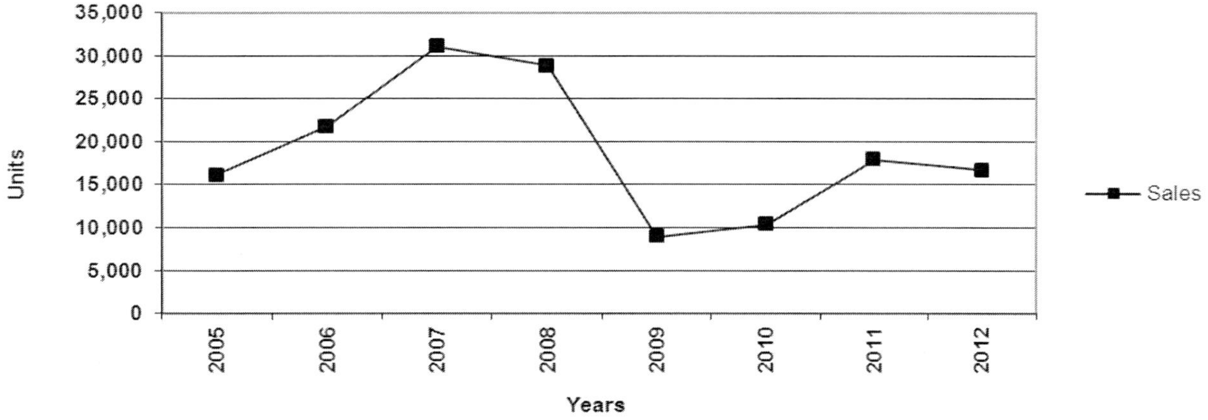

LITHUANIA

SALES

Year	Cars	Commercial Vehicles	Total
2012	12,172	4,502	16,674
2011	13,223	4,690	17,913
2010	7,970	2,399	10,369
2009	7,515	1,403	8,918
2008	22,217	6,668	28,885
2007	21,606	9,484	31,090
2006	14,234	7,510	21,744
2005	10,467	5,668	16,135

SALES BY COMPANY

	2012	% Chg.	2011
Cars			
Aston Martin	2	–	–
Audi	280	-0.7	282
AutoVaz (Lada)	19	-40.6	32
Bentley	4	33.3	3
BMW	*440*	*-10.0*	*489*
Mini	*33*	*10.0*	*30*
BMW Total	473	-8.9	519
Chrysler	–	–	*2*
Dodge	–	–	*2*
Jeep	*39*	*200.0*	*13*
Chrysler Total	39	129.4	17
Dacia	174	-1.7	177
Daihatsu	–	–	28
Daimler (Mercedes-Benz)	159	1.9	156
Alfa Romeo	*71*	–	*71*
Fiat	*873*	*24.0*	*704*
Lancia	*4*	–	–
Fiat Total	948	22.3	775
Ford	431	-30.7	622
Cadillac	–	–	*2*
Chevrolet	*80*	*42.9*	*56*
Opel	*427*	*-43.7*	*759*
GM Total	507	-37.9	817
Honda	215	-16.3	257
Hyundai	533	34.6	396
Jaguar	*6*	*-66.7*	*18*
Land Rover	*96*	*43.3*	*67*
Jaguar Land Rover Total	102	20.0	85
Kia	535	37.5	389
Maserati	1	–	–
Mazda	208	177.3	75
Mitsubishi	165	-28.6	231
Infiniti	*5*	–	*5*
Nissan	*824*	*-49.2*	*1,621*
Nissan Total	829	-49.0	1,626
Citroen	*196*	*-13.7*	*227*
Peugeot	*748*	*-44.7*	*1,352*
Peugeot Citroen Total	944	-40.2	1,579
Porsche	29	3.6	28
Renault	466	13.9	409
Rolls-Royce	1	–	1
Saab	1	-95.2	21
Subaru	315	-33.4	473
Suzuki	81	8.0	75
Tata	–	–	3
Lexus	*102*	*10.9*	*92*
Toyota	*1,129*	*9.2*	*1,034*

LITHUANIA

SALES BY COMPANY

	2012	% Chg.	2011
Toyota Total	1,231	9.3	1,126
SEAT	*40*	*53.8*	*26*
Skoda	*1,074*	*41.9*	*757*
Volkswagen	*2,190*	*7.5*	*2,038*
Volkswagen Total	3,304	17.1	2,821
Volvo	166	-12.6	190
Other	10	–	10
Total Cars	**12,172**	**-7.9**	**13,223**
Light Trucks			
Chrysler (Dodge)	2	–	–
Dacia	19	-36.7	30
Daimler (Mercedes-Benz)	77	16.7	66
Dongfeng	–	–	1
Fiat	158	-36.8	250
Ford	125	8.7	115
GAZ	23	360.0	5
Cadillac	–	–	*2*
Opel	*27*	*-68.6*	*86*
GM Total	27	-69.3	88
Isuzu	17	-54.1	37
Iveco	42	-16.0	50
Kia	10	-83.6	61
Land Rover	1	-75.0	4
Mazda	–	–	4
Mitsubishi	35	-32.7	52
Nissan	62	47.6	42
Citroen	*126*	*21.2*	*104*
Peugeot	*49*	*6.5*	*46*
Peugeot Citroen Total	175	16.7	150
Piaggio	–	–	4
Renault	199	-19.4	247
Tata	–	–	17
Toyota	155	2.0	152
Skoda	*161*	*-30.0*	*230*
Volkswagen	*310*	*41.6*	*219*
Volkswagen Total	471	4.9	449
Total Light Trucks	**1,598**	**-12.4**	**1,824**
Medium/Heavy Trucks			
Other	2,759	-0.1	2,762
Total Med./Hvy Trucks	**2,759**	**-0.1**	**2,762**
Buses			
Other	145	39.4	104
Total Buses	**145**	**39.4**	**104**
Total Comm. Vehs.	**4,502**	**-4.0**	**4,690**
Total Vehicles	**16,674**	**-6.9**	**17,913**

LUXEMBOURG

ECONOMIC SUMMARY

Common Language	Luxembourgish/ French/German
Currency	Euro
Steering	Left Hand Drive

2012 Vehicles in Operation	
Passenger Cars	355,358
Commercial Vehicles	42,303
Total	397,661
Persons per Car	1.4
Persons per Vehicle	1.3

2012 GDP (billions of US $)	
Nominal GDP	56.7
Real GDP	44.6
GDP % annual growth	0.1
GDP per capita	84,289

Current Account Balance (billions of US $)	
2012:	3.4
2011:	4.2

Current Account Balance in % of GDP	
2012:	6.0
2011:	7.1

Total Number of Fatal Road Accidents (most recent data)	
2012:	34

Capital	Luxembourg
Population	514,862
Population Density	515.9
Surface Area sq km/miles	2,586 / 998

Road Infrastructure	
Paved Road Surface %	100.0
Total Road Length in km/miles	5,227 / 3,247
Road Traffic (Vehicles per mile)	122.5

2012 Total Vehicle Percent Market Share by Company

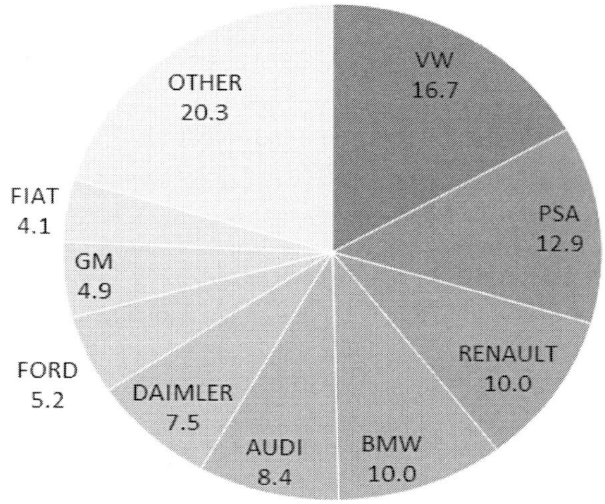

Total Vehicle Sales 2002-2012

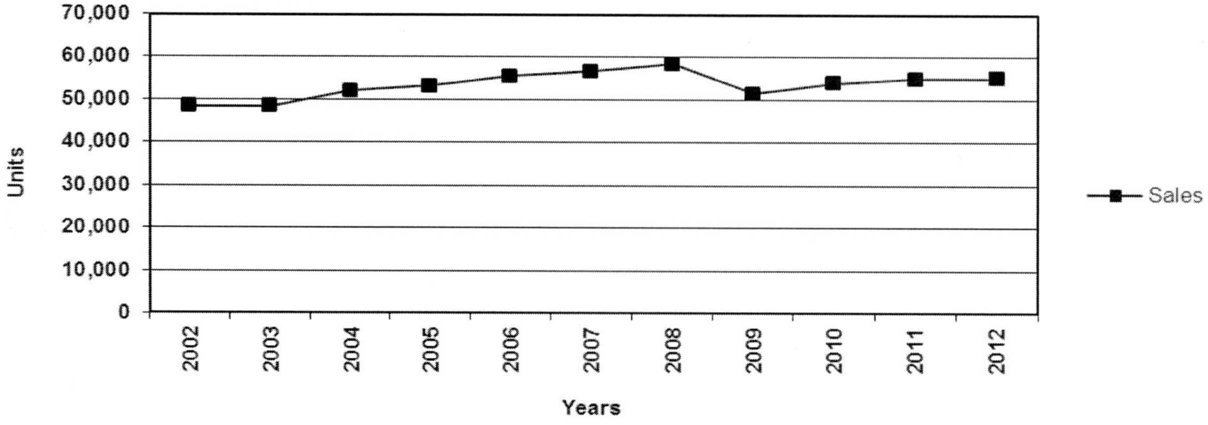

LUXEMBOURG

SALES

Year	Cars	Trucks	Buses	Total
2012	50,398	4,492	159	55,049
2011	49,879	4,939	197	55,015
2010	49,726	4,093	174	53,993
2009	47,265	3,988	209	51,462
2008	52,360	5,810	225	58,395
2007	51,329	5,139	176	56,644
2006	50,833	4,539	168	55,540
2005	48,517	4,478	158	53,153
2004	48,234	3,625	231	52,090
2003	43,624	4,663	136	48,423
2002	43,400	4,944	84	48,428
2001	42,833	5,058	152	48,043
2000	42,304	2,988	116	45,408
1999	40,476	4,262	118	44,856
1998	35,929	3,537	106	39,572
1997	31,418	1,572	*	32,990
1996	30,566	2,617	74	33,257
1995	28,806	1,661	102	30,569
1994	30,120	2,035	51	32,206
1993	25,661	1,887	114	27,662
1992	37,133	2,949	78	40,160
1991	38,777	3,259	90	42,126
1990	34,412	3,006	61	37,479
1989	30,676	2,826	48	33,550
1988	33,847	2,227	30	36,104
1987	32,660	1,895	40	34,595
1986	32,960	1,647	35	34,642
1985	29,235	1,347	37	30,619
1984	28,597	1,148	39	29,784
1983	26,482	1,266	38	27,786
1982	25,796	1,532	33	27,361

* Included with Trucks.

LUXEMBOURG

SALES BY COMPANY

	2012	% Chg.	2011
Cars			
Aston Martin	39	11.4	35
Audi	4,644	–	4,642
AutoVaz (Lada)	8	33.3	6
Bentley	20	42.9	14
BMW	*4,815*	*10.3*	*4,367*
Mini	*681*	*-12.1*	*775*
BMW Total	5,496	6.9	5,142
Caterham	3	–	3
Chrysler	*–*	*–*	*32*
Dodge	*2*	*-96.6*	*58*
Jeep	*248*	*40.1*	*177*
Chrysler Total	250	-6.4	267
Dacia	819	24.8	656
Daihatsu	2	-75.0	8
Mercedes-Benz	*3,260*	*5.1*	*3,102*
Smart	*120*	*-22.6*	*155*
Daimler Total	3,380	3.8	3,257
Ferrari	27	-12.9	31
Alfa Romeo	*563*	*-29.0*	*793*
Fiat	*1,060*	*4.0*	*1,019*
Lancia	*227*	*35.1*	*168*
Fiat Total	1,850	-6.6	1,980
Fisker	2	–	–
Ford	2,536	-13.5	2,932
Cadillac	*7*	*-53.3*	*15*
Chevrolet	*370*	*32.6*	*279*
Opel	*2,146*	*11.0*	*1,933*
GM Total	2,523	13.3	2,227
Honda	233	-9.7	258
Hyundai	1,899	-14.6	2,223
Jaguar	*104*	*11.8*	*93*
Land Rover	*835*	*124.5*	*372*
Jaguar Land Rover Total	939	101.9	465
Kia	328	1.9	322
Lamborghini	6	100.0	3
Maserati	22	-12.0	25
Mazda	124	-31.1	180
Mitsubishi	170	-27.7	235

	2012	% Chg.	2011
Morgan	8	-27.3	11
Infiniti	*36*	*12.5*	*32*
Nissan	*1,378*	*17.2*	*1,176*
Nissan Total	1,414	17.1	1,208
Citroen	*2,894*	*-3.3*	*2,994*
Peugeot	*3,484*	*0.5*	*3,466*
Peugeot Citroen Total	6,378	-1.3	6,460
Porsche	416	5.1	396
Proton (Lotus)	10	-23.1	13
Renault	4,866	23.5	3,940
Rolls-Royce	1	-75.0	4
Saab	3	-94.9	59
Ssangyong	61	-26.5	83
Subaru	110	5.8	104
Suzuki	586	24.4	471
Lexus	*89*	*-13.6*	*103*
Toyota	*1,124*	*31.5*	*855*
Toyota Total	1,213	26.6	958
SEAT	*690*	*-35.5*	*1,069*
Skoda	*1,381*	*-7.9*	*1,499*
Volkswagen	*6,721*	*-6.9*	*7,222*
Volkswagen Total	8,792	-10.2	9,790
Volvo	1,201	-17.3	1,452
Wiesmann	5	66.7	3
Other	14	-12.5	16
Total Cars	**50,398**	**1.0**	**49,879**
Light Trucks			
AutoVaz (Lada)	–	–	3
Dodge	*12*	*500.0*	*2*
Jeep	*1*	*-75.0*	*4*
Chrysler Total	13	116.7	6
Dacia	26	-35.0	40
Mercedes-Benz	*334*	*-1.8*	*340*
Mitsubishi Fuso	*2*	*–*	*–*
Daimler Total	336	-1.2	340
Fiat	379	1.3	374
Ford	292	2.8	284
GAZ (LDV)	–	–	1

SALES BY COMPANY

	2012	% Chg.	2011
GMC	*1*	–	–
Opel	*180*	*7.8*	*167*
GM Total	181	8.4	167
Hyundai	13	8.3	12
Isuzu	61	5.2	58
Iveco	34	61.9	21
Jaguar Land Rover	18	38.5	13
Mazda	–	–	1
Mitsubishi	16	-36.0	25
Nissan	87	-25.0	116
Citroen	*369*	*-7.8*	*400*
Peugeot	*334*	–	*334*
Peugeot Citroen Total	703	-4.2	734
Piaggio	4	33.3	3
Renault	650	-19.9	811
Ssangyong	3	-40.0	5
Suzuki	3	50.0	2
Toyota	24	-22.6	31
Skoda	–	–	*2*
Volkswagen	*412*	*10.5*	*373*
Volkswagen Total	412	9.9	375
Other	7	600.0	1
Total Light Trucks	**3,262**	**-4.7**	**3,423**
Medium/Heavy Trucks			
Mercedes-Benz	*315*	*-29.2*	*445*
Setra	–	–	*1*
Daimler Total	315	-29.4	446
Fiat	20	-35.5	31
Ford	8	-20.0	10
Iveco	137	90.3	72
MAN	*113*	*-45.7*	*208*
Volkswagen Truck & Bus	*18*	*-35.7*	*28*
MAN Total	131	-44.5	236
PACCAR (DAF)	167	10.6	151

	2012	% Chg.	2011
Citroen	–	–	*1*
Peugeot	*2*	–	*2*
Peugeot Citroen Total	2	-33.3	3
Scania	86	-34.8	132
Renault Truck	*72*	*-36.3*	*113*
UD Trucks	*2*	*-60.0*	*5*
Volvo Truck	*161*	*-17.0*	*194*
Volvo Truck Total	235	-24.7	312
Other	129	4.9	123
Total Med./Hvy. Trks.	**1,230**	**-18.9**	**1,516**
Buses			
Evobus	–	–	*4*
Mercedes-Benz	*60*	*15.4*	*52*
Setra	*44*	*2.3*	*43*
Daimler Total	104	5.1	99
Heuliez	*1*	–	*1*
Irisbus	*8*	*-42.9*	*14*
Iveco	–	–	*6*
Iveco Total	9	-57.1	21
MAN	5	-73.7	19
Otokar	–	–	1
Temsa	7	–	7
Van Hool	26	13.0	23
Bova	–	–	*4*
VDL	*5*	–	–
VDL Total	5	25.0	4
Renault Truck	*1*	–	–
Volvo Truck	*2*	*-86.7*	*15*
Volvo Truck Total	3	-80.0	15
Other	–	–	8
Total Buses	**159**	**-19.3**	**197**
Total Comm. Vehs.	**4,651**	**-9.4**	**5,136**
Total Vehicles	**55,049**	**0.1**	**55,015**

NETHERLANDS

ECONOMIC SUMMARY

Common Language	Dutch
Currency	Euro
Steering	Left Hand Drive

2012 Vehicles in Operation

Passenger Cars	8,142,387
Commercial Vehicles	1,071,783
Total	9,214,170
Persons per Car	2.1
Persons per Vehicle	1.8

2012 GDP (billions of US $)

Nominal GDP	773.1
Real GDP	727.3
GDP % annual growth	-0.9
GDP per capita	43,408

Current Account Balance (billions of US $)

2012:	64.6
2011:	81.6

Current Account Balance in % of GDP

2012:	8.3
2011:	9.7

Total Number of Fatal Road Accidents (most recent data)

2010:	640

Capital	Amsterdam
Population	16,805,037
Population Density	1,286.0
Surface Area sq km/miles	33,893 / 13,086

Road Infrastructure

Paved Road Surface %	90.0
Total Road Length in km/miles	136,827 / 85,020
Road Traffic (Vehicles per mile)	108.4

2012 Total Vehicle Percent Market Share by Company

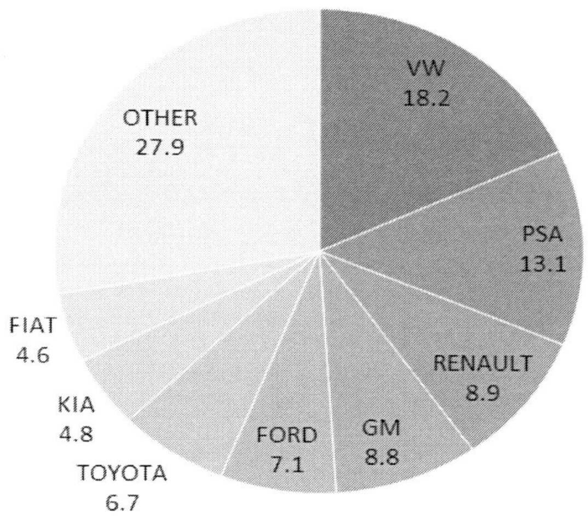

Total Vehicle Production and Sales 2002-2012

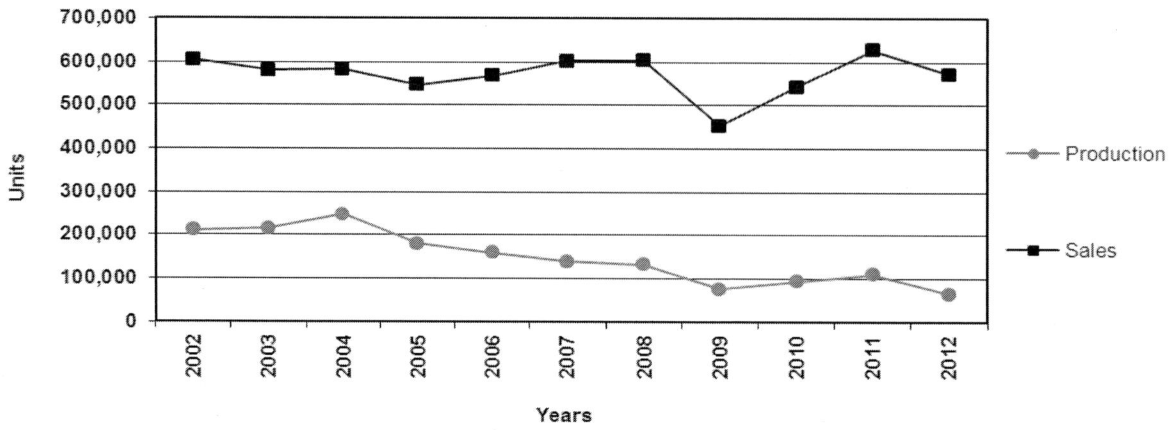

NETHERLANDS

PRODUCTION

Year	Cars	Trucks	Buses	Total
2012	–	64,293	1,461	65,754
2011	40,772	68,176	1,435	110,383
2010	48,025	44,764	1,343	94,132
2009	50,620	24,490	1,641	76,751
2008	59,223	71,326	1,945	132,494
2007	61,912	74,759	2,007	138,678
2006	87,332	70,202	1,920	159,454
2005	115,121	63,463	1,984	180,568
2004	187,600	58,442	1,461	247,503
2003	163,080	50,654	1,547	215,281
2002	182,368	28,058	1,453	211,879
2001	189,261	30,072	1,254	220,587
2000	215,085	28,902	1,632	245,619
1999	262,242	23,330	1,743	287,315
1998	242,989	25,870	1,601	270,460
1997	197,225	19,063	1,365	217,653
1996	145,206	18,750	1,152	165,108
1995	100,434	31,376	1,045	132,855
1994	92,044	22,184	514	114,742
1993	80,246	12,994	880	94,120
1992	94,019	22,146	1,827	117,992
1991	84,709	24,585	1,518	110,812
1990	121,300	28,227	1,605	151,132
1989	134,600	30,583	1,794	166,977
1988	119,786	27,641	1,410	148,837
1987	125,247	25,646	1,031	151,924
1986	118,976	22,411	967	142,354
1985	108,083	19,984	907	128,974
1984	108,598	19,779	744	129,121
1983	105,597	15,652	988	122,237
1982	90,615	17,295	1,095	109,005

NETHERLANDS

PRODUCTION BY MANUFACTURER

	2012	% Chg.	2011		2012	% Chg.	2011
Cars				**Buses**			
Mitsubishi	–	–	40,772	*Bova*	*465*	*-1.9*	*474*
Total Cars	**–**	**–**	**40,772**	*VDL*	*996*	*3.6*	*961*
				VDL Total	1,461	1.8	1,435
Medium/Heavy Trucks				**Total Buses**	**1,461**	**1.8**	**1,435**
DAF	35,000	-6.0	37,232				
Scania	28,000	-6.4	29,906	**Total Comm. Vehs.**	**65,754**	**-5.5**	**69,611**
Other	1,293	24.6	1,038				
Total Med./Hvy. Trks.	**64,293**	**-5.7**	**68,176**	**Total Vehicles**	**65,754**	**-40.4**	**110,383**

NETHERLANDS

SALES

Year	Cars	Trucks	Buses	Total
2012	502,544	68,570	779	571,893
2011	555,920	71,305	640	627,865
2010	483,168	59,209	572	542,949
2009	387,699	63,045	1,163	451,907
2008	499,983	102,825	1,330	604,138
2007	505,643	96,023	1,252	602,918
2006	483,999	83,730	874	568,603
2005	465,196	79,535	1,182	545,913
2004	481,453	100,483	918	582,854
2003	489,952	90,050	864	580,866
2002	510,257	94,771	726	605,754
2001	530,331	100,310	837	631,478
2000	597,182	112,667	907	710,756
1999	611,776	116,097	593	728,466
1998	543,056	113,872	528	657,456
1997	478,318	95,980	372	574,670
1996	473,474	83,222	586	557,282
1995	446,388	64,553	565	511,506
1994	433,913	53,062	622	487,597
1993	391,934	68,715	660	461,309
1992	491,970	92,833	676	585,479
1991	490,755	73,099	952	564,806
1990	502,704	67,606	1,122	571,432
1989	496,115	58,371	834	555,320
1988	482,616	59,048	866	542,530
1987	555,743	83,700	783	640,226
1986	560,510	78,716	770	639,996
1985	495,682	61,808	775	558,265
1984	461,391	46,492	795	508,678
1983	459,129	37,694	743	497,566
1982	406,991	37,862	825	445,678

NETHERLANDS

SALES BY COMPANY

	2012	% Chg.	2011
Cars			
Aston Martin	16	-48.4	31
Audi	20,201	8.9	18,552
AutoVaz (Lada)	3	-50.0	6
Bentley	26	100.0	13
BMW	*18,636*	*16.4*	*16,010*
Mini	*4,039*	*1.7*	*3,972*
BMW Total	22,675	13.5	19,982
Chana (Jiangling)	26	–	–
Chrysler	–	–	*73*
Dodge	*8*	*-95.2*	*167*
Jeep	*387*	*-15.7*	*459*
Chrysler Total	395	-43.5	699
Dacia	2,158	-53.4	4,628
Daihatsu	133	-86.0	947
Mercedes-Benz	*11,264*	*-0.4*	*11,307*
Smart	*469*	*-46.5*	*877*
Daimler Total	11,733	-3.7	12,184
Ferrari	31	34.8	23
Alfa Romeo	*3,653*	*-34.4*	*5,567*
Fiat	*19,031*	*-35.1*	*29,345*
Lancia	*832*	*15.7*	*719*
Fiat Total	23,516	-34.0	35,631
Fisker	140	–	–
Ford	35,861	-12.6	41,050
Cadillac	*15*	–	*15*
Chevrolet	*8,715*	*-6.2*	*9,292*
Opel	*35,527*	*-13.8*	*41,236*
Vauxhall	–	–	*4*
GM Total	44,257	-12.4	50,547
Honda	2,969	-39.4	4,899
Hyundai	21,627	-8.0	23,499
Jaguar	*341*	*0.3*	*340*
Land Rover	*1,342*	*44.6*	*928*
Jaguar Land Rover Total	1,683	32.7	1,268
Kia	27,322	17.7	23,223
Lamborghini	2	–	–
Maserati	1	-92.3	13
Mazda	6,149	30.1	4,727
Micro-Vett	–	–	3
Mitsubishi	3,584	-46.5	6,700
Morgan	3	-25.0	4
Infiniti	*12*	*-71.4*	*42*
Nissan	*9,405*	*-21.8*	*12,032*
Nissan Total	9,417	-22.0	12,074
Citroen	*25,607*	*-12.6*	*29,309*
Peugeot	*40,238*	*-12.6*	*46,038*
Peugeot Citroen Total	65,845	-12.6	75,347
Porsche	944	29.3	730
Proton (Lotus)	1	-85.7	7
Renault	43,821	-1.5	44,510
Rolls-Royce	1	-80.0	5
Saab	55	-90.9	602
Ssangyong	25	-24.2	33
Subaru	967	-12.2	1,101
Suzuki	12,519	-24.3	16,536
Tesla	26	-39.5	43
Lexus	*3,920*	*38.7*	*2,826*
Toyota	*33,726*	*-5.0*	*35,485*
Toyota Total	37,646	-1.7	38,311
SEAT	*12,667*	*-31.7*	*18,559*
Skoda	*15,564*	*-24.6*	*20,647*
Volkswagen	*60,350*	*-0.9*	*60,881*
Volkswagen Total	88,581	-11.5	100,087
Volvo	16,925	2.3	16,538
Other	1,260	-7.8	1,367
Total Cars	**502,544**	**-9.6**	**555,920**
Light Trucks			
Audi	8	–	–
BMW	*7*	–	–
Mini	*5*	–	–
BMW Total	12	–	–
Dodge	*5*	–	–
Jeep	*71*	–	–
Ram	*393*	–	–
Chrysler Total	469	–	–
Dacia	39	–	–
Freightliner	*1*	–	–
Mercedes-Benz	*7,557*	*-9.4*	*8,345*
Mitsubishi Fuso	*113*	–	–
Daimler Total	7,671	-8.1	8,345
Dongfeng	5	–	–
Fiat	*2,978*	*-5.8*	*3,160*
Lancia	*16*	–	–
Fiat Total	2,994	-5.3	3,160

NETHERLANDS

SALES BY COMPANY

	2012	% Chg.	2011
Ford	4,599	-14.7	5,391
GAZ (LDV)	3	–	–
Cadillac	*3*	–	–
Chevrolet	*27*	–	–
GMC	*5*	–	–
Hummer	*1*	–	–
Opel	*5,805*	*2.1*	*5,685*
GM Total	5,841	2.7	5,685
Hyundai	174	–	–
Isuzu	37	–	–
Iveco	1,305	-15.8	1,549
Kia	42	–	–
Land Rover	180	-15.9	214
Mazda	1	–	–
Micro-Vett	2	–	–
Mitsubishi	78	–	–
Nissan	813	–	–
Citroen	*2,835*	*1.4*	*2,796*
Peugeot	*6,089*	*31.2*	*4,642*
Peugeot Citroen Total	8,924	20.0	7,438
Piaggio	73	–	–
Renault	7,225	-0.8	7,281
Ssangyong	55	–	–
Suzuki	3	–	–
Toyota	516	–	–
SEAT	*2*	–	–
Volkswagen	*15,433*	-6.7	*16,533*
Volkswagen Total	15,435	-6.6	16,533
Volvo	38	–	–
Volvo Truck (Renault Truck)	4	-81.0	21
Other	22	-99.3	3,037
Total Light Trucks	**56,568**	**-3.5**	**58,654**

Medium/Heavy Trucks

	2012	% Chg.	2011
Ashok Leyland (Avia)	1	–	–
Astra	1	–	–
Chrysler (Dodge)	5	25.0	4
Freightliner	–	–	*1*
Mercedes-Benz	*1,875*	*17.7*	*1,593*
Mitsubishi Fuso	*8*	*-65.2*	*23*
Daimler Total	1,883	16.5	1,617
Fiat	12	–	12
Ford	8	-11.1	9
GM (Opel)	8	700.0	1
Iveco	541	-14.1	630
MAN	*1,404*	*-3.8*	*1,460*
Volkswagen Truck & Bus	*41*	*-10.9*	*46*
MAN Total	1,445	-4.1	1,506
Modec	–	–	2
Nissan	3	–	–
DAF	*3,759*	*2.6*	*3,663*
Kenworth	*1*	–	–
Peterbilt	*2*	–	–
PACCAR Total	3,762	2.7	3,663
Peugeot Citroen	4	100.0	2
Scania	1,790	-27.1	2,457
Tatra	2	–	–
Toyota	1	–	1
Renault Truck	*406*	*16.0*	*350*
Volvo Truck	*2,060*	*-8.2*	*2,245*
Volvo Truck Total	2,466	-5.0	2,595
Other	70	-53.9	152
Total Med./Hvy. Trks.	**12,002**	**-5.1**	**12,651**

Buses

	2012	% Chg.	2011
BYD	1	–	–
Mercedes-Benz	*65*	*124.1*	*29*
Setra	*6*	*20.0*	*5*
Daimler Total	71	108.8	34
Iveco (Irisbus)	17	183.3	6
MAN	170	136.1	72
Optare	1	-97.1	34
Scania	3	-98.6	208
Temsa	2	-33.3	3
Van Hool	24	-22.6	31
Bova	*11*	*83.3*	*6*
VDL	*298*	*111.3*	*141*
VDL Total	309	110.2	147
Volvo Truck	76	245.5	22
Other	105	26.5	839
Total Buses	**779**	**21.7**	**640**
Total Comm. Vehs.	**69,349**	**-3.6**	**71,945**
Total Vehicles	**571,893**	**-8.9**	**627,865**

NORWAY

ECONOMIC SUMMARY

Common Language	Norwegian
Currency	Norwegian Krone
Steering	Left Hand Drive

2012 Vehicles in Operation

Passenger Cars	2,442,964
Commerical Vehicles	561,420
Total	3,004,384
Persons per Car	1.9
Persons per Vehicle	1.6

2012 GDP (billions of US $)

Nominal GDP	501.1
Real GDP	475.2
GDP % annual growth	3.0
GDP per capita	94,312

Current Account Balance (billions of US $)

2012:	71.2
2011:	62.7

Current Account Balance in % of GDP

2012:	14.2
2011:	12.8

Total Number of Fatal Road Accidents (most recent data)

2012:	145

Capital	Oslo
Population	4,722,701
Population Density	40.2
Surface Area sq km/miles	304,282 / 117,484

Road Infrastructure

Paved Road Surface %	77.5
Total Road Length in km/miles	93,509 / 58,104
Road Traffic (Vehicles per mile)	51.7

2012 Total Vehicle Percent Market Share by Company

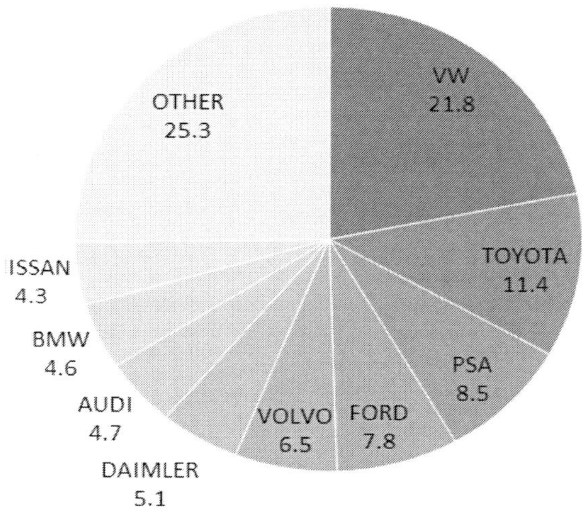

Total Vehicle Sales 2002-2012

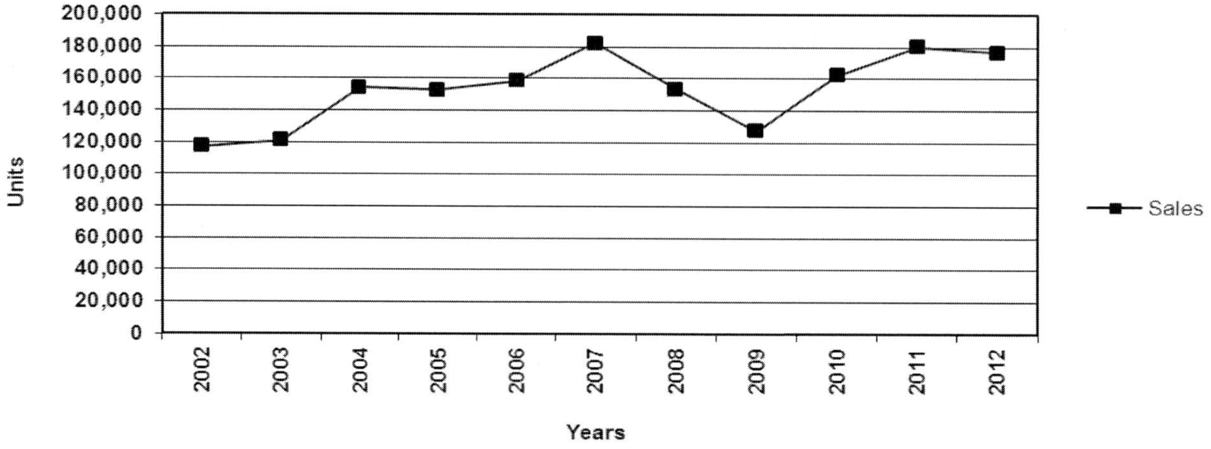

NORWAY

SALES

Year	Cars	Trucks	Buses	Total
2012	137,967	37,711	1,231	176,909
2011	138,345	40,623	1,345	180,313
2010	127,754	33,118	1,482	162,354
2009	98,675	27,602	1,160	127,437
2008	110,617	41,398	1,232	153,247
2007	129,195	51,870	1,138	182,203
2006	109,164	48,400	843	158,407
2005	109,907	41,513	1,168	152,588
2004	115,645	37,471	936	154,052
2003	89,921	30,502	815	121,238
2002	88,721	27,545	977	117,243
2001	91,916	37,378	1,014	130,308
2000	97,378	33,869	1,379	132,626
1999	101,278	27,151	2,150	130,579
1998	118,553	33,569	2,704	154,826
1997	127,733	34,216	2,849	164,798
1996	124,994	33,448	3,380	161,822
1995	90,502	34,543	3,453	128,498
1994	85,067	27,520	2,704	115,291
1993	60,822	20,664	3,021	84,507
1992	59,539	16,703	4,103	80,345
1991	53,382	15,016	2,535	70,933
1990	61,860	21,367	1,668	84,895
1989	54,892	16,495	1,058	72,445
1988	67,822	21,109	1,408	90,339
1987	115,097	29,776	1,718	146,591
1986	167,352	41,376	1,737	210,465
1985	159,079	41,178	1,352	201,609
1984	107,237	26,645	1,322	135,204
1983	109,672	23,710	1,526	134,908
1982	115,573	20,434	1,745	137,752

NORWAY

SALES BY COMPANY

	2012	% Chg.	2011
Cars			
Aston Martin	4	-63.6	11
Audi	8,321	13.2	7,351
Bentley	4	–	–
BMW	*7,325*	*16.5*	*6,288*
Mini	*856*	*5.3*	*813*
BMW Total	8,181	15.2	7,101
Chrysler	*–*	*–*	*3*
Dodge	*1*	*–*	*–*
Jeep	*209*	*–*	*12*
Chrysler Total	210	–	15
Dacia	6	20.0	5
Daihatsu	–	–	34
Mercedes-Benz	*5,616*	*23.6*	*4,544*
Smart	*48*	*-48.9*	*94*
Daimler Total	5,664	22.1	4,638
Ferrari	4	300.0	1
Alfa Romeo	*283*	*-43.8*	*504*
Fiat	*631*	*-10.6*	*706*
Lancia	*–*	*–*	*1*
Fiat Total	914	-24.5	1,211
Fisker	6	–	–
Ford	11,007	-15.3	12,991
Cadillac	*5*	*–*	*–*
Chevrolet	*79*	*243.5*	*23*
Opel	*3,616*	*-23.3*	*4,716*
GM Total	3,700	-21.9	4,739
Honda	1,731	-13.8	2,008
Hyundai	5,557	1.0	5,504
Isuzu	3	–	3
Iveco	2	-90.5	21
Jaguar	*47*	*2.2*	*46*
Land Rover	*684*	*218.1*	*215*
Jaguar Land Rover Total	731	180.1	261
Kia	4,917	88.1	2,614
Mazda	2,927	66.1	1,762
Mitsubishi	4,654	-40.7	7,842
Morgan	5	66.7	3
Infiniti	*–*	*–*	*1*
Nissan	*6,658*	*14.2*	*5,832*
Nissan Total	6,658	14.1	5,833
Citroen	*3,583*	*-2.3*	*3,666*
Peugeot	*6,209*	*-16.4*	*7,427*
Peugeot Citroen Total	9,792	-11.7	11,093
Porsche	328	106.3	159
Proton (Lotus)	–	–	2
Renault	657	-25.3	880
Saab	11	-96.3	298
Ssangyong	36	-67.6	111
Subaru	3,559	50.4	2,367
Suzuki	2,060	-33.2	3,082
Tesla	32	14.3	28
Lexus	*493*	*3.4*	*477*
Toyota	*16,903*	*10.9*	*15,236*
Toyota Total	17,396	10.7	15,713
SEAT	*2*	*–*	*–*
Skoda	*6,793*	*-10.9*	*7,620*
Volkswagen	*20,507*	*-1.9*	*20,898*
Volkswagen Total	27,302	-4.3	28,518
Volvo	11,550	-3.7	11,995
Other	38	-74.8	151
Total Cars	**137,967**	**-0.3**	**138,345**
Light Trucks			
Dodge	*13*	*333.3*	*3*
Jeep	*244*	*–*	*12*
Chrysler Total	257	–	15
Dacia	2	–	–
Daimler (Mercedes-Benz)	2,131	0.1	2,129
Fiat	1,097	20.8	908
Ford	2,728	-22.3	3,510
Cadillac	*7*	*600.0*	*1*
Chevrolet	*62*	*17.0*	*53*
Opel	*994*	*-23.9*	*1,306*
GM Total	1,063	-21.8	1,360
Hyundai	250	-28.4	349
Isuzu	669	15.7	578
Iveco	109	-2.7	112

NORWAY

SALES BY COMPANY

	2012	% Chg.	2011
Kia	109	-40.4	183
Land Rover	223	-14.9	262
Mazda	7	-79.4	34
Mitsubishi	803	-30.5	1,156
Nissan	1,003	-10.5	1,121
Citroen	*2,249*	*-21.2*	*2,854*
Peugeot	*2,986*	*-7.8*	*3,238*
Peugeot Citroen Total	5,235	-14.1	6,092
Piaggio	1	–	–
Renault	519	3.2	503
Ssangyong	142	-37.2	226
Toyota	2,813	-41.6	4,819
Volkswagen	11,180	3.5	10,798
Other	1,509	11.2	1,357
Total Light Trucks	**31,850**	**-10.3**	**35,512**
Medium/Heavy Trucks			
Mercedes-Benz	*740*	*13.3*	*653*
Mitsubishi Fuso	*64*	*-1.5*	*65*
Daimler Total	804	12.0	718
Fiat	2	–	2
Ford	2	-77.8	9
Iveco	202	32.0	153
MAN	*433*	*16.1*	*373*
Volkswagen Truck & Bus	*28*	*115.4*	*13*
MAN Total	461	19.4	386
PACCAR (DAF)	122	16.2	105
Peugeot	1	-92.9	14
Scania	1,484	14.2	1,300
Sisu	4	–	–
Renault Truck	*108*	*50.0*	*72*
UD Trucks	–	–	*2*
Volvo Truck	*1,556*	*24.0*	*1,255*
Volvo Truck Total	1,664	25.2	1,329
Other	1,115	1.8	1,095
Total Med./Hvy. Trks.	**5,861**	**14.7**	**5,111**
Buses			
Mercedes-Benz	*342*	*-23.7*	*448*
Setra	*39*	*129.4*	*17*
Daimler Total	381	-18.1	465
Fiat	16	33.3	12
Ford	64	700.0	8
GM (Opel)	4	–	–
Irisbus	*116*	–	–
Iveco	*46*	*-71.8*	*163*
Iveco Total	162	-0.6	163
MAN	*164*	*1.2*	*162*
Volkswagen Truck & Bus	*8*	*-42.9*	*14*
MAN Total	172	-2.3	176
Peugeot	14	-41.7	24
Scania	39	-15.2	46
Temsa	3	-40.0	5
VDL	13	-18.8	16
Renault Truck	*4*	*100.0*	*2*
Volvo Truck	*299*	*-21.9*	*383*
Volvo Truck Total	303	-21.3	385
Other	60	33.3	45
Total Buses	**1,231**	**-8.5**	**1,345**
Total Comm. Vehs.	**38,942**	**-7.2**	**41,968**
Total Vehicles	**176,909**	**-1.9**	**180,313**

POLAND

ECONOMIC SUMMARY

Common Language	Polish
Currency	Zloty
Steering	Left Hand Drive

2012 Vehicles in Operation

Passenger Cars	18,744,000
Commercial Vehicles	3,278,000
Total	22,022,000
Persons per Car	2.0
Persons per Vehicle	1.7

2012 GDP (billions of US $)

Nominal GDP	487.7
Real GDP	426.8
GDP % annual growth	2.0
GDP per capita	10,974

Current Account Balance (billions of US $)

2012:	-17.4
2011:	-25.0

Current Account Balance in % of GDP

2012:	-3.6
2011:	-4.9

Total Number of Fatal Road Accidents (most recent data)

2012:	3,571

Capital	Warsaw
Population	38,383,809
Population Density	326.7
Surface Area sq km/miles	304,255 / 117,473

Road Infrastructure

Paved Road Surface %	69.7
Total Road Length in km/miles	423,997 / 263,459
Road Traffic (Vehicles per mile)	83.6

2012 Total Vehicle Percent Market Share by Company

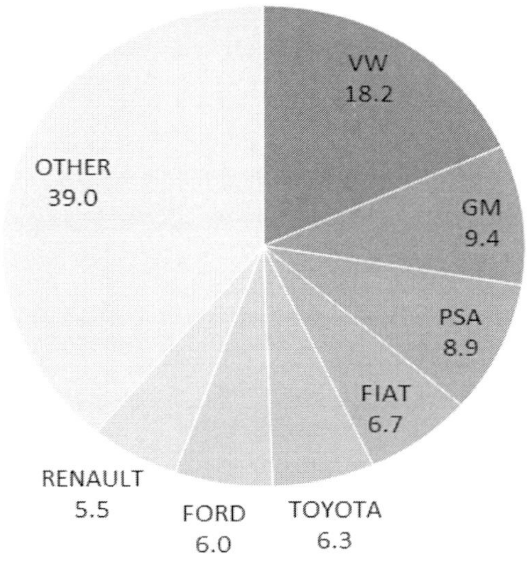

Total Vehicle Production and Sales 2002-2012

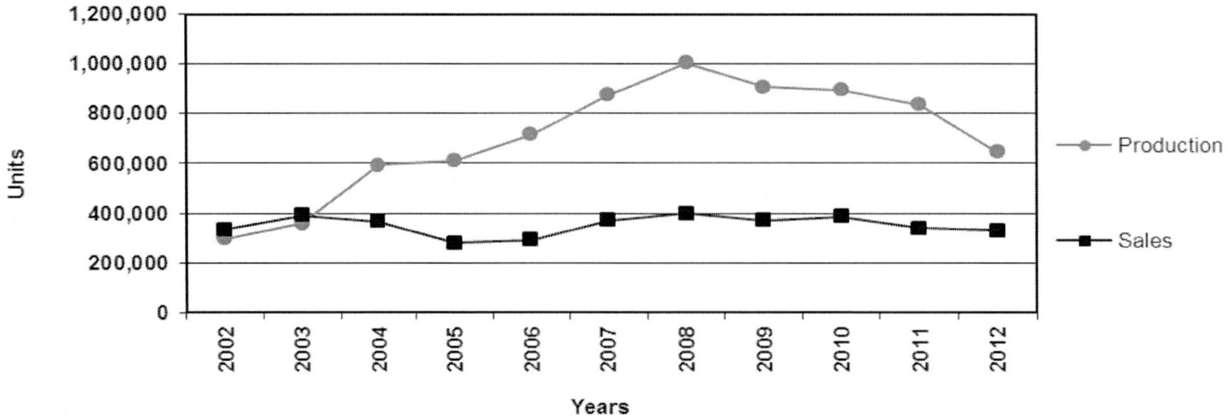

POLAND

PRODUCTION

Year	Cars	Trucks	Total
2012	533,918	110,796	644,714
2011	722,285	113,569	835,854
2010	799,255	95,552	894,807
2009	833,314	73,762	907,076
2008	864,174	141,533	1,005,707
2007	761,993	112,609	874,602
2006	608,919	107,425	716,344
2005	527,020	85,003	612,023
2004	516,152	76,898	593,050
2003	330,176	30,461	360,637
2002	293,796	3,913	297,709
2001	331,153	4,290	335,443
2000	474,674	13,528	488,202
1999	580,241	19,870	600,111
1998	512,140	28,736	540,876
1997	415,884	33,272	449,156
1996	413,251	37,748	450,999
1995	350,124	27,400	377,524

PRODUCTION BY MANUFACTURER

	2012	% Chg.	2011
Cars			
Fiat	330,954	-25.9	446,503
Daewoo	–	–	*6,693*
Opel	*125,469*	*-27.9*	*174,030*
GM Total	125,469	-30.6	180,723
Volkswagen	77,495	-18.5	95,059
Total Cars	**533,918**	**-26.1**	**722,285**
Light Trucks			
Fiat	17,418	-18.0	21,242
Volkswagen	84,501	3.1	81,942
Total Light Trucks	**101,919**	**-1.2**	**103,184**
Medium/Heavy Trucks			
MAN	8,877	-14.5	10,385
Total Med./Hvy. Trks.	**8,877**	**-14.5**	**10,385**
Total Comm. Vehs.	**110,796**	**-2.4**	**113,569**
Total Vehicles	**644,714**	**-22.9**	**835,854**

POLAND

SALES

Year	Cars	Commercial Vehicles	Total
2012	273,589	56,214	329,803
2011	277,430	61,336	338,766
2010	333,599	54,260	387,859
2009	320,119	51,716	371,835
2008	319,922	80,610	400,532
2007	293,305	77,987	371,292
2006	238,982	54,104	293,086
2005	235,522	45,559	281,081
2004	320,156	46,846	367,002
2003	358,252	33,886	392,138
2002	308,158	27,540	335,698
2001	327,251	29,395	356,646
2000	478,752	41,117	519,869
1999	640,183	50,813	690,996
1998	515,256	61,200	576,456
1997	477,960	65,300	543,260
1996	393,478	64,156	457,634
1995	263,881	31,946	295,827
1994	250,282	23,955	274,237
1993	241,613	23,585	265,198
1992	199,279	23,116	222,395

POLAND

SALES BY COMPANY

	2012	% Chg.	2011
Cars			
Audi	5,111	–	–
BMW	5,555	21.9	4,558
Dacia	9,570	96.6	4,867
Daimler (Mercedes-Benz)	5,415	11.4	4,862
Fiat	12,932	-26.7	17,641
Ford	16,923	-26.9	23,151
Chevrolet	*10,474*	*36.7*	*7,660*
Opel	*19,550*	*-10.8*	*21,922*
GM Total	30,024	1.5	29,582
Honda	6,896	1.6	6,790
Hyundai	15,365	8.8	14,119
Kia	15,127	18.8	12,735
Mitsubishi	3,858	-37.5	6,168
Nissan	13,377	-3.9	13,922
Citroen	*10,212*	*-3.1*	*10,544*
Peugeot	*11,797*	*5.6*	*11,170*
Peugeot Citroen Total	22,009	1.4	21,714
Renault	13,587	-19.8	16,936
Suzuki	5,988	-15.7	7,102
Toyota	19,853	5.2	18,866
Skoda	*32,213*	*-0.6*	*32,404*
Volkswagen	*23,394*	*17.3*	*19,947*
Volkswagen Total	55,607	6.2	52,351
Volvo	4,480	-9.6	4,954
Other	11,912	-30.4	17,112
Total Cars	**273,589**	**-1.4**	**277,430**
Light Trucks			
Dacia	713	0.6	709
Daimler (Mercedes-Benz)	2,951	0.1	2,948
Fiat	9,222	-0.6	9,276
Ford	2,912	-25.7	3,921
GM (Opel)	1,016	–	1,016
Hyundai	255	–	–
Isuzu	–	–	176
Iveco	1,785	-15.2	2,104
Mitsubishi	443	-34.6	677

	2012	% Chg.	2011
Nissan	903	-30.6	1,302
Citroen	*3,094*	*-38.3*	*5,017*
Peugeot	*4,173*	*-17.6*	*5,064*
Peugeot Citroen Total	7,267	-27.9	10,081
Renault	4,585	-3.4	4,748
Toyota	827	-16.5	990
Skoda	*1,315*	*56.7*	*839*
Volkswagen	*3,051*	*-4.2*	*3,185*
Volkswagen Total	4,366	8.5	4,024
Other	1,262	75.3	720
Total Light Trucks	**38,507**	**-9.8**	**42,692**
Medium/Heavy Trucks			
Mercedes-Benz	*2,770*	*214.1*	*882*
Mitsubishi Fuso	*28*	*-46.2*	*52*
Daimler Total	2,798	199.6	934
Iveco	1,572	-55.6	3,543
MAN	2,675	34.8	1,984
PACCAR (DAF)	3,545	6.3	3,334
Scania	2,141	-5.8	2,273
Renault Truck	*1,322*	*-34.3*	*2,012*
Volvo Truck	*2,220*	*-16.1*	*2,647*
Volvo Truck Total	3,542	-24.0	4,659
Other	163	-57.0	379
Total Med./Hvy. Trks.	**16,436**	**-3.9**	**17,106**
Buses			
Mercedes-Benz	*407*	*-5.6*	*431*
Setra	*54*	*–*	*–*
Daimler Total	461	7.0	431
Iveco (Irisbus)	34	-59.0	83
MAN	70	-18.6	86
Solaris	257	-47.2	487
Other	449	-0.4	451
Total Buses	**1,271**	**-17.4**	**1,538**
Total Comm. Vehs.	**56,214**	**-8.4**	**61,336**
Total Vehicles	**329,803**	**-2.6**	**338,766**

PORTUGAL

ECONOMIC SUMMARY

Common Language	Portuguese
Currency	Euro
Steering	Left Hand Drive

Capital	Lisbon
Population	10,799,270
Population Density	305.8
Surface Area sq km/miles	91,470 / 35,317

2012 Vehicles in Operation

Passenger Cars	4,497,000
Commercial Vehicles	1,310,000
Total	5,807,000
Persons per Car	2.4
Persons per Vehicle	1.9

Road Infrastructure

Paved Road Surface %	86.0
Total Road Length in km/miles	82,900 / 51,511
Road Traffic (Vehicles per mile)	112.7

2012 GDP (billions of US $)

Nominal GDP	212.7
Real GDP	205.3
GDP % annual growth	-3.2
GDP per capita	19,473

Current Account Balance (billions of US $)

2012:	-3.3
2011:	-16.7

Current Account Balance in % of GDP

2012:	-1.5
2011:	-7.0

Total Number of Fatal Road Accidents (most recent data)

2010:	967

2012 Total Vehicle Percent Market Share by Company

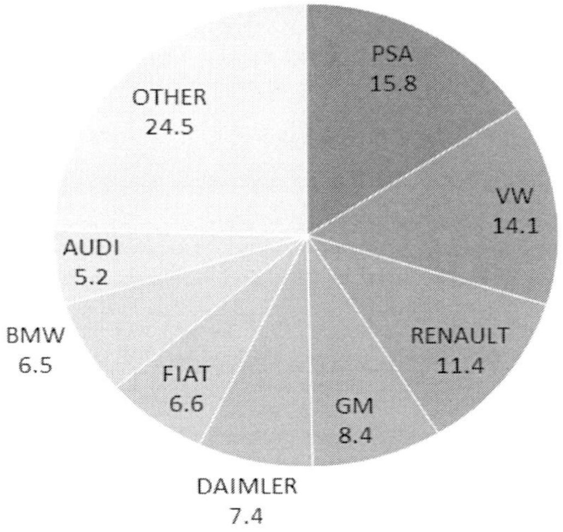

Total Vehicle Production and Sales 2002-2012

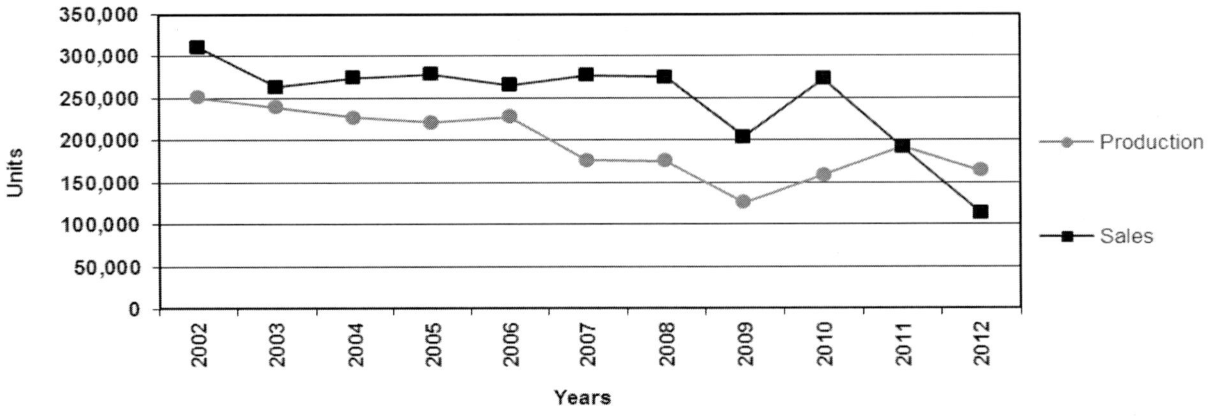

PORTUGAL

PRODUCTION

Year	Cars	Trucks	Buses	Total
2012	115,735	47,821	5	163,561
2011	141,779	50,458	5	192,242
2010	114,563	44,090	70	158,723
2009	101,680	24,251	84	126,015
2008	132,242	42,787	126	175,155
2007	134,047	42,037	158	176,242
2006	143,478	83,720	127	227,325
2005	137,602	83,318	140	221,060
2004	150,781	75,821	126	226,728
2003	165,576	73,785	139	239,500
2002	182,573	68,090	169	250,832
2001	177,357	62,135	227	239,719
2000	178,509	68,037	178	246,724
1999	186,996	65,148	146	252,290
1998	181,388	89,445	197	271,030
1997	186,010	80,974	179	267,163
1996	152,646	80,369	117	233,132
1995	73,185	85,547	163	158,895
1994	37,754	87,278	177	125,209
1993	60,075	61,999	133	122,207
1992	96,179	67,047	*	163,226
1991	71,973	69,404	*	141,377
1990	60,221	77,466	*	137,687
1989	73,181	72,906	*	146,087
1988	71,088	65,436	*	136,524
1987	70,830	53,067	*	123,897
1986	62,113	33,893	*	96,006
1985	60,975	26,548	*	87,523
1984	61,153	23,114	*	84,267
1983	65,903	29,128	*	95,031
1982	64,841	54,117	*	118,958

* Included with Trucks.

PORTUGAL

PRODUCTION BY MANUFACTURER

	2012	% Chg.	2011
Cars			
Citroen	*1,451*	*-60.6*	*3,679*
Peugeot	*1,734*	*-65.3*	*5,000*
Peugeot Citroen Total	3,185	-63.3	8,679
SEAT	*19,393*	*6.9*	*18,140*
Volkswagen	*93,157*	*-19.0*	*114,960*
Volkswagen Total	112,550	-15.4	133,100
Total Cars	**115,735**	**-18.4**	**141,779**
Light Trucks			
Isuzu	294	-35.5	456
Mitsubishi Fuso	1,451	-42.3	2,515
Citroen	*22,123*	*0.5*	*22,002*
Peugeot	*18,632*	*-5.0*	*19,609*
Peugeot Citroen Total	40,755	-2.1	41,611
Toyota	1,331	-26.2	1,803
Total Light Trucks	**43,831**	**-5.5**	**46,385**
Medium/Heavy Trucks			
Isuzu	1,279	46.2	875
Mitsubishi Fuso	2,666	-10.6	2,983
Toyota	45	-79.1	215
Total Med./Hvy. Trucks	**3,990**	**-2.0**	**4,073**
Buses			
Toyota	5	–	5
Total Buses	**5**	**–**	**5**
Total Comm. Vehs.	**47,826**	**-5.2**	**50,463**
Total Vehicles	**163,561**	**-14.9**	**192,242**

PORTUGAL

SALES

Year	Cars	Trucks	Buses	Total
2012	95,290	17,898	223	113,411
2011	153,433	37,552	331	191,316
2010	223,491	48,781	489	272,761
2009	160,996	42,057	629	203,682
2008	213,386	60,899	749	275,034
2007	201,868	74,091	725	276,684
2006	194,684	69,895	579	265,158
2005	203,443	74,388	728	278,559
2004	197,521	75,937	641	274,099
2003	189,792	72,804	558	263,154
2002	226,092	84,037	694	310,823
2001	255,215	105,378	873	361,466
2000	257,836	128,009	927	386,772
1999	272,883	134,143	652	407,678
1998	248,398	125,515	703	374,616
1997	226,593	95,344	493	322,430
1996	269,839	77,153	357	347,349
1995	229,875	62,950	398	293,223
1994	244,379	110,800	352	355,531
1993	249,876	89,749	382	340,007
1992	281,903	87,944	379	370,226
1991	231,397	73,929	431	305,757
1990	212,680	75,831	567	289,078
1989	192,962	70,627	433	264,022
1988	227,064	57,269	372	284,705
1987	129,243	49,673	403	179,319
1986	114,226	32,054	301	146,581
1985	104,195	23,146	294	127,635
1984	85,636	20,670	543	106,849
1983	92,294	27,232	570	120,096
1982	85,194	52,177	871	138,242

PORTUGAL

SALES BY COMPANY

	2012	% Chg.	2011
Cars			
Aston Martin	3	-82.4	17
Audi	5,942	-10.3	6,622
Bentley	3	-25.0	4
BMW	*6,368*	*-13.5*	*7,363*
Mini	*1,054*	*-35.7*	*1,640*
BMW Total	7,422	-17.6	9,003
Chrysler	*–*	*–*	*22*
Dodge	*31*	*-72.8*	*114*
Jeep	*2*	*-77.8*	*9*
Chrysler Total	33	-77.2	145
Dacia	1,109	-50.3	2,231
Mercedes-Benz	*5,554*	*-21.6*	*7,084*
Smart	*1,635*	*-26.7*	*2,230*
Daimler Total	7,189	-22.8	9,314
Ferrari	10	-47.4	19
Alfa Romeo	*933*	*-49.6*	*1,851*
Fiat	*4,741*	*-32.2*	*6,997*
Lancia	*358*	*-7.7*	*388*
Fiat Total	6,032	-34.7	9,236
Ford	4,838	-54.4	10,615
Chevrolet	*2,483*	*-22.6*	*3,206*
Opel	*5,961*	*-47.6*	*11,372*
GM Total	8,444	-42.1	14,578
Honda	1,487	-18.6	1,827
Hyundai	1,173	-58.2	2,807
Jaguar	*111*	*-15.9*	*132*
Land Rover	*420*	*117.6*	*193*
Jaguar Land Rover Total	531	63.4	325
Kia	1,863	-31.0	2,701
Lamborghini	1	-66.7	3
Maserati	–	–	1
Mazda	376	-63.6	1,034
Mitsubishi	868	-62.0	2,283
Nissan	3,921	-41.0	6,650
Citroen	*4,756*	*-47.7*	*9,088*
Peugeot	*8,366*	*-35.0*	*12,875*
Peugeot Citroen Total	13,122	-40.3	21,963
Porsche	298	-15.3	352
Proton (Lotus)	1	–	1
Renault	10,348	-36.7	16,340
Saab	14	-87.7	114
Subaru	3	200.0	1
Suzuki	157	-72.3	566
Lexus	*88*	*-67.8*	*273*
Toyota	*3,791*	*-31.3*	*5,519*
Toyota Total	3,879	-33.0	5,792
SEAT	*2,559*	*-69.7*	*8,435*
Skoda	*1,862*	*-31.3*	*2,711*
Volkswagen	*9,825*	*-33.9*	*14,874*
Volkswagen Total	14,246	-45.2	26,020
Volvo	1,977	-31.1	2,869
Total Cars	**95,290**	**-37.9**	**153,433**

	2012	% Chg.	2011
Light Trucks			
Chrysler (Jeep)	150	-53.3	321
Dacia	31	-71.8	110
Daimler (Mercedes-Benz)	752	-44.5	1,355
Fiat	1,438	-48.7	2,805
Ford	750	-71.1	2,594
GM (Opel)	1,075	-54.6	2,367
Hyundai	29	-83.8	179
Isuzu	170	-56.6	392
Iveco	314	-41.7	539
Kia	29	-61.3	75
Land Rover	41	-48.1	79
Mazda	54	-78.6	252
Mitsubishi	649	-48.5	1,259
Nissan	314	-48.9	614
Citroen	*2,257*	*-61.3*	*5,829*
Peugeot	*2,558*	*-44.2*	*4,582*
Peugeot Citroen Total	4,815	-53.8	10,411

PORTUGAL

SALES BY COMPANY

	2012	% Chg.	2011
Renault	2,599	-58.5	6,259
Suzuki	12	–	1
Toyota	1,076	-54.3	2,354
SEAT	*46*	*-92.5*	*612*
Skoda	*18*	*350.0*	*4*
Volkswagen	*1,647*	*-28.6*	*2,306*
Volkswagen Total	1,711	-41.4	2,922
Total Light Trucks	**16,009**	**-54.1**	**34,888**
Medium/Heavy Trucks			
Mercedes-Benz	*268*	*-31.8*	*393*
Mitsubishi Fuso	*81*	*-30.8*	*117*
Daimler Total	349	-31.6	510
Ford	–	–	1
GM (Opel)	2	–	–
Isuzu	13	-71.7	46
Iveco	191	-0.5	192
MAN	*275*	*10.0*	*250*
Volkswagen Truck & Bus	–	–	*5*
MAN Total	275	7.8	255
PACCAR (DAF)	82	-74.7	324
Scania	170	-48.6	331
Toyota	74	-25.3	99
Renault Truck	*490*	*-9.8*	*543*
UD Trucks	*5*	*150.0*	*2*
Volvo Truck	*238*	*-34.1*	*361*
Volvo Truck Total	733	-19.1	906
Total Med./Hvy. Trucks	**1,889**	**-29.1**	**2,664**
Buses			
Citroen	2	100.0	1
Mercedes-Benz	*77*	*-32.5*	*114*
Setra	*1*	*-50.0*	*2*
Daimler Total	78	-32.8	116
Fiat	1	-85.7	7
Ford	1	–	1
GM (Opel)	3	–	–
Iveco	22	-50.0	44
MAN	*38*	*-56.3*	*87*
Neoplan	*2*	–	–
Volkswagen Truck & Bus	*10*	*150.0*	*4*
MAN Total	50	-45.1	91
Scania	9	-40.0	15
Toyota	11	-31.3	16
Renault Truck	*19*	*216.7*	*6*
Volvo Truck	*27*	*-20.6*	*34*
Volvo Truck Total	46	15.0	40
Total Buses	**223**	**-32.6**	**331**
Total Comm. Vehs.	**18,121**	**-52.2**	**37,883**
Total Vehicles	**113,411**	**-40.7**	**191,316**

ROMANIA

ECONOMIC SUMMARY

Common Language	Romanian
Currency	New Leu
Steering	Left Hand Drive

2012 Vehicles in Operation

Passenger Cars	4,487,000
Commercial Vehicles	762,000
Total	5,249,000
Persons per Car	4.9
Persons per Vehicle	4.2

2012 GDP (billions of US $)

Nominal GDP	169.4
Real GDP	100.2
GDP % annual growth	0.3
GDP per capita	4,696

Current Account Balance (billions of US $)

2012:	-6.5
2011:	-8.3

Current Account Balance in % of GDP

2012:	-3.8
2011:	-4.5

Total Number of Fatal Road Accidents (most recent data)

2010:	2,377

Capital	Bucharest
Population	21,790,479
Population Density	245.5
Surface Area sq km/miles	229,891 / 88,761

Road Infrastructure

Paved Road Surface %	30.2
Total Road Length in km/miles	198,817 / 123,539
Road Traffic (Vehicles per mile)	42.5

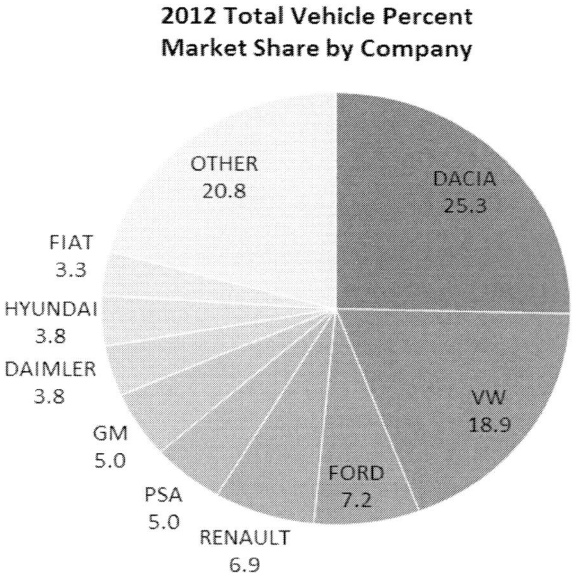

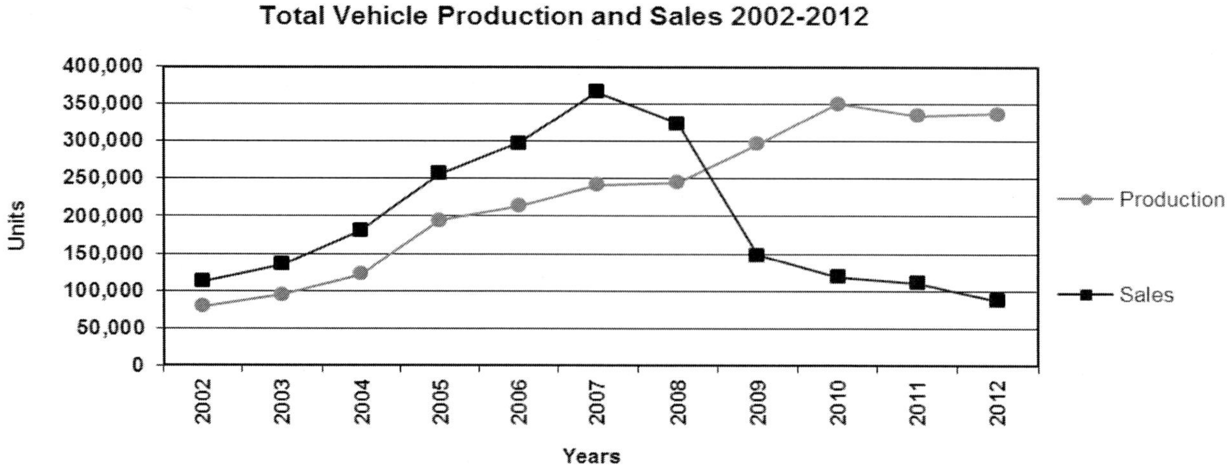

ROMANIA

PRODUCTION

Year	Cars	Commercial Vehicles	Total
2012	326,556	11,209	337,765
2011	310,243	24,989	335,232
2010	323,587	27,325	350,912
2009	279,320	17,178	296,498
2008	231,056	14,252	245,308
2007	234,103	7,609	241,712
2006	201,663	11,934	213,597
2005	174,538	20,264	194,802
2004	98,997	23,188	122,185
2003	75,706	19,546	95,252
2002	65,266	14,190	79,456
2001	56,774	11,987	68,761
2000	64,181	13,984	78,165
1999	88,312	18,612	106,924
1998	103,931	22,973	126,904
1997	109,189	20,529	129,718
1996	97,923	24,637	122,560
1995	87,708	4,907	92,615
1994	85,858	5,187	91,045
1993	93,330	7,169	100,499
1992	74,152	7,391	81,543
1991	83,574	10,223	93,797
1990	99,252	11,337	110,589

ROMANIA

PRODUCTION BY MANUFACTURER

	2012	% Chg.	2011
Cars			
Ford	30,591	–	–
Dacia	*226,954*	*-9.1*	*249,656*
Renault	*69,011*	*13.9*	*60,587*
Renault Total	295,965	-4.6	310,243
Total Cars	**326,556**	**5.3**	**310,243**
Light Trucks			
Ford	–	–	7,547
Renault (Dacia)	11,187	-35.6	17,377
Total Light Trucks	**11,187**	**-55.1**	**24,924**
Medium/Heavy Trucks			
Roman	22	-66.2	65
Total Med./Hvy. Trks.	**22**	**-66.2**	**65**
Total Comm. Vehs.	**11,209**	**-55.1**	**24,989**
Total Vehicles	**337,765**	**0.8**	**335,232**

ROMANIA

SALES

Year	Cars	Commercial Vehicles	Total
2012	72,179	15,326	87,505
2011	94,624	16,044	110,668
2010	106,328	13,089	119,417
2009	130,193	17,769	147,962
2008	270,995	53,085	324,080
2007	315,621	51,198	366,819
2006	256,364	40,798	297,162
2005	215,532	40,882	256,414
2004	145,106	35,821	180,927
2003	106,763	28,542	135,305
2002	88,804	23,196	112,000
2001	72,157	19,363	91,520
2000	66,276	17,894	84,170
1999	96,765	21,299	118,064
1998	115,833	27,095	142,928
1997	93,724	20,878	114,602
1996	96,754	21,370	118,124
1995	81,814	12,300	94,114
1994	60,000	28,000	88,000
1993	57,000	22,000	79,000
1992	31,000	9,000	40,000
1991	57,000	14,200	71,200
1990	37,666	12,469	50,135

ROMANIA

SALES BY COMPANY

	2012	% Chg.	2011		2012	% Chg.	2011
Cars				Peugeot Citroen Total	2,707	-23.8	3,554
Audi	1,900	18.8	1,600	Porsche	138	35.3	102
AutoVaz (Lada)	–	–	1	Renault	4,785	-33.8	7,229
BMW	*2,161*	*6.6*	*2,028*	Saab	–	–	13
Mini	*168*	*15.9*	*145*	Ssangyong	6	-68.4	19
BMW Total	2,329	7.2	2,173	Subaru	175	-22.9	227
Chrysler (Jeep)	183	577.8	27	Suzuki	1,638	-22.4	2,111
Dacia	19,948	-29.6	28,337	*Lexus*	*63*	*21.2*	*52*
Mercedes-Benz	*1,574*	*1.9*	*1,544*	*Toyota*	*2,315*	*-20.2*	*2,902*
Smart	*12*	*-29.4*	*17*	Toyota Total	2,378	-19.5	2,954
Daimler Total	1,586	1.6	1,561	*SEAT*	*650*	*-51.9*	*1,351*
Alfa Romeo	*159*	*-35.1*	*245*	*Skoda*	*6,291*	*-23.8*	*8,260*
Fiat	*1,507*	*-19.3*	*1,868*	*Volkswagen*	*8,046*	*-19.9*	*10,043*
Lancia	*39*	*-44.3*	*70*	Volkswagen Total	14,987	-23.7	19,654
Fiat Total	1,705	-21.9	2,183	Volvo	455	8.3	420
Ford	4,682	-15.6	5,547	Other	34	88.9	18
Chevrolet	*1,661*	*-45.1*	*3,027*	**Total Cars**	**72,179**	**-23.7**	**94,624**
Opel	*2,446*	*-49.8*	*4,872*				
GM Total	4,107	-48.0	7,899	**Light Trucks**			
Great Wall	–	–	17	Dacia	2,200	-13.0	2,530
Honda	553	-14.8	649	Daimler (Mercedes-Benz)	1,131	17.8	960
Hyundai	3,300	-3.1	3,405	Fiat	1,176	-13.8	1,364
Jaguar	*97*	*162.2*	*37*	Ford	1,660	26.9	1,308
Land Rover	*697*	*74.3*	*400*	GM (Opel)	248	31.9	188
Jaguar Land Rover Total	794	81.7	437	Great Wall	–	–	3
Kia	1,010	-9.2	1,112	Hyundai	–	–	4
Maserati	6	200.0	2	*Irisbus*	*69*	*-2.8*	*71*
Mazda	432	-13.4	499	*Iveco*	*341*	*20.5*	*283*
Mitsubishi	766	-30.4	1,101	Iveco Total	410	15.8	354
Infiniti	*35*	*25.0*	*28*	Kia	–	–	1
Nissan	*1,540*	*-11.7*	*1,745*	Mazda	4	-85.7	28
Nissan Total	1,575	-11.2	1,773	Mitsubishi	254	11.9	227
Citroen	*797*	*-29.4*	*1,129*	Nissan	273	-48.6	531
Peugeot	*1,910*	*-21.2*	*2,425*	*Citroen*	*937*	*-1.0*	*946*

ROMANIA

SALES BY COMPANY

	2012	% Chg.	2011
Peugeot	*[illegible]*	*[illegible]*	*[illegible]*
Peugeot Citroen Total	1,686	7.3	1,572
Renault	1,280	-13.8	1,485
Toyota	238	161.5	91
Skoda	*12*	*-47.8*	*23*
Volkswagen	*1,559*	*7.7*	*1,447*
Volkswagen Total	1,571	6.9	1,470
Volvo Truck (Renault Truck)	7	75.0	4
Total Light Trucks	**12,138**	**0.1**	**12,120**
Medium/Heavy Trucks			
Mercedes-Benz	*548*	*-13.3*	*632*
Mitsubishi Fuso	–	–	*47*
Daimler Total	548	-19.3	679
Iveco	353	-39.9	587
MAN	705	-23.0	915
PACCAR (DAF)	420	-30.9	608
Roman	23	-42.5	40
Scania	354	22.5	289
Renault Truck	*178*	*-34.1*	*270*
Volvo Truck	*495*	*11.2*	*445*
Volvo Truck Total	673	-5.9	715
Total Med./Hvy. Trks.	**3,076**	**-19.7**	**3,833**
Buses			
BMC	–	–	16
Mercedes-Benz	*37*	*54.2*	*24*
Setra	–	–	*2*
Daimler Total	37	42.3	26
Isuzu	38	850.0	4
Iveco (Irisbus)	19	-42.4	33
MAN	–	–	1
Otokar	18	80.0	10
Volvo Truck	–	–	1
Total Buses	**112**	**23.1**	**91**
Total Comm. Vehs.	**15,326**	**-4.5**	**16,044**
Total Vehicles	**87,505**	**-20.9**	**110,668**

RUSSIA

ECONOMIC SUMMARY

Common Language	Russian
Currency	Ruble
Steering	Left Hand Drive

2012 Vehicles in Operation

Passenger Cars	38,482,00
Commercial Vehicles	6,901,000
Total	45,383,000
Persons per Car	3.7
Persons per Vehicle	3.1

2012 GDP (billions of US $)

Nominal GDP	2,022.0
Real GDP	1,406.9
GDP % annual growth	3.4
GDP per capita	9,913

Current Account Balance (billions of US $)

2012:	81.3
2011:	98.8

Current Account Balance in % of GDP

2012:	4.0
2011:	5.2

Total Number of Fatal Road Accidents (most recent data)

2010:	26,567

Capital	Moscow
Population	142,500,482
Population Density	22.5
Surface Area sq km/miles	16,377,742 / 6,323,481

Road Infrastructure

Paved Road Surface %	79.0
Total Road Length in km/miles	982,000 / 610,808
Road Traffic (Vehicles per mile)	74.3

2012 Total Vehicle Percent Market Share by Company

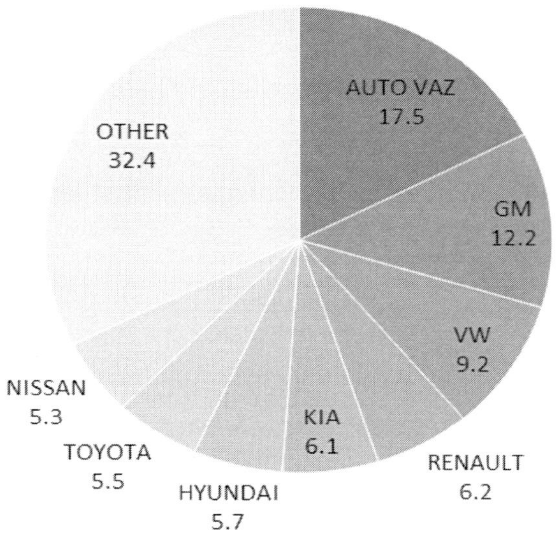

Total Vehicle Production and Sales 2005-2012

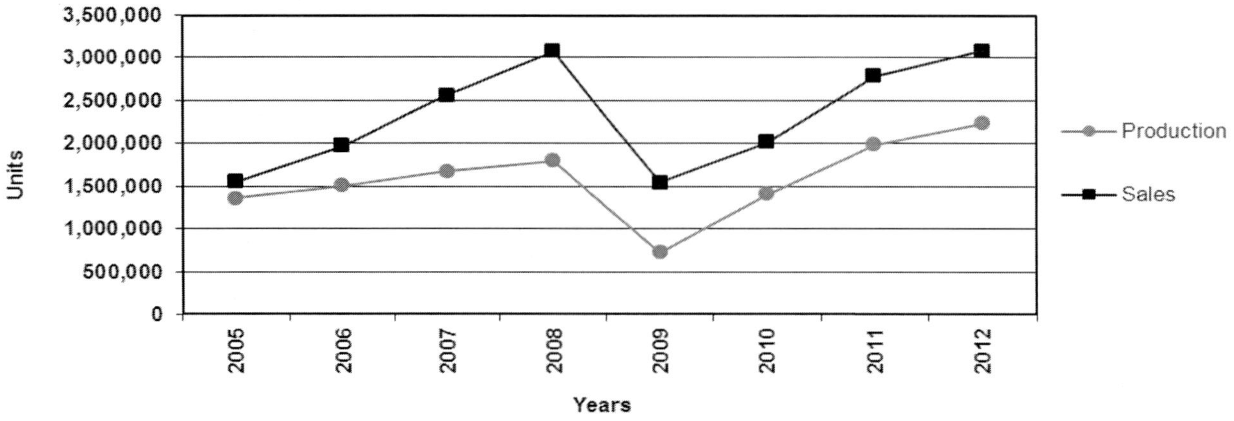

RUSSIA

PRODUCTION

Year	Cars	Trucks	Buses	Total
2012	1,968,788	206,136	57,247	2,232,171
2011	1,738,162	203,947	46,810	1,988,919
2010	1,208,362	150,377	45,112	1,403,851
2009	595,807	91,090	35,494	722,391
2008	1,469,898	256,709	67,011	1,793,618
2007	1,293,094	290,442	88,578	1,672,114
2006	1,176,337	248,741	81,764	1,506,842
2005	1,068,141	210,275	75,016	1,353,432
2004	1,110,293	202,212	76,193	1,388,698
2003	1,010,027	194,551	76,578	1,281,156
2002	980,061	174,250	67,308	1,221,619
2001	1,021,981	174,393	56,981	1,253,355
2000	969,235	187,111	53,952	1,210,298
1999	954,701	179,050	50,395	1,184,146
1998	832,807	143,144	44,097	1,020,048
1997	970,424	148,150	43,416	1,161,990
1996	868,132	136,025	*	1,004,157
1995	838,280	156,480	*	994,760
1994	795,903	205,935	*	1,001,838
1993	1,065,439	553,494	*	1,618,933
1992	1,052,573	640,829	*	1,693,402
1991	1,307,724	744,493	*	2,052,217
1990	1,259,000	774,000	84,380	2,117,380
1989	1,217,000	755,200	88,800	2,061,000
1988	1,262,000	800,000	85,000	2,147,000
1987	1,332,000	780,000	90,000	2,202,000
1986	1,326,000	816,000	84,000	2,226,000
1985	1,332,000	813,900	86,100	2,232,000
1984	1,327,000	794,000	85,400	2,206,400
1983	1,315,700	776,300	86,000	2,178,000
1982	1,307,000	780,000	85,700	2,172,700

* Included with Trucks.

RUSSIA

PRODUCTION BY MANUFACTURER

	2012	% Chg.	2011
Cars			
AutoVaz (Lada)	607,369	6.8	568,458
Dacia	*160,369*	*15.2*	*139,219*
Renault	*7,730*	*150.6*	*3,085*
AvtoFramos Total	168,099	18.1	142,304
BMW	*23,001*	*28.0*	*17,965*
Cadillac	*1,377*	*-59.2*	*3,379*
Chevrolet	*81,856*	*15.3*	*70,968*
Kia	*91,285*	*-4.0*	*95,041*
Opel	*52,116*	*50.1*	*34,728*
Avtotor Total	249,635	12.4	222,081
Geely	*23,272*	*333.7*	*5,366*
Haima	*595*	*-78.4*	*2,754*
Lifan	*22,779*	*-2.6*	*23,383*
Derways Total	46,646	48.1	31,503
Ford	107,724	9.0	98,807
GAZ	–	–	1
Chevrolet	*62,979*	*13.7*	*55,403*
Other GM	*106,644*	*50.3*	*70,942*
GM Total	169,623	34.3	126,345
Great Wall	23,904	56.9	15,235
Hyundai	58,400	50.3	38,848
Kia	–	–	*2,480*
Lada	–	–	*63,753*
IzhAvto Total	–	–	66,233
Nissan	126,058	50.3	83,859
Peugeot Citroen	40,258	4.9	38,361
Fiat	*207*	*-98.1*	*10,638*
Ford	*6,992*	–	–
Mazda	*3,108*	–	–
Ssangyong	*32,773*	*-3.3*	*33,886*
Sollers Total	43,080	-3.2	44,524
TagAZ (Hyundai)	17,740	-31.9	26,043
Toyota	98,229	50.3	65,342
UAZ	32,469	6.8	30,394
Volkswagen	179,554	28.4	139,824
Total Cars	**1,968,788**	**13.3**	**1,738,162**
Light Trucks			
AutoVaz (IzhAvto)	3,755	–	–
BAW	2,073	52.8	1,357
GAZ	71,558	-2.4	73,328
IzhAvto	–	–	5,241
Fiat	*986*	*-88.3*	*8,401*
Ford	*11,019*	–	–
Sollers Total	12,005	42.9	8,401
TagAZ (Hyundai)	1,378	-51.6	2,849
UAZ	24,253	1.5	23,892
VAZ	3,610	15.9	3,114
Total Light Trucks	**118,632**	**0.4**	**118,182**
Medium/Heavy Trucks			
Amur	8	-95.6	181
Avtotor (Hyundai)	378	–	10
Bryansk	126	22.3	103
GAZ	16,959	-2.4	17,377
Hyundai	243	-36.7	384
Iveco	1,027	9.5	938
KamAZ	47,221	4.4	45,227
Mercedes-Benz	*2,479*	*108.7*	*1,188*
Mitsubishi Fuso	*2,206*	*75.1*	*1,260*
Mercedes-Benz Total	4,685	91.4	2,448
Scania	1,288	17.4	1,097
Severstal (Isuzu)	1,567	58.4	989
Tonar	77	–	–
Ural	6,494	-39.6	10,752
Volvo Truck	6,252	23.6	5,057
ZIL	1,154	-4.0	1,202
Other	25	–	–
Total Med./Hvy. Trks.	**87,504**	**2.0**	**85,765**
Buses			
GAZ	15,449	5.8	14,604
GolAZ	110	-52.2	230
KAVZ	696	-22.8	901
LiAZ	2,002	-11.8	2,269
MZPA	–	–	11
NefAZ	743	40.2	530
Nizhegorod (Ford)	7,404	80.5	4,102
PAZ	11,523	14.9	10,026
Fiat	*5*	*-99.7*	*1,965*
Ford	*254*	–	–
Sollers Total	259	-86.8	1,965
TagAZ (Hyundai)	295	-73.8	1,125
UAZ	13,712	42.8	9,604
Ural	528	-40.2	883
Volzhanin	666	100.0	333
ZIL	6	–	–
Other	3,854	–	227
Total Buses	**57,247**	**22.3**	**46,810**
Total Comm. Vehs.	**263,383**	**5.0**	**250,757**
Total Vehicles	**2,232,171**	**21.6**	**1,988,919**

RUSSIA

SALES

Year	Light Vehicles	Med./Hvy. Trucks	Buses	Total
2012	2,935,111	87,360	57,200	3,079,671
2011	2,653,408	87,040	46,760	2,787,208
2010	1,910,573	59,820	44,980	2,015,373
2009	1,465,917	44,375	35,405	1,545,697
2008	2,897,459	110,125	67,055	3,074,639
2007	2,360,949	111,435	88,715	2,561,099
2006	1,803,163	87,610	81,615	1,972,388
2005	1,401,069	71,440	74,910	1,547,419

RUSSIA

SALES BY COMPANY

	2012	% Chg.	2011
Light Vehicles			
Audi	33,512	44.1	23,250
AutoVaz (Lada)	537,625	-7.0	578,387
BAW	1,845	–	–
BMW	*37,515*	*33.2*	*28,165*
Mini	*2,629*	*31.3*	*2,002*
BMW Total	40,144	33.1	30,167
Bogdan	7,512	-13.5	8,683
BYD	2,630	31.0	2,007
Chery	19,004	161.1	7,278
Chrysler	*194*	*155.3*	*76*
Dodge	*206*	*30.4*	*158*
Jeep	*4,704*	*124.7*	*2,093*
Chrysler Total	5,104	119.3	2,327
Mercedes-Benz	*41,410*	*30.4*	*31,761*
Smart	*28*	–	–
Daimler Total	41,438	30.5	31,761
FAW	560	–	–
Fiat	8,623	-69.5	28,254
Ford	130,809	10.8	118,031
Foton	141	–	–
GAZ	90,247	0.5	89,773
Cadillac	*2,024*	*-9.1*	*2,226*
Chevrolet	*205,042*	*18.2*	*173,484*
Daewoo	*88,232*	*-4.9*	*92,778*
Opel	*81,242*	*20.3*	*67,555*
GM Total	376,540	12.1	336,043
Great Wall	14,373	112.1	6,777
Haima	591	-81.0	3,114
Honda	21,512	12.6	19,101
Hyundai	174,286	6.6	163,447
Isuzu	156	-58.0	371
IzhAvto	2,465	-52.3	5,163
Jaguar	*1,506*	*27.4*	*1,182*
Land Rover	*19,043*	*44.5*	*13,183*
Jaguar Land Rover Total	20,549	43.0	14,365
Kia	187,330	22.5	152,873
Lifan	20,544	14.7	17,910

	2012	% Chg.	2011
Mazda	44,443	11.9	39,718
Mitsubishi	74,294	0.2	74,166
Infiniti	*9,209*	*30.8*	*7,042*
Nissan	*153,747*	*10.7*	*138,827*
Nissan Total	162,956	11.7	145,869
Citroen	*32,859*	*18.9*	*27,633*
Peugeot	*44,420*	*0.2*	*44,311*
Peugeot Citroen Total	77,279	7.4	71,944
Porsche	3,613	64.1	2,202
Renault	189,852	22.7	154,734
Ssangyong	31,198	37.5	22,692
Subaru	14,296	15.6	12,371
Suzuki	32,684	-7.9	35,469
TagAZ	*1,331*	*-27.7*	*1,840*
Vortex	*5,953*	*-50.8*	*12,103*
TagAZ Total	7,284	-47.8	13,943
Lexus	*15,653*	*14.3*	*13,698*
Toyota	*153,047*	*28.1*	*119,505*
Toyota Total	168,700	26.6	133,203
UAZ	60,653	6.1	57,148
SEAT	*2,500*	*121.8*	*1,127*
Skoda	*99,062*	*33.7*	*74,074*
Volkswagen	*180,863*	*38.8*	*130,348*
Volkswagen Total	282,425	37.4	205,549
Volvo	20,364	6.0	19,209
ZAZ	9,964	-50.3	20,049
Zhejiang Geely	17,566	189.9	6,060
Total Light Vehicles	**2,935,111**	**10.6**	**2,653,408**
Medium/Heavy Trucks			
Amur	10	-94.4	180
Beijing AIC	–	–	1,320
Bryansk	110	10.0	100
Mercedes-Benz	*2,470*	*105.8*	*1,200*
Mitsubishi Fuso	*2,190*	*73.8*	*1,260*
Daimler Total	4,660	89.4	2,460
GAZ	16,950	-2.4	17,370
Hyundai	620	59.0	390

SALES BY COMPANY

	2012	% Chg.	2011		2012	% Chg.	2011
Isuzu	1,560	57.6	990	GolAZ	100	-56.5	230
Iveco	1,030	9.6	940	Hyundai	300	-73.0	1,110
KamAZ	47,180	4.4	45,180	KAVZ	680	-23.6	890
Scania	1,280	16.4	1,100	LiAZ	2,010	-11.5	2,270
Tonar	70	–	–	NefAZ	750	36.4	550
Ural	6,490	-39.6	10,750	PAZ	11,520	15.0	10,020
Volvo Truck	6,250	23.5	5,060	UAZ	13,720	43.1	9,590
ZIL	1,150	-4.2	1,200	Ural	530	-39.1	870
Total Med./Hvy. Trks.	**87,360**	**0.4**	**87,040**	Volzhanin	660	94.1	340
				Other	3,830	–	230
Buses				**Total Buses**	**57,200**	**22.3**	**46,760**
Fiat	–	–	1,960				
Ford	7,670	87.1	4,100	**Total Comm. Vehs.**	**144,560**	**8.0**	**133,800**
GAZ	15,430	5.7	14,600	**Total Vehicles**	**3,079,671**	**10.5**	**2,787,208**

SERBIA

ECONOMIC SUMMARY

Common Language	Serbian
Currency	Dinar
Steering	Left Hand Drive

2012 Vehicles in Operation

Passenger Cars	1,792,000
Commercial Vehicles	190,000
Total	1,982,000
Persons per Car	4.3
Persons per Vehicle	3.9

2012 GDP (billions of US $)

Nominal GDP	37.4
Real GDP	21.6
GDP % annual growth	-1.8
GDP per capita	2,854

Current Account Balance (billions of US $)

2012:	-4.1
2011:	-4.0

Current Account Balance in % of GDP

2012:	-10.9
2011:	-9.2

Total Number of Fatal Road Accidents (most recent data)

2010:	656

Capital	Belgrade
Population	7,243,007
Population Density	316.1
Surface Area sq km/miles	77,474 / 29,913

Road Infrastructure

Paved Road Surface %	62.0
Total Road Length in km/miles	41,913 / 26,043
Road Traffic (Vehicles per mile)	76.1

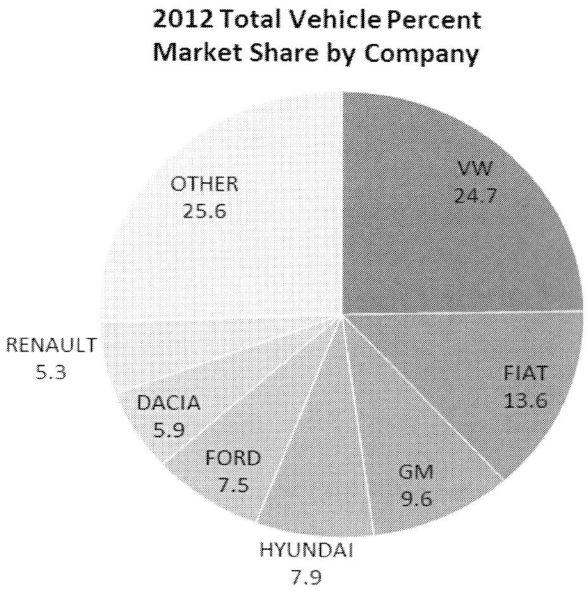

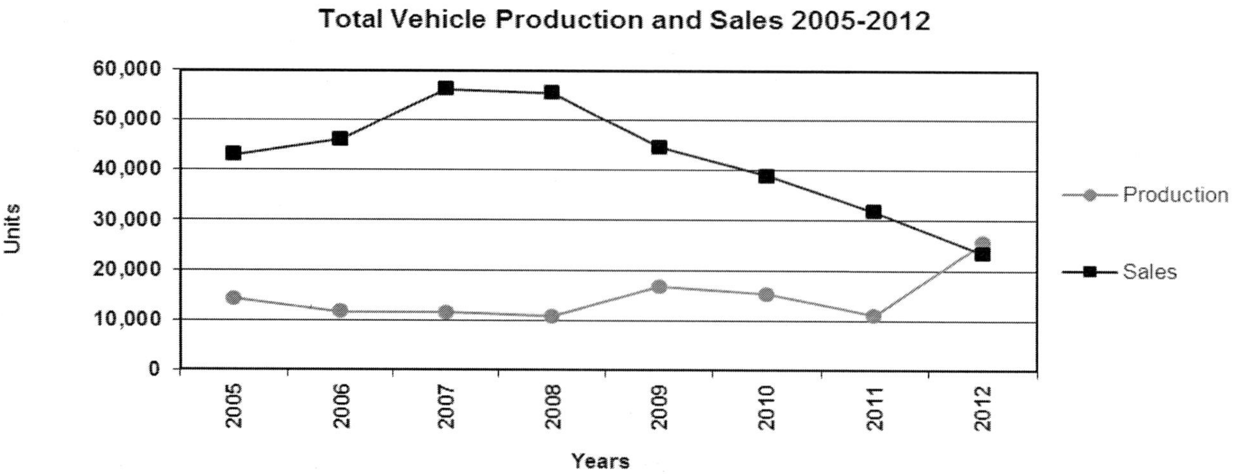

SERBIA

PRODUCTION

Year	Cars	Trucks	Buses	Total
2012	25,281	144	89	25,514
2011	10,227	661	135	11,023
2010	14,551	619	30	15,200
2009	16,337	300	101	16,738
2008	9,818	541	360	10,719
2007	10,658	721	145	11,524
2006	10,252	1,142	173	11,567
2005	12,575	1,208	397	14,180
2004	13,355	1,601	238	15,194
2003	13,101	929	186	14,216
2002	10,272	1,437	317	12,026
2001	7,668	1,032	279	8,979
2000	11,175	1,327	238	12,740
1999	3,816	1,357	96	5,269
1998	14,011	1,845	182	16,038
1997	11,124	2,143	236	13,503
1996	8,848	1,338	212	10,398
1995	7,502	1,668	139	9,309
1994	7,648	1,387	170	9,205
1993	7,359	873	81	8,313
1992	25,271	4,982	874	31,127

PRODUCTION BY MANUFACTURER

	2012	% Chg.	2011		2012	% Chg.	2011
Cars				**Buses**			
Fiat	25,281	147.2	10,227	FAP	16	166.7	6
Total Cars	**25,281**	**147.2**	**10,227**	Ikarbus	72	-1.4	73
				Neobus	–	–	53
Trucks				Zastava	1	-66.7	3
FAP	62	-36.1	97	**Total Buses**	**89**	**-34.1**	**135**
Fiat	–	–	366				
Zastava	82	-58.6	198	**Total Comm. Vehs.**	**233**	**-70.7**	**796**
Total Trucks	**144**	**-78.2**	**661**	**Total Vehicles**	**25,514**	**131.5**	**11,023**

SERBIA

SALES

Year	Light Vehicles	Year	Light Vehicles
2012	23,510	2008	55,464
2011	31,780	2007	56,231
2010	38,909	2006	46,105
2009	44,682	2005	43,000

SALES BY COMPANY

	2012	% Chg.	2011
Light Vehicles			
Audi	520	6.6	488
AutoVaz (Lada)	509	117.5	234
BMW	*392*	*-3.0*	*404*
Mini	*118*	*59.5*	*74*
BMW Total	510	6.7	478
Chery	92	-24.6	122
Chrysler	*–*	*–*	*12*
Dodge	*4*	*-75.0*	*16*
Jeep	*55*	*14.6*	*48*
Chrysler Total	59	-22.4	76
Dacia	1,389	-45.8	2,562
Mercedes-Benz	*503*	*-32.1*	*741*
Smart	*49*	*2.1*	*48*
Daimler Total	552	-30.0	789
Alfa Romeo	*51*	*-7.3*	*55*
Fiat	*3,114*	*-49.2*	*6,130*
Lancia	*32*	*60.0*	*20*
Fiat Total	3,197	-48.5	6,205
Ford	1,763	-14.5	2,063
Chevrolet	*852*	*-42.5*	*1,482*
Opel	*1,399*	*-34.4*	*2,134*
GM Total	2,251	-37.7	3,616
Honda	391	27.4	307
Hyundai	1,847	-22.8	2,393
Isuzu	–	–	6
Kia	832	18.7	701
Land Rover	107	21.6	88
Mahindra & Mahindra	–	–	11
Mazda	243	-22.9	315
Mitsubishi	88	-65.1	252
Infiniti	*9*	*800.0*	*1*
Nissan	*420*	*-14.5*	*491*
Nissan Total	429	-12.8	492
Citroen	*22*	*-95.8*	*519*
Peugeot	*273*	*-63.6*	*749*
Peugeot Citroen Total	295	-76.7	1,268
Porsche	16	-59.0	39
Renault	1,249	-29.8	1,779
Saab	–	–	9
Subaru	23	-42.5	40
Suzuki	460	16.8	394
Lexus	*3*	*-83.3*	*18*
Toyota	*801*	*-16.2*	*956*
Toyota Total	804	-17.5	974
SEAT	*221*	*-17.5*	*268*
Skoda	*3,665*	*-1.5*	*3,719*
Volkswagen	*1,919*	*-4.7*	*2,014*
Volkswagen Total	5,805	-3.3	6,001
Volvo	79	1.3	78
Total Light Vehicles	**23,510**	**-26.0**	**31,780**
Total Vehicles	**23,510**	**-26.0**	**31,780**

SLOVAKIA

ECONOMIC SUMMARY

Common Language	Slovak/Hungarian
Currency	Euro
Steering	Left Hand Drive

2012 Vehicles in Operation

Passenger Cars	1,771,000
Commercial Vehicles	315,000
Total	2,086,000
Persons per Car	3.1
Persons per Vehicle	2.6

2012 GDP (billions of US $)

Nominal GDP	91.9
Real GDP	86.2
GDP % annual growth	2.0
GDP per capita	15,852

Current Account Balance (billions of US $)

2012:	2.1
2011:	-2.0

Current Account Balance in % of GDP

2012:	2.3
2011:	-2.1

Total Number of Fatal Road Accidents (most recent data)

2010:	345

Capital	Bratislava
Population	5,488,339
Population Density	295.5
Surface Area sq km/miles	48,105 / 18,573

Road Infrastructure

Paved Road Surface %	87.0
Total Road Length in km/miles	43,761 / 27,192
Road Traffic (Vehicles per mile)	76.7

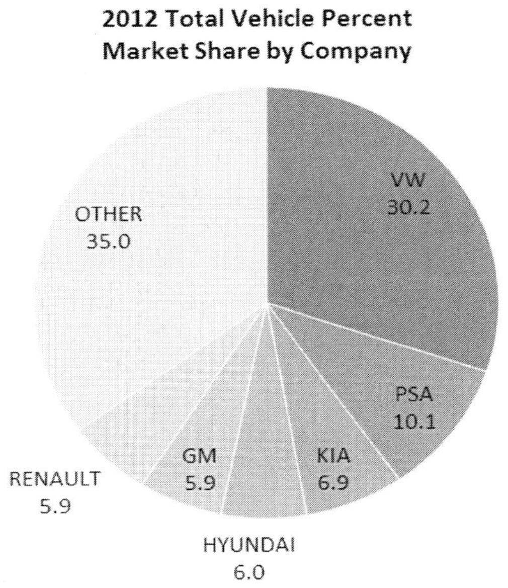

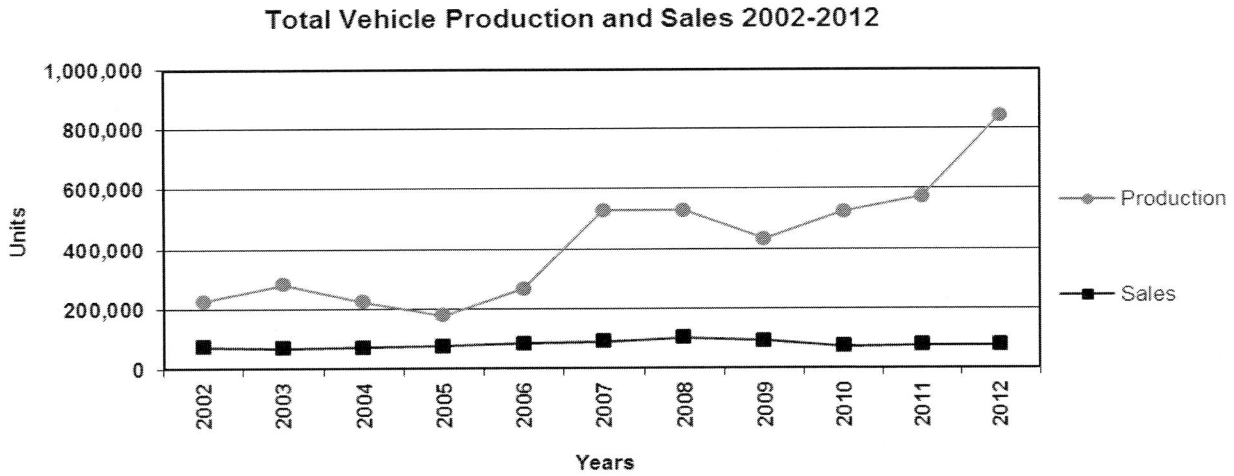

SLOVAKIA

PRODUCTION

Year	Cars	Trucks	Buses	Total
2012	844,110	–	–	844,110
2011	573,660	–	–	573,660
2010	523,019	–	–	523,019
2009	433,314	–	–	433,314
2008	527,534	–	–	527,534
2007	528,163	–	–	528,163
2006	267,088	–	–	267,088
2005	176,189	–	–	176,189
2004	223,542	–	–	223,542
2003	281,160	181	6	281,347
2002	225,442	259	17	225,718
2001	181,644	323	36	182,003
2000	181,333	246	186	181,765
1999	126,503	300	43	126,846
1998	125,089	582	70	125,741
1997	40,855	1,071	10	41,936
1996	30,147	1,880	–	32,027
1995	19,688	3,136	–	22,824
1994	6,010	3,019	–	9,029
1993	2,958	2,774	–	5,732
1992	1,630	9,094	35	10,759
1991	1	11,758	–	11,759
1990	1	18,203	–	18,204
1989	1	13,316	–	13,317
1988	1	17,706	–	17,707
1987	53	16,224	–	16,277
1986	554	17,930	–	18,484
1985	738	17,530	–	18,268
1984	1,038	16,373	–	17,411
1983	840	15,062	–	15,902
1982	257	15,481	–	15,738

PRODUCTION BY MANUFACTURER

	2012	% Chg.	2011		2012	% Chg.	2011
<u>**Cars**</u>				*Volkswagen*	*282,004*	*240.8*	*82,753*
Kia	292,000	15.8	252,200	Volkswagen Total	337,110	147.0	136,460
Peugeot	215,000	16.2	185,000	**Total Cars**	**844,110**	**47.1**	**573,660**
Audi	*55,106*	*2.6*	*53,707*	**Total Vehicles**	**844,110**	**47.1**	**573,660**

SLOVAKIA

SALES

Year	Cars	Light Trucks	Medium/Heavy Trucks	Buses	Total
2012	69,268	5,113	3,511	297	78,189
2011	68,203	5,743	3,664	294	77,904
2010	64,033	6,934	2,378	488	73,833
2009	74,717	15,731	1,664	649	92,761
2008	70,040	26,889	4,885	564	102,378
2007	59,700	23,573	5,506	315	89,094
2006	59,084	19,510	4,433	492	83,519
2005	56,916	14,415	3,460	242	75,033
2004	57,430	10,196	2,857	258	70,741
2003	57,513	9,899	2,560	212	70,184
2002	65,312	5,621	2,422	150	73,505

SALES BY COMPANY

	2012	% Chg.	2011
Cars			
Audi	1,730	7.6	1,608
AutoVaz (Lada)	87	1.2	86
BMW	*2,204*	*-11.4*	*2,487*
Mini	*214*	*-8.5*	*234*
BMW Total	2,418	-11.1	2,721
Chrysler	–	–	*2*
Dodge	–	–	*6*
Jeep	*184*	*275.5*	*49*
Chrysler Total	184	222.8	57
Dacia	1,517	-11.7	1,718
Mercedes-Benz	*1,754*	*12.1*	*1,564*
Smart	*10*	*-37.5*	*16*
Daimler Total	1,764	11.6	1,580
Alfa Romeo	*106*	*-29.8*	*151*
Fiat	*1,511*	*-6.7*	*1,619*
Lancia	*106*	*37.7*	*77*
Fiat Total	1,723	-6.7	1,847
Ford	2,161	-23.6	2,827
Chevrolet	*1,569*	*3.2*	*1,520*
Opel	*2,855*	*-4.2*	*2,980*
GM Total	4,424	-1.7	4,500
Honda	1,100	38.9	792
Hyundai	4,683	31.1	3,572
Jaguar	*45*	*-16.7*	*54*
Land Rover	*261*	*32.5*	*197*
Jaguar Land Rover Total	306	21.9	251
Kia	5,398	9.7	4,919
Mazda	741	23.5	600
Mitsubishi	477	-39.2	784
Nissan	1,773	31.5	1,348
Citroen	*2,397*	*-14.1*	*2,792*
Peugeot	*3,944*	*-1.8*	*4,015*
Peugeot Citroen Total	6,341	-6.8	6,807
Porsche	111	2.8	108
Renault	4,028	-8.4	4,395
Saab	–	–	17
Ssangyong	8	33.3	6
Subaru	237	-11.2	267
Suzuki	2,431	4.1	2,336
Lexus	*78*	*-13.3*	*90*
Toyota	*1,985*	*-9.9*	*2,203*
Toyota Total	2,063	-10.0	2,293
SEAT	*1,373*	*-40.8*	*2,321*

SALES BY COMPANY

	2012	% Chg.	2011
Skoda	*15,337*	*11.8*	*13,720*
Volkswagen	*6,397*	*4.7*	*6,108*
Volkswagen Total	23,107	4.3	22,149
Volvo	456	-25.7	614
Other	–	–	1
Total Cars	**69,268**	**1.6**	**68,203**
Light Trucks			
BMW	–	–	5
Dacia	53	-3.6	55
Mercedes-Benz	*259*	*-5.8*	*275*
Mitsubishi Fuso	*2*	*-80.0*	*10*
Daimler Total	261	-8.4	285
Dongfeng	1	-75.0	4
Alfa Romeo	–	–	*1*
Fiat	*1,216*	*-13.5*	*1,405*
Fiat Total	1,216	-13.5	1,406
Ford	228	6.0	215
GM (Opel)	204	-7.7	221
GoNow	2	-83.3	12
Hyundai	3	–	3
Isuzu	–	–	2
Iveco	214	-4.0	223
Kia	–	–	1
Land Rover	5	–	5
Mahindra & Mahindra	5	-66.7	15
Mitsubishi	20	-70.6	68
Multicar	6	20.0	5
Nissan	144	9.1	132
Citroen	*705*	*-12.3*	*804*
Peugeot	*818*	*-23.8*	*1,073*
Peugeot Citroen Total	1,523	-18.9	1,877
Piaggio	4	-81.8	22
Renault	595	16.7	510
Toyota	89	-32.1	131
Skoda	*120*	*-17.2*	*145*
Volkswagen	*410*	*11.7*	*367*
Volkswagen Total	530	3.5	512
Volvo Truck (Renault Truck)	10	-70.6	34
Total Light Trucks	**5,113**	**-11.0**	**5,743**

	2012	% Chg.	2011
Medium/Heavy Trucks			
Ashok Leyland (Avia)	30	42.9	21
Citroen	1	–	–
Mercedes-Benz	*523*	*-15.8*	*621*
Mitsubishi Fuso	*16*	*-33.3*	*24*
Daimler Total	539	-16.4	645
GM (Opel)	1	–	1
Isuzu	3	-66.7	9
Iveco	519	0.6	516
MAN	*429*	*-17.3*	*519*
Volkswagen Truck & Bus	*4*	*100.0*	*2*
MAN Total	433	-16.9	521
Multicar	3	-40.0	5
PACCAR (DAF)	415	-1.7	422
Renault	4	–	–
Scania	479	-11.9	544
Tatra	8	-46.7	15
Renault Truck	*266*	*12.7*	*236*
Volvo Truck	*808*	*11.3*	*726*
Volvo Truck Total	1,074	11.6	962
Other	2	-33.3	3
Total Med./Hvy. Trks.	**3,511**	**-4.2**	**3,664**
Buses			
Mercedes-Benz	*5*	*-54.5*	*11*
Setra	–	–	*1*
Daimler Total	5	-58.3	12
Isuzu	–	–	1
Iveco (Irisbus)	159	-25.4	213
MAN	*10*	*42.9*	*7*
Neoplan	*1*	–	–
MAN Total	11	57.1	7
Scania	3	–	–
Solaris	–	–	2
Yutong	4	300.0	1
Other	115	98.3	58
Total Buses	**297**	**1.0**	**294**
Total Comm. Vehs.	**8,921**	**-8.0**	**9,701**
Total Vehicles	**78,189**	**0.4**	**77,904**

SLOVENIA

ECONOMIC SUMMARY

Common Language	Slovenian
Currency	Euro
Steering	Left Hand Drive

2012 Vehicles in Operation

Passenger Cars	1,066,028
Commercial Vehicles	85,445
Total	1,151,473
Persons per Car	1.9
Persons per Vehicle	1.7

2012 GDP (billions of US $)

Nominal GDP	45.6
Real GDP	31.4
GDP % annual growth	-2.3
GDP per capita	15,298

Current Account Balance (billions of US $)

2012:	1.0
2011:	0.0

Current Account Balance in % of GDP

2012:	2.3
2011:	0.0

Total Number of Fatal Road Accidents (most recent data)

2010:	138

Capital	Ljubljana
Population	1,992,690
Population Density	256.1
Surface Area sq km/miles	20,151 / 7,780

Road Infrastructure

Paved Road Surface %	71.4
Total Road Length in km/miles	38,925 / 24,187
Road Traffic (Vehicles per mile)	47.6

2012 Total Vehicle Percent Market Share by Company

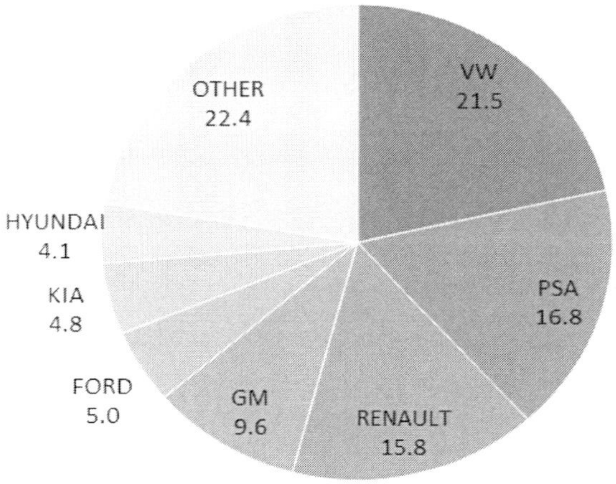

Total Vehicle Production and Sales 2005-2012

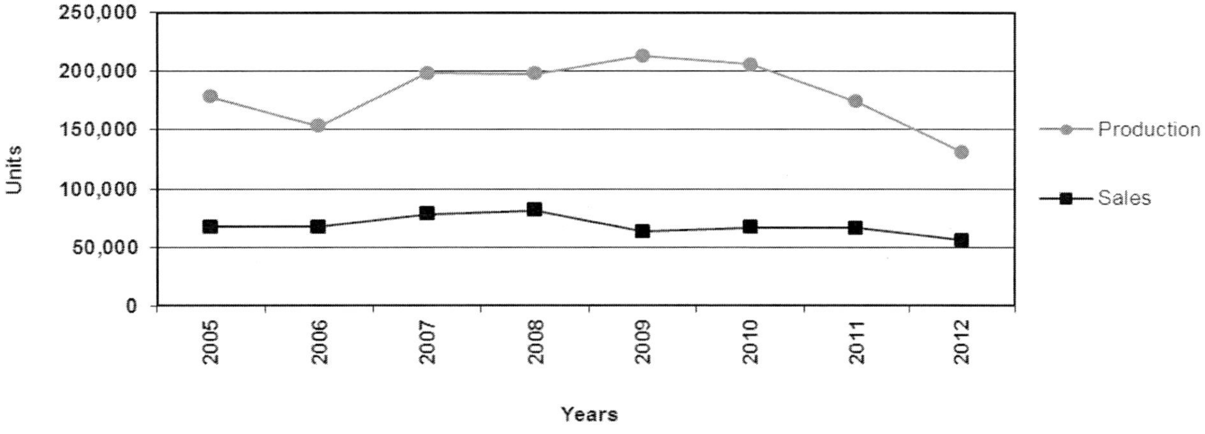

SLOVENIA

PRODUCTION

Year	Cars	Commercial Vehicles	Total
2012	126,836	4,113	130,949
2011	168,955	5,164	174,119
2010	195,207	10,504	205,711
2009	202,570	10,179	212,749
2008	180,233	17,610	197,843
2007	174,209	24,193	198,402
2006	119,212	33,915	153,127
2005	138,393	39,558	177,951
2004	131,646	15,037	146,683
2003	118,172	–	118,172
2002	126,661	–	126,661
2001	116,082	–	116,082
2000	122,949	–	122,949
1999	118,132	–	118,132
1998	126,397	–	126,397
1997	95,876	50	95,926
1996	89,229	185	89,414

PRODUCTION BY MANUFACTURER

	2012	% Chg.	2011		2012	% Chg.	2011
Cars				Commercial Vehicles			
Revoz (Renault)	126,836	-24.9	168,955	Revoz (Renault)	4,113	-20.4	5,164
Total Cars	**126,836**	**-24.9**	**168,955**	**Total Comm. Vehs.**	**4,113**	**-20.4**	**5,164**
				Total Vehicles	**130,949**	**-24.8**	**174,119**

SLOVENIA

SALES

Year	Cars	Trucks	Buses	Total
2012	33,882	22,275	92	56,249
2011	40,816	25,465	123	66,404
2010	44,555	22,174	142	66,871
2009	42,081	21,079	126	63,286
2008	50,608	30,804	189	81,601
2007	48,188	30,070	140	78,398
2006	47,852	19,852	132	67,836
2005	52,108	15,650	98	67,856

SALES BY COMPANY

	2012	% Chg.	2011
Cars			
Aston Martin	–	–	1
Audi	956	-10.2	1,064
BMW	*1,031*	*-25.7*	*1,387*
Mini	*186*	*-57.7*	*440*
BMW Total	1,217	-33.4	1,827
Chrysler	–	–	21
Dacia	482	15.3	418
Mercedes-Benz	*225*	*-23.5*	*294*
Smart	*11*	*-31.3*	*16*
Daimler Total	236	-23.9	310
Alfa Romeo	*283*	*-28.7*	*397*
Fiat	*1,294*	*-24.1*	*1,704*
Lancia	*35*	*45.8*	*24*
Fiat Total	1,612	-24.1	2,125
Ford	1,671	-26.2	2,265
Chevrolet	*1,473*	*-14.9*	*1,731*
Opel	*2,729*	*-28.1*	*3,797*
GM Total	4,202	-24.0	5,528
Honda	215	-42.4	373
Hyundai	1,414	-27.7	1,956
Jaguar	13	8.3	12
Kia	1,938	42.2	1,363
Maserati	–	–	1
Mazda	216	-45.5	396
Mitsubishi	12	-73.9	46
Nissan	523	-20.8	660
Citroen	*1,614*	*-11.2*	*1,818*
Peugeot	*2,921*	*-7.8*	*3,168*
Peugeot Citroen Total	4,535	-9.0	4,986
Porsche	15	–	15
Proton (Lotus)	–	–	1
Renault	5,642	-21.9	7,220
Saab	–	–	11
Subaru	23	-61.7	60
Suzuki	47	-19.0	58
Tata	–	–	2
Lexus	*13*	*-60.6*	*33*
Toyota	*801*	*-15.2*	*945*
Toyota Total	814	-16.8	978
SEAT	*703*	*-26.2*	*952*
Skoda	*2,520*	*-1.9*	*2,570*
Volkswagen	*4,800*	*-12.3*	*5,471*
Volkswagen Total	8,023	-10.8	8,993

SLOVENIA

SALES BY COMPANY

	2012	% Chg.	2011
Volvo	76	-39.7	126
Total Cars	**33,882**	**-17.0**	**40,816**
Light Trucks			
Audi	431	27.9	337
AutoVaz (Lada)	–	–	3
BMW	275	-0.4	276
Chrysler (Jeep)	37	236.4	11
Dacia	530	-1.3	537
Mercedes-Benz	*725*	*36.3*	*532*
Mitsubishi Fuso	*7*	*75.0*	*4*
Daimler Total	732	36.6	536
Fiat	*605*	*-2.3*	*619*
Lancia	*3*	*50.0*	*2*
Fiat Total	608	-2.1	621
Ford	1,152	-20.1	1,442
Chevrolet	*156*	*-38.1*	*252*
Opel	*1,058*	*-33.8*	*1,597*
GM Total	1,214	-34.3	1,849
Honda	85	-25.4	114
Hyundai	866	-18.4	1,061
Isuzu	–	–	2
Iveco	77	-19.8	96
Kia	785	-17.1	947
Land Rover	121	70.4	71
Mazda	265	47.2	180
Mitsubishi	100	-41.9	172
Infiniti	*1*	*–*	*1*
Nissan	*993*	*-7.9*	*1,078*
Nissan Total	994	-7.9	1,079
Citroen	*2,348*	*-21.9*	*3,007*

	2012	% Chg.	2011
Peugeot	*2,589*	*-3.7*	*2,688*
Peugeot Citroen Total	4,937	-13.3	5,695
Piaggio	2	-66.7	6
Porsche	31	6.9	29
Renault	3,248	-8.5	3,551
Ssangyong	23	666.7	3
Subaru	52	2.0	51
Suzuki	157	-54.9	348
Lexus	*6*	*-57.1*	*14*
Toyota	*267*	*-31.0*	*387*
Toyota Total	273	-31.9	401
SEAT	*325*	*-26.1*	*440*
Skoda	*630*	*16.7*	*540*
Volkswagen	*3,089*	*-12.5*	*3,530*
Volkswagen Total	4,044	-10.3	4,510
Volvo	47	-35.6	73
Volvo Truck (Renault Truck)	4	33.3	3
Other	88	11.4	79
Total Light Trucks	**21,178**	**-12.1**	**24,083**
Medium/Heavy Trucks			
Other	1,097	-20.6	1,382
Total Med./Hvy Trucks	**1,097**	**-20.6**	**1,382**
Buses			
Other	92	-25.2	123
Total Buses	**92**	**-25.2**	**123**
Total Comm. Vehs.	**22,367**	**-12.6**	**25,588**
Total Vehicles	**56,249**	**-15.3**	**66,404**

SPAIN

ECONOMIC SUMMARY

Common Language	Spanish
Currency	Euro
Steering	Left Hand Drive

2012 Vehicles in Operation	
Passenger Cars	22,247,528
Commercial Vehicles	5,232,813
Total	27,480,341
Persons per Car	2.1
Persons per Vehicle	1.7

2012 GDP (billions of US $)	
Nominal GDP	1,352.1
Real GDP	1,365.3
GDP % annual growth	-1.4
GDP per capita	29,576

Current Account Balance (billions of US $)	
2012:	-14.5
2011:	-55.4

Current Account Balance in % of GDP	
2012:	-1.1
2011:	-3.7

Total Number of Fatal Road Accidents (most recent data)	
2010:	2,479

Capital	Madrid
Population	47,370,542
Population Density	245.9
Surface Area sq km/miles	498,980 / 192,657

Road Infrastructure	
Paved Road Surface %	100.0
Total Road Length in km/miles	681,298 / 423,339
Road Traffic (Vehicles per mile)	64.9

2012 Total Vehicle Percent Market Share by Company

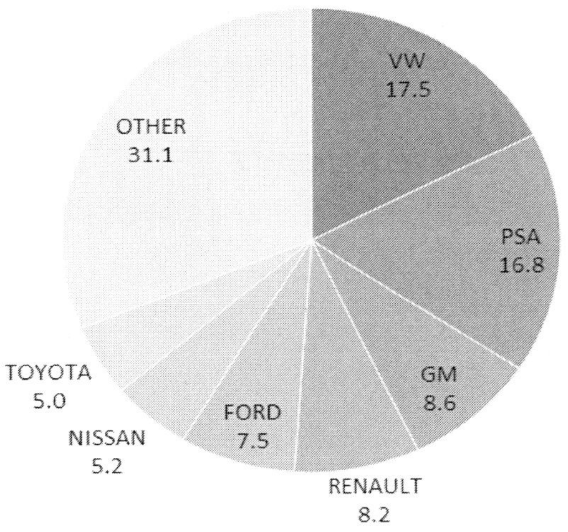

Total Vehicle Production and Sales 2002-2012

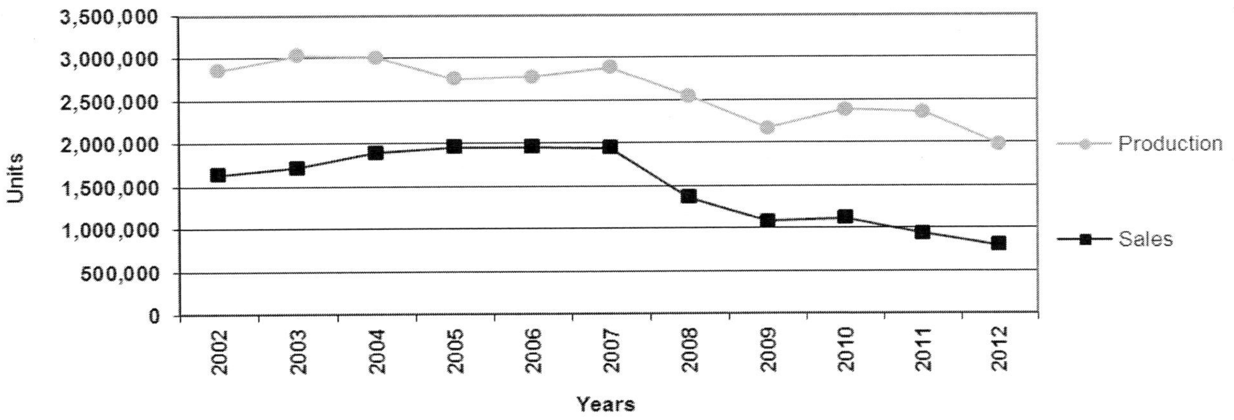

SPAIN

PRODUCTION

Year	Cars	Trucks	Buses	Total
2012	1,595,073	384,030	--	1,979,103
2011	1,867,657	485,644	381	2,353,682
2010	1,951,380	436,265	254	2,387,899
2009	1,827,423	336,500	552	2,164,475
2008	2,013,861	526,399	1,384	2,541,644
2007	2,308,774	579,383	1,546	2,889,703
2006	2,187,363	588,602	1,470	2,777,435
2005	2,174,012	577,226	1,262	2,752,500
2004	2,402,501	608,238	1,435	3,012,174
2003	2,399,374	630,452	1,502	3,031,328
2002	2,266,902	588,337	1,428	2,856,667
2001	2,211,172	637,308	1,408	2,849,888
2000	2,366,359	665,008	1,507	3,032,874
1999	2,208,708	642,093	1,588	2,852,389
1998	2,216,386	608,411	1,245	2,826,042
1997	2,010,267	550,457	1,353	2,562,077
1996	2,213,102	198,197	1,010	2,412,309
1995	2,130,540	202,743	504	2,333,787
1994	1,974,488	167,531	243	2,142,262
1993	1,622,378	144,796	466	1,767,640
1992	1,972,431	150,094	587	2,123,112
1991	1,943,181	137,868	662	2,081,711
1990	1,915,856	136,706	788	2,053,350
1989	1,896,973	147,519	1,065	2,045,557
1988	1,722,261	142,845	1,358	1,866,464
1987	1,593,794	109,104	1,415	1,704,313
1986	1,439,275	91,568	1,780	1,532,623
1985	1,344,736	71,113	1,755	1,417,604
1984	1,254,515	52,623	1,631	1,308,769
1983	1,224,798	61,880	2,043	1,288,721
1982	1,008,874	57,656	2,977	1,069,507

SPAIN

PRODUCTION BY MANUFACTURER

	2012	% Chg.	2011
Cars			
Ford	149,745	-34.9	229,919
GM (Opel)	264,858	-21.5	337,478
Mercedes-Benz	19,290	-24.2	25,432
Nissan	55,469	15.1	48,204
Citroen	*87,700*	*-24.1*	*115,622*
Peugeot	*88,800*	*1.8*	*87,225*
Peugeot Citroen Total	176,500	-13.0	202,847
Renault	264,698	-16.5	316,957
Audi	*106,919*	*445.1*	*19,613*
SEAT	*270,306*	*-19.0*	*333,807*
Volkswagen	*287,288*	*-18.7*	*353,400*
Volkswagen Total	664,513	-6.0	706,820
Total Cars	**1,595,073**	**-14.6**	**1,867,657**
Light Trucks			
GM (Opel)	–	–	27,941
Iveco	17,564	-0.6	17,677
Mercedes-Benz	57,150	-7.2	61,568
Nissan	*35,091*	*-30.6*	*50,544*
Opel	*800*	*33.3*	*600*
Renault	*51,805*	*-6.5*	*55,404*
Nissan Total	87,696	-17.7	106,548
Citroen	*101,000*	*-21.9*	*129,356*
Peugeot	*96,000*	*-19.6*	*119,403*
Peugeot Citroen Total	197,000	-20.8	248,759
Renault	5,060	-13.7	5,864
Total Light Trucks	**364,470**	**-22.2**	**468,357**
Medium/Heavy Trucks			
Iveco	19,560	13.1	17,287
Total Med./Hvy. Trks.	**19,560**	**13.1**	**17,287**
Buses			
Iveco	–	–	381
Total Buses	**–**	**–**	**381**
Total Comm. Vehs.	**384,030**	**-21.0**	**486,025**
Total Vehicles	**1,979,103**	**-15.9**	**2,353,682**

Note: Includes Ward's estimates.

SPAIN

SALES

Year	Cars	Trucks	Buses	Total
2012	699,559	89,530	1,872	790,961
2011	808,036	120,530	2,823	931,389
2010	981,976	129,546	2,558	1,114,080
2009	952,771	118,546	2,904	1,074,221
2008	1,161,176	197,558	3,809	1,362,543
2007	1,614,835	320,359	4,102	1,939,296
2006	1,634,608	314,712	3,814	1,953,134
2005	1,649,320	306,045	4,123	1,959,488
2004	1,517,518	370,304	3,635	1,891,457
2003	1,383,098	329,828	3,278	1,716,204
2002	1,331,877	301,924	3,142	1,636,943
2001	1,425,573	320,537	4,233	1,750,343
2000	1,381,256	331,462	4,222	1,716,940
1999	1,406,246	340,101	4,519	1,750,866
1998	1,192,530	195,228	3,288	1,391,046
1997	1,016,383	239,984	*	1,256,367
1996	977,855	107,731	6,207	1,091,793
1995	893,556	96,201	7,160	996,917
1994	972,993	97,512	1,573	1,072,078
1993	805,876	92,640	1,676	900,192
1992	1,076,528	139,422	2,211	1,218,161
1991	819,894	98,979	1,252	920,125
1990	926,773	105,924	1,223	1,033,920
1989	1,068,285	113,699	1,561	1,183,545
1988	987,852	101,671	1,185	1,090,708
1987	745,605	68,190	1,219	815,014
1986	616,901	54,574	1,262	672,737
1985	540,038	49,997	1,451	591,486
1984	503,982	38,315	1,409	543,706
1983	526,276	42,599	1,634	570,509
1982	504,997	46,577	1,638	553,212

* Included with Trucks.

SPAIN

SALES BY COMPANY

	2012	% Chg.	2011
Cars			
Aston Martin	22	-4.3	23
Audi	36,128	-5.3	38,140
AutoVaz (Lada)	37	-2.6	38
Bentley	23	-8.0	25
BMW	*25,703*	*-9.3*	*28,350*
Mini	*7,286*	*-13.0*	*8,379*
BMW Total	32,989	-10.2	36,729
BYD	2	-75.0	8
Caterham	1	–	–
Chrysler	*14*	*-97.4*	*549*
Dodge	*7*	*-96.8*	*222*
Eagle	–	–	*1*
Jeep	*1,652*	*-5.5*	*1,748*
Chrysler Total	1,673	-33.6	2,520
Dacia	17,747	15.1	15,422
Daihatsu	5	-54.5	11
Mercedes-Benz	*21,271*	*-8.6*	*23,280*
Smart	*2,358*	*-19.8*	*2,939*
Daimler Total	23,629	-9.9	26,219
Ferrari	50	-3.8	52
Alfa Romeo	*3,799*	*-25.7*	*5,111*
Fiat	*16,017*	*1.0*	*15,861*
Lancia	*1,928*	*17.3*	*1,643*
Fiat Total	21,744	-3.9	22,615
Ford	*51,058*	*-21.4*	*64,928*
Lincoln	–	–	*2*
Mercury	*1*	–	–
Ford Total	51,059	-21.4	64,930
Buick	–	–	*2*
Cadillac	*5*	*-54.5*	*11*
Chevrolet	*15,165*	*-13.3*	*17,482*
GMC	–	–	*3*
Hummer	*3*	*-66.7*	*9*
Opel	*49,586*	*-14.9*	*58,236*
Saturn	–	–	*2*
GM Total	64,759	-14.5	75,745
Acura	–	–	*1*
Honda	*5,917*	*-28.2*	*8,241*
Honda Total	5,917	-28.2	8,242
Hyundai	29,409	4.6	28,114
Iveco	5	-70.6	17
Jaguar	*830*	*5.1*	*790*
Land Rover	*6,415*	*28.4*	*4,997*
Jaguar Land Rover Total	7,245	25.2	5,787
Kia	18,416	-6.4	19,668
Lamborghini	7	133.3	3
Mahindra & Mahindra	3	-93.9	49
Maserati	19	-42.4	33
Mazda	4,599	-21.4	5,848
Mitsubishi	4,006	-47.7	7,656
Morgan	3	-50.0	6
Infiniti	*302*	*-25.2*	*404*
Nissan	*35,588*	*-10.3*	*39,666*
Nissan Total	35,890	-10.4	40,070
Citroen	*52,450*	*-11.4*	*59,198*
Peugeot	*55,644*	*-13.1*	*64,058*
Peugeot Citroen Total	108,094	-12.3	123,256
Porsche	1,414	-19.8	1,763
Proton (Lotus)	2	-75.0	8
Renault	52,561	-17.4	63,670
Rolls-Royce	2	-33.3	3
Saab	35	-92.1	444
Santana	1	-50.0	2
Ssangyong	1,424	-54.6	3,135
Subaru	993	1.2	981
Suzuki	3,717	-29.3	5,258
Tata	310	-11.4	350
Tesla	–	–	2
Lexus	*2,031*	*-21.5*	*2,587*
Scion	*2*	–	*2*
Toyota	*35,936*	*-7.8*	*38,967*
Toyota Total	37,969	-8.6	41,556
SEAT	*54,856*	*-25.4*	*73,524*
Skoda	*13,974*	*-12.1*	*15,895*
Volkswagen	*62,490*	*-12.7*	*71,608*
Volkswagen Total	131,320	-18.4	161,027
Volvo	6,318	-26.5	8,593
Wiesmann	–	–	1
Other	12	-29.4	17
Total Cars	**699,559**	**-13.4**	**808,036**
Light Trucks			
Chrysler	*1*	*-75.0*	*4*
Dodge	–	–	*2*
Chrysler Total	1	-83.3	6
Dacia	100	-54.3	219
Daimler (Mercedes-Benz)	4,152	-11.3	4,680
Fiat	6,464	-30.5	9,296
Ford	8,496	-27.8	11,762

SPAIN

SALES BY COMPANY

	2012	% Chg.	2011
GAZ (LDV)	2	-83.3	12
Chevrolet	–	–	*7*
GMC	–	–	*2*
Opel	*3,111*	*-22.8*	*4,029*
GM Total	3,111	-23.0	4,038
Hyundai	–	–	5
Isuzu	163	-34.8	250
Iveco	2,282	-30.4	3,277
Kia	2	-90.0	20
Mahindra & Mahindra	19	-67.8	59
Mazda	1	-50.0	2
Mitsubishi	313	-44.3	562
Modec	–	–	1
Multicar	2	–	–
Nissan	4,955	-13.8	5,745
Citroen	*13,193*	*-31.2*	*19,171*
Peugeot	*11,848*	*-23.8*	*15,545*
Peugeot Citroen Total	25,041	-27.9	34,716
Piaggio	122	-68.0	381
Renault	12,320	-34.6	18,831
Santana	–	–	8
Ssangyong	22	-46.3	41
Tata	27	-79.4	131
Toyota	1,666	-18.3	2,039
SEAT	*131*	*107.9*	*63*
Skoda	*13*	*-38.1*	*21*
Volkswagen	*7,206*	*-8.9*	*7,911*
Volkswagen Total	7,350	-8.1	7,995
Volvo Truck (Renault Truck)	602	-37.4	961
Other	8	-74.2	31
Total Light Trucks	**77,221**	**-26.5**	**105,068**
Medium/Heavy Trucks			
BMC	–	–	4
Mercedes-Benz	*1,768*	*-19.4*	*2,193*
Mitsubishi Fuso	*36*	*-45.5*	*66*
Daimler Total	1,804	-20.1	2,259
Dennis Eagle	–	–	6
Hyundai	–	–	1
Isuzu	60	-4.8	63
Iveco	2,666	-16.8	3,204
ERF	–	–	*2*
MAN	*1,544*	*-24.5*	*2,044*
MAN Total	1,544	-24.5	2,046
Nissan	78	-33.3	117
PACCAR (DAF)	1,178	-34.2	1,791
Scania	1,187	-23.3	1,547
Renault Truck	*2,087*	*-9.4*	*2,304*
Volvo Truck	*1,702*	*-19.4*	*2,112*
Volvo Truck Total	3,789	-14.2	4,416
Other	3	-62.5	8
Total Med./Hvy. Trks.	**12,309**	**-20.4**	**15,462**
Buses			
BMC	11	-47.6	21
Breda	16	-70.9	55
Mercedes-Benz	*472*	*-24.7*	*627*
Setra	*66*	*-39.4*	*109*
Daimler Total	538	-26.9	736
Dennis Eagle	4	–	–
Fiat	64	-22.0	82
Ford	33	725.0	4
Chevrolet	–	–	*1*
Opel	*1*	–	–
GM Total	1	–	1
Iveco	313	-42.5	544
MAN	*296*	*-33.5*	*445*
Neoplan	–	–	*1*
Volkswagen Truck & Bus	*11*	*-31.3*	*16*
MAN Total	307	-33.5	462
PACCAR (DAF)	11	-60.7	28
Citroen	*27*	*200.0*	*9*
Peugeot	*20*	*-70.1*	*67*
Peugeot Citroen Total	47	-38.2	76
Renault	38	442.9	7
Scania	185	-48.9	362
Tata	11	450.0	2
Van Hool	1	-50.0	2
Renault Truck	*2*	–	–
Volvo Truck	*233*	*-36.5*	*367*
Volvo Truck Total	235	-36.0	367
Other	57	-23.0	74
Total Buses	**1,872**	**-33.7**	**2,823**
Total Comm. Vehs.	**91,402**	**-25.9**	**123,353**
Total Vehicles	**790,961**	**-15.1**	**931,389**

SWEDEN

ECONOMIC SUMMARY

Common Language	Swedish
Currency	Swedish Krona
Steering	Left Hand Drive

2012 Vehicles in Operation

Passenger Cars	4,457,145
Commercial Vehicles	572,416
Total	5,029,561
Persons per Car	2.0
Persons per Vehicle	1.8

2012 GDP (billions of US $)

Nominal GDP	526.2
Real GDP	543.3
GDP % annual growth	1.2
GDP per capita	56,955

Current Account Balance (billions of US $)

2012:	37.6
2011:	37.9

Current Account Balance in % of GDP

2012:	7.1
2011:	7.0

Total Number of Fatal Road Accidents (most recent data)

2010:	270

Capital	Stockholm
Population	9,119,423
Population Density	57.6
Surface Area sq km/miles	410,335 / 158,431

Road Infrastructure

Paved Road Surface %	79.0
Total Road Length in km/miles	572,900 / 355,984
Road Traffic (Vehicles per mile)	14.1

2012 Total Vehicle Percent Market Share by Company

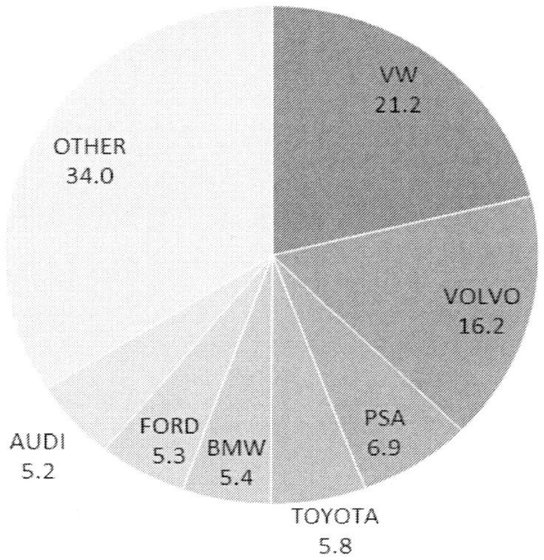

Total Vehicle Production and Sales 2002-2012

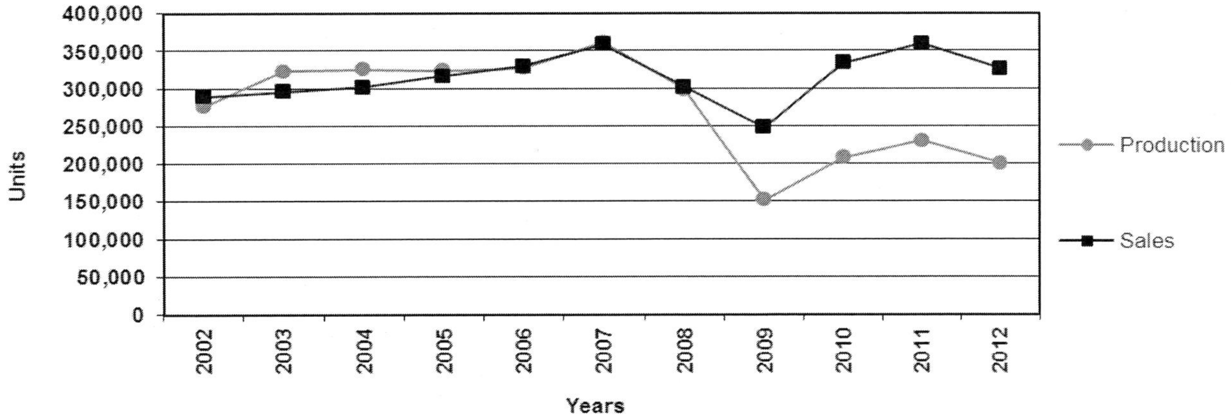

SWEDEN

PRODUCTION

Year	Cars	Trucks	Buses	Total
2012	162,814	28,600	8,900	200,314
2011	188,969	30,960	10,500	230,429
2010	177,084	21,322	10,021	208,427
2009	128,738	13,549	9,839	152,126
2008	252,287	37,156	9,154	298,597
2007	316,850	35,099	9,288	361,237
2006	288,583	29,755	9,004	327,342
2005	288,659	25,697	9,224	323,580
2004	290,383	27,257	8,132	325,772
2003	280,394	34,588	8,050	323,032
2002	237,975	29,931	8,287	276,193
2001	251,035	25,223	7,940	284,198
2000	259,959	27,896	7,779	295,634
1999	213,900	24,600	7,150	245,650
1998	191,600	24,500	7,180	223,280
1997	200,800	22,900	7,240	230,940
1996	193,600	24,500	5,500	223,600
1995	215,100	24,900	5,850	245,850
1994	352,951	73,488	8,556	434,995
1993	279,002	50,775	7,609	337,386
1992	293,499	53,349	9,747	356,595
1991	269,431	67,266	7,993	344,690
1990	335,853	67,451	6,964	410,268
1989	384,206	73,443	8,227	465,876
1988	407,117	68,417	8,090	483,624
1987	431,777	62,660	7,385	501,822
1986	421,255	59,354	6,543	487,152
1985	400,748	54,186	6,138	461,072
1984	372,866	53,722	5,264	431,852
1983	362,527	46,981	5,014	414,522
1982	311,078	48,376	5,898	365,352

SWEDEN

PRODUCTION BY MANUFACTURER

	2012	% Chg.	2011		2012	% Chg.	2011
Cars				**Buses**			
Saab	–	–	12,160	Scania	3,500	-30.0	5,000
Volvo	162,814	-7.9	176,809	Volvo Truck	5,400	-1.8	5,500
Total Cars	**162,814**	**-13.8**	**188,969**	**Total Buses**	**8,900**	**-15.2**	**10,500**
Trucks				**Total Comm. Vehs.**	**37,500**	**-9.6**	**41,460**
Scania	10,100	-8.2	11,000				
Volvo Truck	18,500	-7.3	19,960	**Total Vehicles**	**200,314**	**-13.1**	**230,429**
Total Trucks	**28,600**	**-7.6**	**30,960**				

SWEDEN

SALES

Year	Cars	Trucks	Buses	Total
2012	279,478	44,945	1,597	326,020
2011	304,984	52,474	1,608	359,066
2010	289,477	42,793	1,657	333,927
2009	213,408	32,946	1,159	247,513
2008	253,982	46,303	1,174	301,459
2007	306,799	50,932	991	358,722
2006	282,766	45,846	1,348	329,960
2005	274,301	40,635	1,172	316,108
2004	264,213	36,244	1,127	301,584
2003	261,206	33,160	1,118	295,484
2002	254,589	33,395	1,181	289,165
2001	246,581	34,140	1,126	281,847
2000	290,529	37,161	1,313	329,003
1999	295,249	34,044	1,253	330,546
1998	253,430	30,584	1,000	285,014
1997	225,263	24,631	1,166	251,060
1996	183,820	20,137	1,062	205,019
1995	169,756	13,952	866	184,574
1994	156,375	9,650	642	166,667
1993	124,434	7,673	618	132,725
1992	154,173	11,186	785	166,144
1991	188,288	19,798	1,000	209,086
1990	229,899	32,051	1,082	263,032
1989	307,104	43,115	1,301	351,520
1988	343,963	36,015	959	380,937
1987	315,950	29,550	949	346,449
1986	270,070	23,610	1,031	294,711
1985	263,033	21,945	988	285,966
1984	231,121	19,282	1,020	251,423
1983	217,123	16,865	1,229	235,217
1982	218,428	17,377	1,384	237,189

SWEDEN

SALES BY COMPANY

	2012	% Chg.	2011
Cars			
Aston Martin	23	27.8	18
Audi	17,079	-9.5	18,872
AutoVaz (Lada)	24	118.2	11
Bentley	18	80.0	10
BMW	*16,035*	*–*	*16,036*
Mini	*1,492*	*5.4*	*1,415*
BMW Total	17,527	0.4	17,451
Chrysler	*3*	*-99.0*	*302*
Dodge	*2*	*-99.0*	*193*
Jeep	*528*	*4.1*	*507*
Chrysler Total	533	-46.8	1,002
Dacia	624	-61.5	1,620
Mercedes-Benz	*11,399*	*-3.9*	*11,861*
Smart	*55*	*-24.7*	*73*
Daimler Total	11,454	-4.0	11,934
Ferrari	64	-11.1	72
Alfa Romeo	*271*	*-50.4*	*546*
Fiat	*4,376*	*4.9*	*4,171*
Lancia	*467*	*357.8*	*102*
Fiat Total	5,114	6.1	4,819
Fisker	19	–	–
Ford	13,852	-18.3	16,954
Cadillac	*11*	*-15.4*	*13*
Chevrolet	*2,776*	*60.9*	*1,725*
Opel	*5,811*	*-19.5*	*7,215*
GM Total	8,598	-4.0	8,953
Honda	2,821	-16.5	3,377
Hyundai	11,847	-5.6	12,546
Iveco	43	-23.2	56
Jaguar	*116*	*-46.5*	*217*
Land Rover	*822*	*28.6*	*639*
Jaguar Land Rover Total	938	9.6	856
Kia	15,914	7.8	14,764
Lamborghini	5	-44.4	9
Maserati	23	9.5	21
Mazda	2,149	-20.9	2,716
Mitsubishi	3,525	-33.4	5,296
Morgan	23	-17.9	28
Nissan	6,462	-10.1	7,191
Citroen	*5,537*	*-12.8*	*6,347*
Peugeot	*10,630*	*14.3*	*9,298*
Peugeot Citroen Total	16,167	3.3	15,645
Porsche	522	6.5	490
Proton (Lotus)	2	–	2
Renault	9,474	-25.0	12,634
Saab	694	-80.1	3,484
Ssangyong	4	–	–
Subaru	4,305	23.8	3,476
Suzuki	2,691	-32.1	3,964
Lexus	*720*	*-27.5*	*993*
Toyota	*16,594*	*–*	*16,595*
Toyota Total	17,314	-1.6	17,588
SEAT	*2,086*	*-23.4*	*2,724*
Skoda	*11,937*	*-14.3*	*13,931*
Volkswagen	*42,712*	*-1.8*	*43,492*
Volkswagen Total	56,735	-5.7	60,147
Volvo	52,755	-10.4	58,862
Other	136	17.2	116
Total Cars	**279,478**	**-8.4**	**304,984**
Light Trucks			
Chrysler (Ram)	251	-7.4	271
Dacia	447	-30.6	644
Daimler (Mercedes-Benz)	3,191	-2.2	3,263
Fiat	2,023	-11.8	2,294
Ford	3,508	-21.3	4,458
Chevrolet	*125*	*-16.7*	*150*
GMC	*5*	*-16.7*	*6*
Opel	*1,392*	*-27.0*	*1,907*
GM Total	1,522	-26.2	2,063
Hyundai	90	-37.9	145

SWEDEN

SALES BY COMPANY

	2012	% Chg.	2011
Isuzu	355	-1.4	360
Iveco	413	-2.1	422
Jaguar Land Rover	18	-50.0	36
Mazda	38	-72.1	136
Mitsubishi	769	-25.7	1,035
Nissan	2,843	-25.4	3,812
Citroen	*2,999*	*-18.3*	*3,669*
Peugeot	*3,265*	*9.3*	*2,986*
Peugeot Citroen Total	6,264	-5.9	6,655
Renault	3,726	-21.2	4,728
Ssangyong	20	81.8	11
Toyota	1,673	-16.5	2,003
Skoda	*182*	*-9.0*	*200*
Volkswagen	*12,199*	*-13.2*	*14,050*
Volkswagen Total	12,381	-13.1	14,250
Other	49	58.1	31
Total Light Trucks	**39,581**	**-15.1**	**46,617**
Medium/Heavy Trucks			
Mercedes-Benz	*557*	*-10.3*	*621*
Mitsubishi Fuso	*1*	–	*1*
Daimler Total	558	-10.3	622
Iveco	120	13.2	106
MAN	319	59.5	200
PACCAR (DAF)	129	-30.3	185
Scania	2,088	-3.4	2,162
Renault Truck	*1*	–	–
Volvo Truck	*2,142*	*-16.5*	*2,566*
Volvo Truck Total	2,143	-16.5	2,566
Other	7	-56.3	16
Total Med./Hvy. Trks.	**5,364**	**-8.4**	**5,857**
Buses			
Mercedes-Benz	*313*	*27.8*	*245*
Setra	*51*	*-25.0*	*68*
Daimler Total	364	16.3	313
Fiat	7	250.0	2
Ford	36	–	–
Iveco	38	11.8	34
MAN	*132*	*-35.6*	*205*
Neoplan	*13*	*-31.6*	*19*
MAN Total	145	-35.3	224
Renault	4	-42.9	7
Scania	345	-32.6	512
Volvo Truck	408	14.3	357
Other	250	57.2	159
Total Buses	**1,597**	**-0.7**	**1,608**
Total Comm. Vehs.	**46,542**	**-13.9**	**54,082**
Total Vehicles	**326,020**	**-9.2**	**359,066**

SWITZERLAND

ECONOMIC SUMMARY

Common Language	German/French/Italian
Currency	Swiss Franc/Franken
Steering	Left Hand Drive

2012 Vehicles in Operation

Passenger Cars	4,254,725
Commercial Vehicles	420,204
Total	4,674,929
Persons per Car	1.9
Persons per Vehicle	1.7

2012 GDP (billions of US $)

Nominal GDP	632.4
Real GDP	600.1
GDP % annual growth	1.0
GDP per capita	74,991

Current Account Balance (billions of US $)

2012:	84.7
2011:	55.7

Current Account Balance in % of GDP

2012	13.4
2011:	8.4

Total Number of Fatal Road Accidents (most recent data)

2012:	339

Capital	Bern
Population	7,996,026
Population Density	517.8
Surface Area sq km/miles	39,997 / 15,443

Road Infrastructure

Paved Road Surface %	100.0
Total Road Length in km/miles	71,454 / 44,399
Road Traffic (Vehicles per mile)	105.3

2012 Total Vehicle Percent Market Share by Company

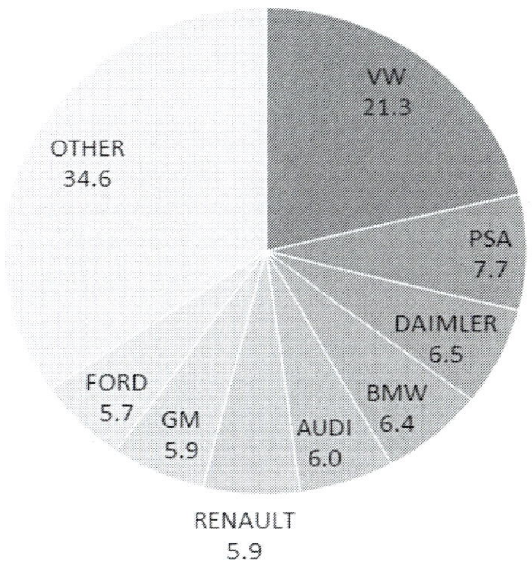

Total Vehicle Sales 2002-2012

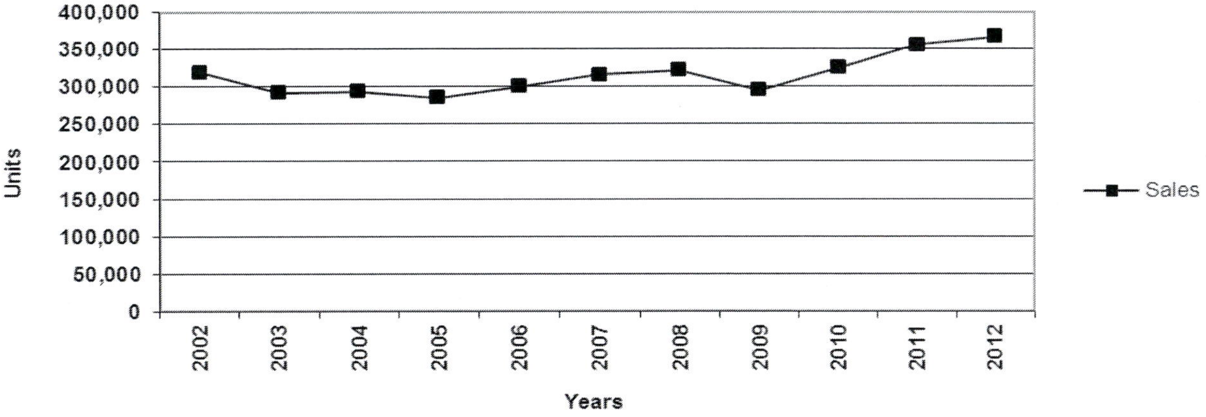

SWITZERLAND

SALES

Year	Cars	Trucks	Buses	Total
2012	328,139	34,325	3,809	366,273
2011	318,958	32,783	3,515	355,256
2010	294,239	27,691	2,849	324,779
2009	266,018	25,964	2,717	294,699
2008	288,525	29,760	3,041	321,326
2007	284,674	28,158	2,570	315,402
2006	269,421	27,560	2,598	299,579
2005	259,426	24,347	1,261	285,034
2004	269,211	23,853	*	293,064
2003	270,359	21,112	*	291,471
2002	295,065	23,995	*	319,060
2001	317,126	28,690	*	345,816
2000	314,482	26,687	*	341,169
1999	314,691	25,595	*	340,286
1998	295,165	23,667	*	318,832
1997	270,625	21,636	*	292,261
1996	272,214	20,549	*	292,763
1995	267,975	20,123	*	288,098
1994	265,892	17,914	*	283,806
1993	256,917	15,135	*	272,052
1992	286,289	19,046	*	305,335
1991	310,193	23,353	*	333,546
1990	322,974	28,877	*	351,851
1989	320,054	26,864	*	346,918
1988	319,406	28,092	*	347,498
1987	303,302	25,225	*	328,527
1986	300,170	23,578	*	323,748
1985	265,467	20,486	*	285,953
1984	267,488	19,495	*	286,983
1983	273,886	18,618	*	292,504
1982	290,890	20,303	*	311,193

* Included with Trucks.

SWITZERLAND

SALES BY COMPANY

	2012	% Chg.	2011
Cars			
Aston Martin	224	-8.2	244
Audi	22,083	19.4	18,502
BMW	*18,968*	*8.0*	*17,564*
Mini	*4,593*	*0.3*	*4,581*
BMW Total	23,561	6.4	22,145
Chrysler	*87*	*-56.1*	*198*
Dodge	*624*	*51.5*	*412*
Jeep	*2,377*	*56.6*	*1,518*
Chrysler Total	3,088	45.1	2,128
Dacia	5,309	-8.8	5,823
Daihatsu	869	-34.6	1,328
Mercedes-Benz	*16,638*	*18.5*	*14,037*
Smart	*1,888*	*-12.1*	*2,148*
Daimler Total	18,526	14.5	16,185
Alfa Romeo	*2,818*	*-31.4*	*4,106*
Fiat	*10,261*	*4.2*	*9,849*
Lancia	*1,092*	*30.9*	*834*
Fiat Total	14,171	-4.2	14,789
Ford	18,179	-1.2	18,400
Chevrolet	*3,997*	*16.4*	*3,435*
Opel	*15,307*	*-4.2*	*15,978*
GM Total	19,304	-0.6	19,413
Honda	5,540	-5.3	5,851
Hyundai	11,456	14.8	9,976
Jaguar	*511*	*10.8*	*461*
Land Rover	*3,789*	*110.1*	*1,803*
Jaguar Land Rover Total	4,300	89.9	2,264
Kia	4,677	20.7	3,876
Maserati	161	-12.0	183
Mazda	7,152	-3.2	7,387
Mitsubishi	4,611	-17.4	5,580
Infiniti	*153*	*–*	*–*
Nissan	*9,063*	*-4.5*	*9,486*
Nissan Total	9,216	-2.8	9,486
Citroen	*11,678*	*-10.1*	*12,984*
Peugeot	*12,298*	*-19.0*	*15,186*
Peugeot Citroen Total	23,976	-14.9	28,170
Porsche	2,310	6.1	2,178
Renault	17,084	-6.8	18,323
Saab	–	–	299
Ssangyong	476	5.8	450
Subaru	8,483	16.5	7,283
Suzuki	7,702	-6.5	8,240
Lexus	*1,022*	*1.7*	*1,005*
Toyota	*13,038*	*-2.2*	*13,335*
Toyota Total	14,060	-2.0	14,340
SEAT	*8,934*	*7.5*	*8,314*
Skoda	*19,132*	*9.5*	*17,470*
Volkswagen	*44,258*	*9.0*	*40,594*
Volkswagen Total	72,324	9.0	66,378
Volvo	8,059	-4.9	8,478
Other	1,238	-1.7	1,259
Total Cars	**328,139**	**2.9**	**318,958**
Light Trucks			
Dacia	327	-14.4	382
Mercedes-Benz	*3,240*	*0.2*	*3,232*
Mitsubishi Fuso	*89*	*-25.2*	*119*
Daimler Total	3,329	-0.7	3,351
Fiat	1,696	1.1	1,678
Ford	2,663	33.9	1,989
GM (Opel)	2,111	-2.6	2,167
Hyundai	99	-8.3	108
Iveco	1,779	8.4	1,641
Jaguar Land Rover	267	37.6	194
Kia	60	-64.7	170
Mazda	59	-50.0	118

SWITZERLAND

SALES BY COMPANY

	2012	% Chg.	2011
Mitsubishi	291	4.7	278
Nissan	1,437	-4.4	1,503
Citroen	*2,280*	*-0.1*	*2,282*
Peugeot	*1,740*	*-1.4*	*1,764*
Peugeot Citroen Total	4,020	-0.6	4,046
Renault	4,525	20.4	3,759
Toyota	1,354	-5.0	1,426
Skoda	*109*	*31.3*	*83*
Volkswagen	*5,429*	*14.5*	*4,740*
Volkswagen Total	5,538	14.8	4,823
Other	917	3.6	885
Total Light Trucks	**30,472**	**6.9**	**28,518**
Medium/Heavy Trucks			
Mercedes-Benz	*1,196*	*2.2*	*1,170*
Mitsubishi Fuso	*40*	–	*40*
Daimler Total	1,236	2.1	1,210
Fiat	–	–	1
Ford	1	–	–
GM (Opel)	3	200.0	1
Hino	8	700.0	1
Iveco	474	18.5	400
MAN	*556*	*-24.1*	*733*
Volkswagen Truck & Bus	*3*	*-57.1*	*7*
MAN Total	559	-24.5	740
PACCAR (DAF)	161	-20.7	203
Peugeot Citroen	1	–	1
Scania	525	-9.0	577
Renault Truck	*316*	*-11.5*	*357*
UD Trucks	*4*	*-20.0*	*5*

	2012	% Chg.	2011
Volvo Truck	*408*	*-30.3*	*585*
Volvo Truck Total	728	-23.1	947
Other	157	-14.7	184
Total Med./Hvy. Trks.	**3,853**	**-9.7**	**4,265**
Buses			
Dacia	–	–	2
Daimler (Mercedes-Benz)	792	-6.7	849
Fiat	1,485	15.7	1,284
Ford	148	-40.6	249
GM (Opel)	42	100.0	21
Iveco	57	–	57
MAN	*74*	*-15.9*	*88*
Volkswagen Truck & Bus	*620*	*73.7*	*357*
MAN Total	694	56.0	445
Nissan	3	200.0	1
Citroen	*49*	*16.7*	*42*
Peugeot	*6*	*-40.0*	*10*
Peugeot Citroen Total	55	5.8	52
Renault	135	95.7	69
Scania	9	-52.6	19
Toyota	7	-12.5	8
Renault Truck	*11*	*-21.4*	*14*
Volvo Truck	*22*	*-50.0*	*44*
Volvo Truck Total	33	-43.1	58
Other	349	-13.0	401
Total Buses	**3,809**	**8.4**	**3,515**
Total Comm. Vehs.	**38,134**	**5.1**	**36,298**
Total Vehicles	**366,273**	**3.1**	**355,256**

TURKEY

ECONOMIC SUMMARY

Common Language	Turkish
Currency	Turkish Lira
Steering	Left Hand Drive
Capital	Ankara
Population	80,694,485
Population density	271.6
Surface Area sq km/miles	769,632 / 297,156

2012 Vehicles in Operation

Passenger Cars	8,648,875
Commercial Vehicles	4,178,324
Total	12,827,199
Persons per Car	9.3
Persons per Vehicle	6.3

Road Infrastructure

Paved Road Surface %	89.7
Total Road Length in km/miles	367,263 / 228,207
Road Traffic (Vehicles per mile)	56.2

2012 GDP (billions of US $)

Nominal GDP	794.5
Real GDP	65.8
GDP % annual growth	2.6
GDP per capita	879

Current Account Balance (billions of US $)

2012: -46.9
2011: -75.1

Current Account Balance in % of GDP

2012: -5.9
2011: -9.7

Total Number of Fatal Road Accidents (most recent data)

2010: 4,045

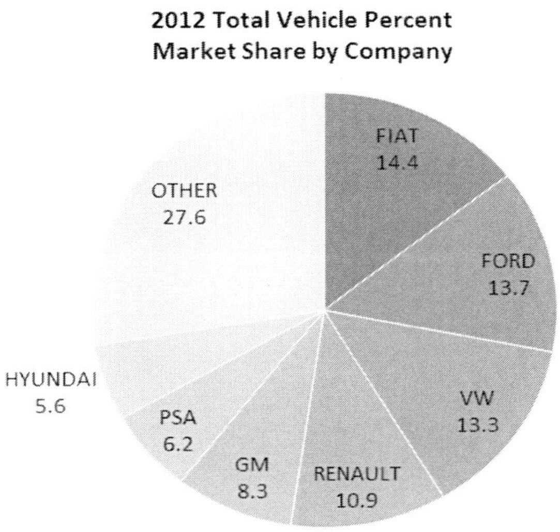

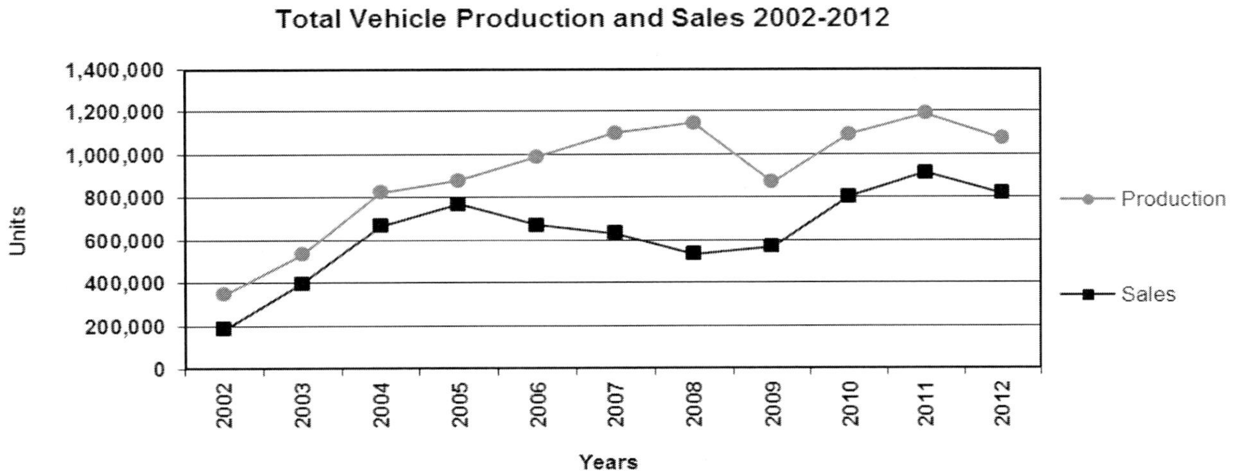

TURKEY

PRODUCTION

Year	Cars	Trucks	Buses	Total
2012	577,296	485,097	10,585	1,072,978
2011	639,734	538,981	10,416	1,189,131
2010	603,394	483,237	7,926	1,094,557
2009	510,931	350,119	8,555	869,605
2008	621,567	507,357	18,186	1,147,110
2007	634,883	448,280	16,251	1,099,414
2006	545,682	427,616	14,282	987,580
2005	453,663	413,274	12,515	879,452
2004	447,152	333,353	42,903	823,408
2003	294,116	214,647	24,909	533,672
2002	204,198	129,167	13,200	346,565
2001	175,343	83,355	11,987	270,685
2000	297,476	97,155	36,316	430,947
1999	222,041	50,647	23,542	296,230
1998	239,937	77,340	27,225	344,502
1997	242,780	76,128	25,444	344,352
1996	207,757	50,464	18,526	276,747
1995	233,412	36,567	12,461	282,440
1994	212,651	21,710	8,813	243,174
1993	348,095	51,109	21,452	420,656
1992	265,245	38,250	19,264	322,759
1991	195,574	30,554	15,710	241,838
1990	167,556	27,512	13,647	208,715
1989	118,314	19,191	8,772	146,277
1988	120,796	20,162	8,717	149,675
1987	107,185	21,125	10,597	138,907
1986	82,032	20,251	10,639	112,922
1985	60,353	26,050	11,225	97,628
1984	54,832	24,797	11,976	91,605
1983	42,509	24,336	10,058	76,903
1982	31,195	20,122	7,140	58,457

Note: Minibus included with trucks. Prior to 2005, minibus was included with buses.

TURKEY

PRODUCTION BY MANUFACTURER

	2012	% Chg.	2011
Cars			
Fiat	81,695	-28.7	114,529
Honda	24,378	97.5	12,341
Hyundai	90,857	0.7	90,231
Renault	292,066	-11.8	330,994
Toyota	88,300	-3.6	91,639
Total Cars	**577,296**	**-9.8**	**639,734**
Light Trucks			
Fiat	174,414	-9.8	193,259
Ford	264,790	-7.6	286,611
Isuzu	764	23.6	618
Otokar	440	-47.8	843
Peugeot Citroen	15,188	-22.8	19,678
Temsa (Mitsubishi)	372	-35.4	576
Total Light Trucks	**455,968**	**-9.1**	**501,585**
Medium/Heavy Trucks			
BMC	1,210	-62.6	3,234
Ford	7,210	-22.0	9,239
Isuzu	2,145	-15.3	2,531
Mercedes-Benz	16,666	-9.1	18,344
Peugeot Citroen	1,185	-50.5	2,392
Temsa (Mitsubishi Fuso)	713	-56.9	1,656
Total Med./Hvy. Trks.	**29,129**	**-22.1**	**37,396**
Buses			
BMC	303	-38.2	490
Isuzu	1,832	55.9	1,175
MAN	1,134	-29.6	1,610
Mercedes-Benz	3,336	10.5	3,018
Otokar	2,479	11.7	2,219
Peugeot Citroen	–	–	76
Temsa	1,501	-17.9	1,828
Total Buses	**10,585**	**1.6**	**10,416**
Total Comm. Vehs.	**495,682**	**-9.8**	**549,397**
Total Vehicles	**1,072,978**	**-9.8**	**1,189,131**

TURKEY

SALES

Year	Cars	Trucks	Buses	Total
2012	556,280	255,846	5,494	817,620
2011	594,307	313,424	5,636	913,367
2010	515,595	283,485	4,050	803,130
2009	365,052	199,398	3,542	567,992
2008	308,813	215,352	7,982	532,147
2007	357,946	264,046	8,602	630,594
2006	371,175	290,766	7,068	669,009
2005	437,474	323,076	6,916	767,466
2004	451,347	215,195	*	666,542
2003	260,683	134,446	*	395,129
2002	104,599	80,153	*	184,752
2001	148,121	70,028	*	218,149
2000	473,378	188,569	*	661,947
1999	284,402	110,799	*	395,201
1998	280,487	121,423	*	401,910

*Included with Trucks.

TURKEY

SALES BY COMPANY

	2012	% Chg.	2011
Cars			
Aston Martin	10	-50.0	20
Audi	13,720	13.7	12,064
AutoVaz (Lada)	–	–	10
Bentley	18	80.0	10
BMW	*15,247*	*1.5*	*15,018*
Mini	*1,758*	*1.3*	*1,735*
BMW Total	17,005	1.5	16,753
Chery	810	-58.1	1,932
Chrysler	–	–	*47*
Dodge	–	–	*1*
Jeep	*799*	*-10.0*	*888*
Chrysler Total	799	-14.6	936
Dacia	27,666	54.2	17,937
Daihatsu	–	–	379
Mercedes-Benz	*12,730*	*0.8*	*12,630*
Smart	*96*	*60.0*	*60*
Daimler Total	12,826	1.1	12,690
Ferrari	21	-8.7	23
Alfa Romeo	*1,052*	*17.1*	*898*
Fiat	*44,295*	*-22.7*	*57,280*
Lancia	*276*	*55.1*	*178*
Fiat Total	45,623	-21.8	58,356
Ford	47,030	-20.0	58,807
Chevrolet	*18,492*	*-21.7*	*23,610*
Opel	*49,143*	*-6.6*	*52,624*
GM Total	67,635	-11.3	76,234
Honda	16,516	25.1	13,197
Hyundai	43,665	-1.8	44,449
Jaguar	*84*	*-50.0*	*168*
Land Rover	*1,590*	*-25.7*	*2,139*
Jaguar Land Rover Total	1,674	-27.4	2,307
Kia	11,870	22.2	9,714
Lamborghini	2	-66.7	6
Maserati	18	-25.0	24
Mazda	810	-50.0	1,620
Mitsubishi	1,289	-57.1	3,005
Infiniti	*41*	*-59.8*	*102*
Nissan	*18,873*	*1.6*	*18,573*
Nissan Total	18,914	1.3	18,675
Citroen	*14,711*	*1.7*	*14,469*
Peugeot	*14,519*	*-6.2*	*15,480*
Peugeot Citroen Total	29,230	-2.4	29,949
Porsche	497	12.4	442
Proton	760	-31.7	1,113
Renault	72,874	-25.4	97,651
Ssangyong	946	7.4	881
Subaru	640	-15.9	761
Suzuki	1,780	32.8	1,340
Tata	684	-33.2	1,024
Toyota	31,961	-13.0	36,733
SEAT	*5,811*	*-4.1*	*6,059*
Skoda	*10,118*	*33.3*	*7,589*
Volkswagen	*66,792*	*20.2*	*55,550*
Volkswagen Total	82,721	19.5	69,198
Volvo	5,247	12.5	4,662
Zhejiang Geely	1,019	-27.5	1,405
Total Cars	**556,280**	**-6.4**	**594,307**
Light Trucks			
BMC	–	–	4
Chery	–	–	113
Dacia	1,298	-61.8	3,402
Daimler (Mercedes-Benz)	9,234	10.0	8,395
Dongfeng	921	-73.0	3,407
Fiat	72,119	-16.4	86,300
Ford	58,741	-21.4	74,745
GM (Opel)	247	-86.4	1,813

TURKEY

SALES BY COMPANY

	2012	% Chg.	2011
Hyundai	2,454	-4.0	2,557
Isuzu	2,602	-10.4	2,904
Iveco	1,673	-8.7	1,833
Kia	845	8.3	780
Mazda	203	-11.0	228
Mitsubishi	3,115	-37.8	5,008
Nissan	1,247	-28.6	1,746
Otokar	304	-52.0	633
Citroen	*7,672*	*-17.1*	*9,258*
Peugeot	*12,309*	*-16.9*	*14,817*
Peugeot Citroen Total	19,981	-17.0	24,075
Renault	16,331	-35.0	25,128
Ssangyong	1,143	12.6	1,015
Tata	360	-56.8	833
Toyota	2,615	93.6	1,351
Skoda	–	–	*1*
Volkswagen	*26,048*	*-1.2*	*26,361*
Volkswagen Total	26,048	-1.2	26,362
Total Light Trucks	**221,481**	**-18.8**	**272,632**
Medium/Heavy Trucks			
BMC	870	-75.2	3,511
Daimler (Mercedes-Benz)	13,605	-9.1	14,970
Ford	6,429	-22.9	8,342
Isuzu	1,802	-29.2	2,544
Mitsubishi	738	-39.1	1,212
Peugeot Citroen	1,104	-54.1	2,404
Other	9,817	25.7	7,809
Total Med./Hvy. Trks.	**34,365**	**-15.8**	**40,792**
Buses			
BMC	216	-20.0	270
Daimler (Mercedes-Benz)	1,132	9.2	1,037
Isuzu	890	-13.7	1,031
MAN	274	-25.1	366
Otokar	2,198	23.1	1,786
Peugeot Citroen	–	–	1
Temsa	665	-36.3	1,044
Other	119	17.8	101
Total Buses	**5,494**	**-2.5**	**5,636**
Total Comm. Vehs.	**261,340**	**-18.1**	**319,060**
Total Vehicles	**817,620**	**-10.5**	**913,367**

UKRAINE

ECONOMIC SUMMARY

Common Language	Ukrainian/Russian
Currency	Hryvnia
Steering	Left Hand Drive

2012 Vehicles in Operation

Passenger Cars	8,301,000
Commercial Vehicles	1,642,000
Total	9,943,000
Persons per Car	5.4
Persons per Vehicle	4.5

2012 GDP (billions of US $)

Nominal GDP	176.2
Real GDP	85.4
GDP % annual growth	0.2
GDP per capita	1,879

Current Account Balance (billions of US $)

2012: -14.4
2011: -10.2

Current Account Balance in % of GDP

2012: -8.2
2011: -6.3

Total Number of Fatal Road Accidents (most recent data)

2010: 4,709

Capital	Kiev
Population	44,573,205
Population Density	199.3
Surface Area sq km/miles	579,330 / 223,680

Road Infrastructure

Paved Road Surface %	97.8
Total Road Length in km/miles	169,495 / 105,319
Road Traffic (Vehicles per mile)	94.4

2012 Total Vehicle Percent Market Share by Company

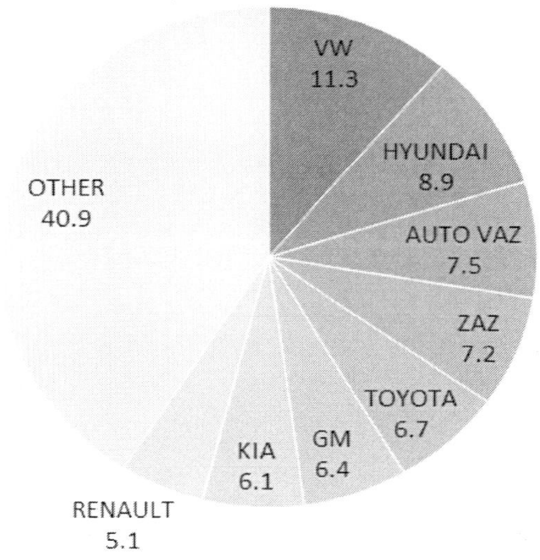

Total Vehicle Production and Sales 2005-2012

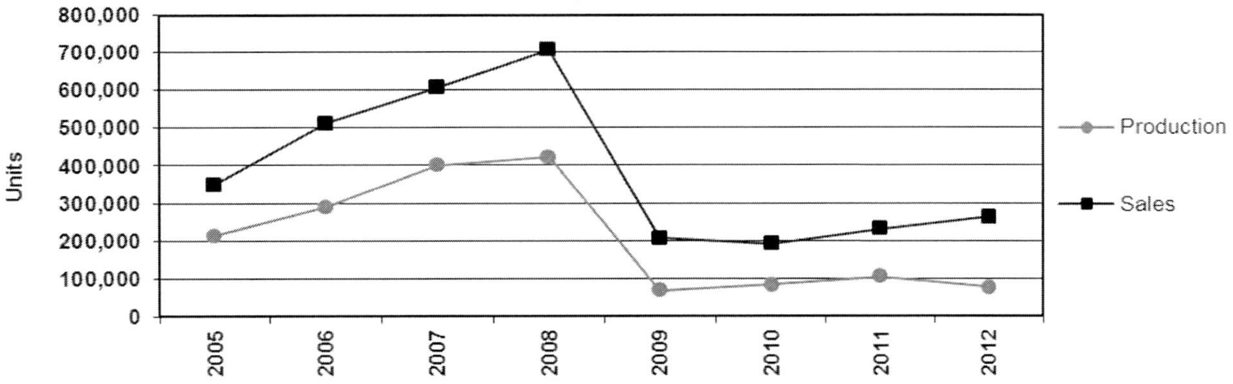

UKRAINE

PRODUCTION

Year	Cars	Trucks	Buses	Total
2012	69,687	3,235	3,359	76,281
2011	97,585	3,330	3,738	104,653
2010	74,992	5,129	2,693	82,814
2009	65,646	2,331	1,318	69,295
2008	400,799	12,551	9,783	423,133
2007	379,561	12,210	8,980	400,751
2006	267,500	14,648	7,639	289,787
2005	192,000	15,642	4,874	212,516
2004	171,700	12,897	2,762	187,359
2003	101,300	1,265	2,673	105,238
2002	50,393	3,430	*	53,823
2001	27,004	6,587	*	33,591
2000	18,124	13,131	*	31,255
1999	9,433	9,204	*	18,637
1998	25,523	4,205	*	29,728
1997	1,704	5,544	*	7,248
1996	8,916	4,346	*	13,262
1995	58,984	8,553	*	67,537

* Included with Trucks.

UKRAINE

PRODUCTION BY MANUFACTURER

	2012	% Chg.	2011
Cars			
Hyundai	*4,210*	*-40.6*	*7,085*
Kia	*3,612*	*-40.5*	*6,071*
Lada	*4,212*	*-40.5*	*7,084*
Bogdan Total	12,034	-40.5	20,240
Eurocar (Volkswagen)	14,556	24.9	11,656
KrASZ	3,180	-49.8	6,329
Chery	*7,960*	*35.4*	*5,880*
Chevrolet	*16,428*	*326.7*	*3,850*
Kia	*1,440*	*-52.1*	*3,006*
Lada	–	–	*2,342*
ZAZ	*14,089*	*-68.2*	*44,282*
ZAZ Total	39,917	-32.8	59,360
Total Cars	**69,687**	**-28.6**	**97,585**
Light Trucks			
KrASZ	100	-50.0	200
ZAZ	721	-20.8	910
Total Light Trucks	**821**	**-26.0**	**1,110**

	2012	% Chg.	2011
Medium/Heavy Trucks			
BAZ	277	19.4	232
Bogdan (Isuzu)	876	35.0	649
KRAZ	901	-11.5	1,018
ZAZ (Tata)	236	-19.2	292
Other	124	327.6	29
Total Med./Hvy. Trks.	**2,414**	**8.7**	**2,220**
Buses			
BAZ	1,116	-4.0	1,163
Bogdan	455	216.0	144
LAZ	–	–	97
ZAZ	669	123.0	300
Other	1,119	-45.0	2,034
Total Buses	**3,359**	**-10.1**	**3,738**
Total Comm. Vehs.	**6,594**	**-6.7**	**7,068**
Total Vehicles	**76,281**	**-27.1**	**104,653**

UKRAINE

SALES

Year	Light Vehicles	Medium/Heavy Trucks	Buses	Total
2012	237,602	21,984	4,018	263,604
2011	207,453	19,797	3,640	230,890
2010	169,540	18,868	2,701	191,109
2009	175,165	29,428	2,665	207,258
2008	610,200	86,800	8,900	705,900
2007	514,100	84,400	8,800	607,300
2006	378,000	122,500	11,300	511,800
2005	273,600	65,900	9,300	348,800

UKRAINE

SALES BY MANUFACTURER

	2012	% Chg.	2011
Light Vehicles			
Audi	2,752	87.1	1,471
AutoVaz (Lada)	19,718	-30.7	28,436
BMW	2,674	60.3	1,668
Bogdan	2,115	–	–
Chery	3,519	-19.5	4,374
Daimler (Mercedes-Benz)	2,791	69.3	1,649
Fiat	2,118	56.9	1,350
Ford	10,238	-2.0	10,446
Chevrolet	*9,616*	*-11.5*	*10,869*
Daewoo	*4,549*	*-10.3*	*5,072*
Opel	*2,460*	*98.4*	*1,240*
GM Total	16,625	-3.2	17,181
Honda	3,752	19.0	3,154
Hyundai	22,868	4.8	21,829
Jaguar Land Rover	1,472	67.5	879
Kia	16,111	73.1	9,307
Mazda	1,714	0.9	1,698
Mitsubishi	5,501	-20.7	6,938
Nissan	12,609	36.4	9,245
Citroen	*3,149*	*124.8*	*1,401*
Peugeot	*3,718*	*14.1*	*3,259*
Peugeot Citroen Total	6,867	47.4	4,660
Renault	13,567	18.2	11,478
Ssangyong	3,017	10.8	2,723
Subaru	1,142	-15.7	1,355
Suzuki	1,687	-1.5	1,712
Lexus	*1,887*	*2.7*	*1,838*
Toyota	*15,710*	*21.7*	*12,906*
Toyota Total	17,597	19.4	14,744
Skoda	*13,948*	*31.6*	*10,596*
Volkswagen	*15,786*	*56.3*	*10,099*
Volkswagen Total	29,734	43.7	20,695
Volvo	1,045	95.0	536
ZAZ	18,124	-7.2	19,529
Zhejiang Geely	12,099	87.6	6,449
Other	6,146	55.7	3,947
Total Light Vehicles	**237,602**	**14.5**	**207,453**
Medium/Heavy Trucks			
BAZ	177	-30.0	253
Bogdan	207	8.9	190
Daimler (Mercedes-Benz)	897	80.8	496
FAW	130	-13.3	150
Fiat	3,061	46.2	2,094
Ford	1,403	49.7	937
GAZ	3,069	-12.3	3,499
GM (Opel)	198	18.6	167
Hyundai	496	-12.7	568
Iveco	180	2.3	176
Kamaz	922	1.9	905
MAN	*173*	*-3.4*	*179*
Volkswagen Truck & Bus	*3,031*	*6.6*	*2,843*
MAN Total	3,204	6.0	3,022
MAZ	760	6.0	717
PACCAR (DAF)	150	-46.0	278
Citroen	*947*	*-14.8*	*1,111*
Peugeot	*1,386*	*42.9*	*970*
Peugeot Citroen Total	2,333	12.1	2,081
Scania	473	118.0	217
Tata	226	-0.4	227
UAZ	387	-14.2	451
Renault Truck	*2,154*	*35.6*	*1,588*
Volvo Truck	*260*	*30.0*	*200*
Volvo Truck Total	2,414	35.0	1,788
ZAZ	747	-32.5	1,106
Other	550	15.8	475
Total Med./Hvy Trucks	**21,984**	**11.0**	**19,797**
Buses			
BAZ	1,153	-3.8	1,199
Bogdan	244	-20.3	306
Daimler (Mercedes-Benz)	8	-42.9	14
Ford	73	82.5	40
GAZ	494	12.0	441
Isuzu	517	104.3	253
MAZ	221	370.2	47
PAZ	541	63.9	330
UAZ	196	100.0	98
Volvo Truck (Renault Truck)	6	–	–
Other	565	-38.0	912
Total Buses	**4,018**	**10.4**	**3,640**
Total Comm. Vehs.	**26,002**	**10.9**	**23,437**
Total Vehicles	**263,604**	**14.2**	**230,890**

UNITED KINGDOM

ECONOMIC SUMMARY

Common Language	English
Currency	Pound
Steering	Right Hand Drive

2012 Vehicles in Operation

Passenger Cars	31,481,823
Commercial Vehicles	4,279,078
Total	35,760,901
Persons per Car	2.0
Persons per Vehicle	1.8

2012 GDP (billions of US $)

Nominal GDP	2,440.5
Real GDP	1,906.3
GDP % annual growth	0.2
GDP per capita	30,142

Current Account Balance (billions of US $)

2012:	-85.5
2011:	-32.8

Current Account Balance in % of GDP

2012:	-3.5
2011:	-1.3

Total Number of Fatal Road Accidents (most recent data)

2011:	1,901

Capital	London
Population	63,395,574
Population Density	678.7
Surface Area sq km/miles	241,930 / 93,405

Road Infrastructure

Paved Road Surface %	100.0
Total Road Length in km/miles	394,428 / 245,086
Road Traffic (Vehicles per mile)	145.9

2012 Total Vehicle Percent Market Share by Company

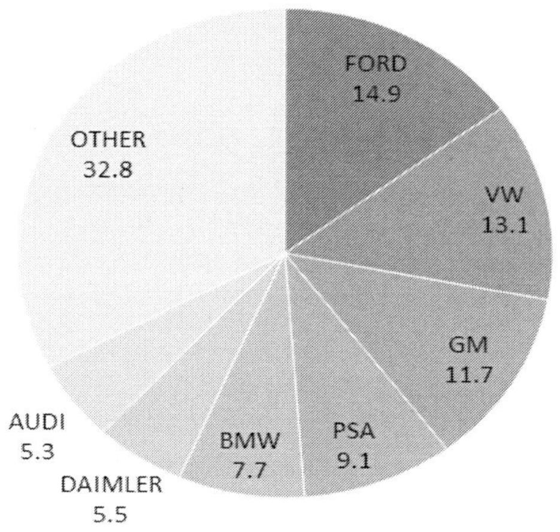

Total Vehicle Production and Sales 2002-2012

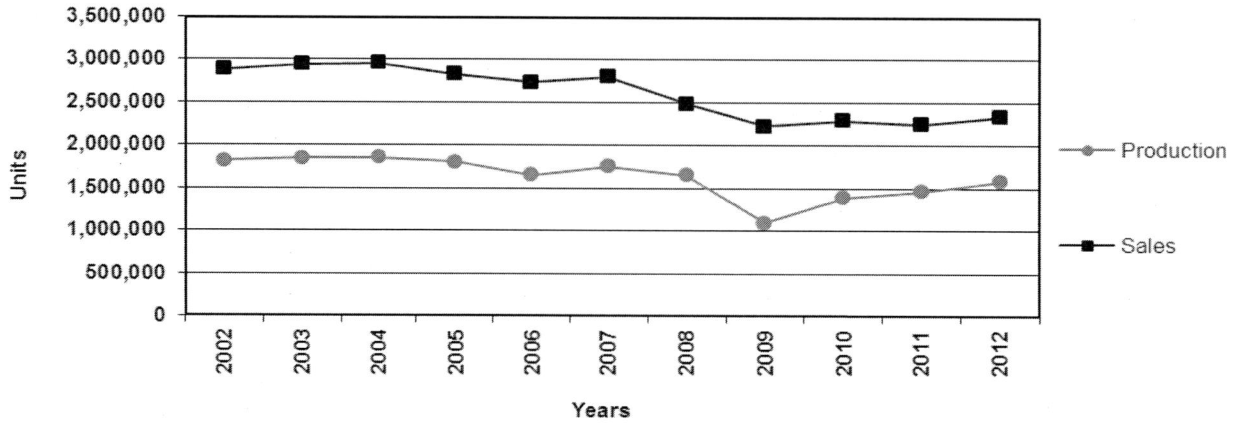

UNITED KINGDOM

PRODUCTION

Year	Cars	Trucks	Buses	Total
2012	1,463,810	109,734	2,305	1,575,849
2011	1,343,794	118,362	1,827	1,463,983
2010	1,274,291	117,692	1,508	1,393,491
2009	999,435	89,030	1,470	1,089,935
2008	1,447,777	199,964	1,752	1,649,493
2007	1,534,676	214,403	1,355	1,750,434
2006	1,441,879	206,677	1,267	1,649,823
2005	1,596,088	205,322	1,431	1,802,841
2004	1,646,881	207,877	1,291	1,856,049
2003	1,657,558	187,134	1,737	1,846,429
2002	1,627,972	189,178	2,069	1,819,219
2001	1,492,365	190,922	1,951	1,685,238
2000	1,641,317	170,257	2,185	1,813,759
1999	1,786,624	183,095	2,812	1,972,531
1998	1,748,258	224,943	2,436	1,975,637
1997	1,698,015	235,352	2,351	1,935,718
1996	1,686,134	227,044	11,270	1,924,448
1995	1,532,084	224,062	8,939	1,765,085
1994	1,466,823	218,249	9,566	1,694,638
1993	1,375,524	186,448	7,019	1,568,991
1992	1,291,880	239,936	8,517	1,540,333
1991	1,236,900	207,304	9,837	1,454,041
1990	1,295,611	258,026	12,320	1,565,957
1989	1,299,082	311,732	14,858	1,625,672
1988	1,226,835	300,843	16,500	1,544,178
1987	1,142,683	232,732	13,995	1,389,410
1986	1,018,962	216,345	12,340	1,247,647
1985	1,047,973	249,946	16,027	1,313,946
1984	908,906	208,326	16,499	1,133,731
1983	1,044,597	228,898	15,616	1,289,111
1982	887,679	255,691	13,107	1,156,477

UNITED KINGDOM

PRODUCTION BY MANUFACTURER

	2012	% Chg.	2011
Cars			
Aston Martin	3,276	-8.2	3,568
Bentley	9,108	21.0	7,528
BMW (Mini)	206,989	8.1	191,474
Caterham	396	-15.4	468
GM (Vauxhall)	90,187	-34.6	137,971
Honda	165,630	69.9	97,459
Jaguar	*55,916*	*12.0*	*49,932*
Land Rover	*305,467*	*28.2*	*238,237*
Jaguar Land Rover Total	361,383	25.4	288,169
Lotus	1,043	-28.5	1,458
LTI	955	-30.5	1,375
McLaren	1,035	–	–
Morgan	576	-16.9	693
Nissan	510,572	6.3	480,485
Rolls-Royce	3,231	-12.3	3,685
Shanghai AIC (MG)	–	–	1,315
Toyota	109,429	-14.6	128,146
Total Cars	**1,463,810**	**8.9**	**1,343,794**
Light Trucks			
BMW (Mini)	261	–	–
Ford	28,031	-0.5	28,170
Nissan	*609*	–	–
Renault	*16,482*	*9.9*	*15,001*
Vauxhall	*42,381*	*-25.6*	*56,981*
GM Total	59,472	-17.4	71,982
Jaguar Land Rover	6,281	119.5	2,862
Total Light Trucks	**94,045**	**-8.7**	**103,014**
Medium/Heavy Trucks			
Dennis Eagle	1,004	28.7	780
Leyland (DAF)	14,685	0.8	14,568
Total Med./Hvy. Trks.	**15,689**	**2.2**	**15,348**
Buses			
Alexander Dennis	1,762	14.5	1,539
Optare	543	88.5	288
Total Buses	**2,305**	**26.2**	**1,827**
Total Comm. Vehs.	**112,039**	**-6.8**	**120,189**
Total Vehicles	**1,575,849**	**7.6**	**1,463,983**

UNITED KINGDOM

SALES

Year	Cars	Trucks	Buses	Total
2012	2,046,184	285,343	3,811	2,335,338
2011	1,942,995	303,097	3,391	2,249,483
2010	2,032,977	257,373	3,226	2,293,576
2009	1,997,087	221,132	4,323	2,222,542
2008	2,133,874	346,873	4,511	2,485,258
2007	2,404,007	388,488	3,993	2,796,488
2006	2,344,864	382,734	4,234	2,731,832
2005	2,439,717	381,771	4,198	2,825,686
2004	2,567,269	385,911	4,012	2,957,192
2003	2,579,048	359,345	4,342	2,942,735
2002	2,563,596	318,266	3,992	2,885,854
2001	2,458,769	309,648	3,763	2,772,180
2000	2,221,647	293,662	4,381	2,519,690
1999	2,197,615	283,552	4,548	2,485,715
1998	2,247,402	290,285	4,225	2,541,912
1997	2,170,725	270,346	3,895	2,444,966
1996	2,025,450	253,213	7,039	2,285,702
1995	1,945,366	246,428	6,235	2,198,029
1994	1,910,933	223,251	4,832	2,139,016
1993	1,778,426	192,965	4,141	1,975,532
1992	1,593,601	196,101	5,085	1,794,787
1991	1,592,326	203,820	5,201	1,801,347
1990	2,005,106	218,873	7,398	2,231,377
1989	2,304,405	295,042	8,005	2,607,452
1988	2,210,309	296,468	9,154	2,515,931
1987	2,016,231	250,614	8,661	2,275,506
1986	1,883,154	238,821	8,924	2,130,899
1985	1,842,075	238,901	6,830	2,087,806
1984	1,759,256	222,895	7,158	1,989,309
1983	1,806,052	229,711	7,271	2,043,034
1982	1,584,328	227,137	7,062	1,818,527

UNITED KINGDOM

SALES BY COMPANY

	2012	% Chg.	2011
Cars			
Aston Martin	932	-9.0	1,024
Audi	123,622	8.6	113,797
Bentley	1,267	22.2	1,037
BMW	*127,530*	*9.3*	*116,642*
Mini	*51,324*	*2.4*	*50,138*
BMW Total	178,854	7.2	166,780
Chrysler	*3,333*	*182.0*	*1,182*
Jeep	*2,306*	*7.5*	*2,146*
Chrysler Total	5,639	69.4	3,328
Daihatsu	–	–	3
Mercedes-Benz	*92,433*	*12.5*	*82,164*
Smart	*5,616*	*11.3*	*5,044*
Daimler Total	98,049	12.4	87,208
Alfa Romeo	*7,253*	*-37.3*	*11,563*
Fiat	*51,199*	*19.3*	*42,903*
Fiat Total	58,452	7.3	54,466
Ford	281,917	6.0	265,894
Chevrolet	*13,476*	*7.6*	*12,524*
Vauxhall	*232,255*	*-1.0*	*234,710*
GM Total	245,731	-0.6	247,234
Honda	54,208	7.2	50,577
Hyundai	74,285	18.1	62,900
Jaguar	*14,109*	*2.3*	*13,787*
Land Rover	*48,626*	*29.2*	*37,637*
Jaguar Land Rover Total	62,735	22.0	51,424
Kia	66,629	24.3	53,615
LTI	997	-31.3	1,451
Maserati	309	-20.4	388
Mazda	26,183	-16.1	31,219
MG Motors	782	117.2	360
Mitsubishi	6,549	-33.5	9,843
Infiniti	*530*	*38.7*	*382*
Nissan	*105,835*	*9.9*	*96,269*
Nissan Total	106,365	10.1	96,651
Perodua	416	-23.4	543
Citroen	*73,656*	*7.6*	*68,464*
Peugeot	*99,486*	*4.7*	*94,989*
Peugeot Citroen Total	173,142	5.9	163,453
Porsche	7,998	25.3	6,382
Lotus	*137*	*-58.4*	*329*
Proton	*208*	*-53.4*	*446*
Proton Total	345	-55.5	775
Renault	40,760	-40.5	68,449
Saab	232	-94.4	4,138
Ssangyong	875	351.0	194
Subaru	2,023	-23.2	2,634
Suzuki	24,893	22.7	20,295
Lexus	*8,404*	*1.6*	*8,269*
Toyota	*84,563*	*14.9*	*73,589*
Toyota Total	92,967	13.6	81,858
SEAT	*38,798*	*7.5*	*36,089*
Skoda	*53,602*	*19.0*	*45,061*
Volkswagen	*183,098*	*2.1*	*179,290*
Volkswagen Total	275,498	5.8	260,440
Volvo	31,790	-2.7	32,657
Other	1,740	-12.0	1,978
Total Cars	**2,046,184**	**5.3**	**1,942,995**
Light Trucks			
BMW (Mini)	86	–	–
Mercedes-Benz	*22,422*	*7.0*	*20,953*
Mitsubishi Fuso	*155*	*-18.4*	*190*
Daimler Total	22,577	6.8	21,143
Fiat	8,476	-8.9	9,301
Ford	65,251	-8.9	71,607
GAZ (LDV)	–	–	1
GM (Vauxhall)	26,597	-20.8	33,565
Great Wall	476	–	7
Hino	–	–	4
Hyundai	379	-44.0	677
Isuzu	3,031	13.5	2,670
Iveco	4,037	-3.8	4,195
Land Rover	5,917	-4.7	6,209
Mitsubishi	4,853	-33.9	7,341
Nissan	10,142	-6.7	10,865
Citroen	*18,383*	*6.3*	*17,293*
Peugeot	*21,631*	*9.9*	*19,682*

UNITED KINGDOM

SALES BY COMPANY

	2012	% Chg.	2011
Peugeot Citroen Total	40,014	8.2	36,975
Renault	14,925	-23.4	19,495
Ssangyong	58	–	4
Toyota	7,747	-7.7	8,391
Volkswagen	31,207	-2.3	31,937
Volvo Truck (Renault Truck)	484	-18.4	593
Other	510	-27.9	707
Total Light Trucks	**246,767**	**-7.1**	**265,687**
Medium/Heavy Trucks			
BMC	1	-90.9	11
Mercedes-Benz	*6,422*	*1.5*	*6,326*
Mitsubishi Fuso	*735*	*23.1*	*597*
Daimler Total	7,157	3.4	6,923
Dennis Eagle	748	12.7	664
Hino	198	40.4	141
Isuzu	873	21.3	720
Iveco	2,908	2.6	2,834
MAN	4,324	-9.4	4,772
PACCAR (DAF)	11,153	13.1	9,863
Scania	4,652	14.3	4,071
Renault Truck	*2,555*	*-7.5*	*2,763*
Volvo Truck	*3,976*	*-14.0*	*4,624*
Volvo Truck Total	6,531	-11.6	7,387
Other	31	29.2	24
Total Med./Hvy. Trks.	**38,576**	**3.1**	**37,410**

	2012	% Chg.	2011
Buses			
Alexander Dennis	1,437	14.4	1,256
BMC	–	–	6
Mercedes-Benz	*141*	*69.9*	*83*
Setra	*10*	*-83.1*	*59*
Daimler Total	151	6.3	142
Iveco (Irisbus)	113	-1.7	115
King Long	21	61.5	13
MAN	*16*	*-5.9*	*17*
Neoplan	*88*	*35.4*	*65*
MAN Total	104	26.8	82
Optare	366	36.1	269
Scania	279	0.4	278
Van Hool	64	25.5	51
Bova	*21*	*-44.7*	*38*
VDL	*112*	*12.0*	*100*
VDL Total	133	-3.6	138
Renault Truck	*–*	*–*	*3*
Volvo Truck	*953*	*11.9*	*852*
Volvo Truck Total	953	11.5	855
Other	190	2.2	186
Total Buses	**3,811**	**12.4**	**3,391**
Total Comm. Vehs.	**289,154**	**-5.7**	**306,488**
Total Vehicles	**2,335,338**	**3.8**	**2,249,483**

UZBEKISTAN

ECONOMIC SUMMARY

Common Language	Uzbek
Currency	Som
Steering	Left Hand Drive

2012 Vehicles in Operation

Passenger Cars	1,453,000
Commercial Vehicles	510,000
Total	1,963,000
Persons per Car	19.7
Persons per Vehicle	14.6

2012 GDP (billions of US $)

Nominal GDP	51.2
Real GDP	0.4
GDP % annual growth	8.0
GDP per capita	14

Total Number of Fatal Road Accidents (most recent data)

2007: 2,731

Capital	Tashkent
Population	28,661,637
Population Density	174.5
Surface Area sq km/miles	425,400 / 164,247

Road Infrastructure

Paved Road Surface %	87.3
Total Road Length in km/miles	86,496 / 53,746
Road Traffic (Vehicles per mile)	36.5

Current Account Balance (billions of US $)

2012: 1.4
2011: 2.6

Current Account Balance in % of GDP

2012: 2.7
2011: 5.8

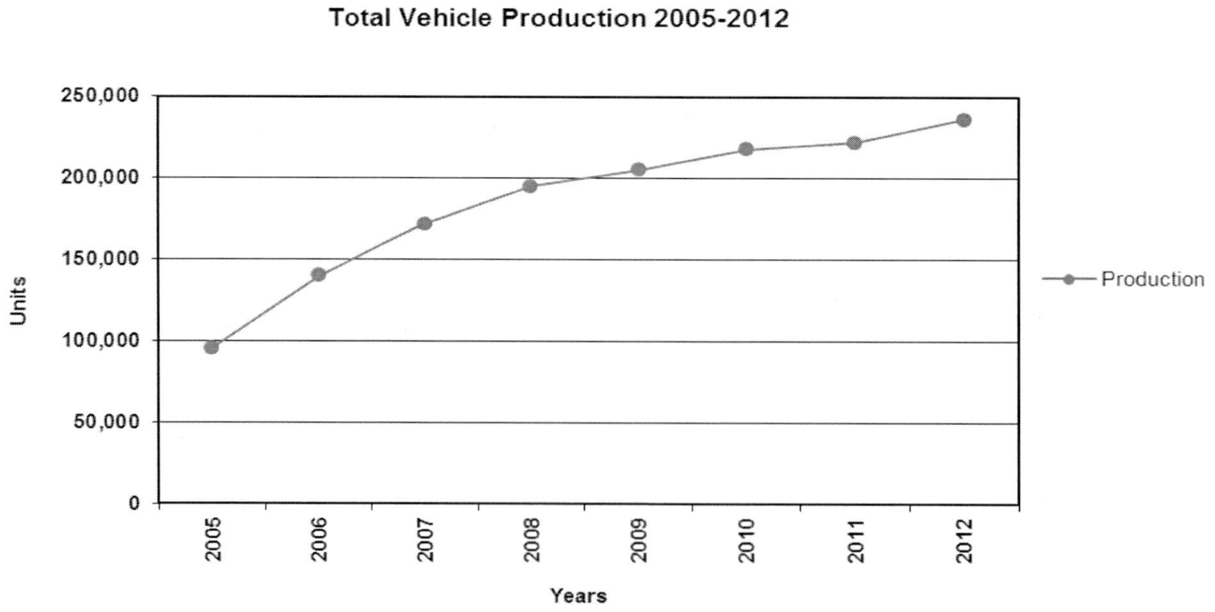

UZBEKISTAN

PRODUCTION

Year	Cars	Commercial Vehicles	Total
2012	210,921	25,280	236,201
2011	201,251	20,771	222,022
2010	196,833	20,900	217,733
2009	181,378	23,633	205,011
2008	170,528	24,510	195,038
2007	150,966	20,843	171,809
2006	124,348	15,731	140,079
2005	87,512	8,302	95,814

PRODUCTION BY MANUFACTURER

	2012	% Chg.	2011
Cars			
GM (Chevrolet)	210,921	4.8	201,251
Total Cars	**210,921**	**4.8**	**201,251**

	2012	% Chg.	2011
Light Trucks			
GM (Chevrolet)	25,280	21.7	20,771
Total Light Trucks	**25,280**	**21.7**	**20,771**
Total Vehicles	**236,201**	**6.4**	**222,022**

Ward's World Motor Vehicle Data 2013

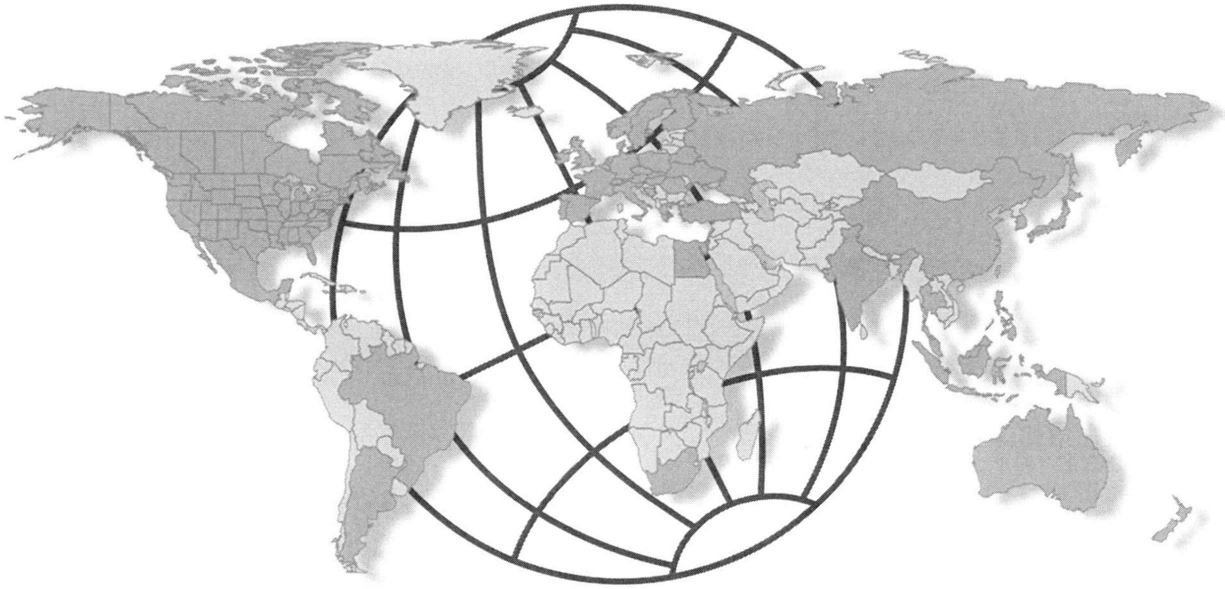

North America Data

CANADA

ECONOMIC SUMMARY

Common Language	English/French
Currency	Canadian Dollar
Steering	Left Hand Drive

2012 Vehicles in Operation

Passenger Cars	20,750,000
Commercial Vehicles	995,000
Total	21,745,000
Persons per Car	1.7
Persons per Vehicle	1.6

2012 GDP (billions of US $)

Nominal GDP	1,819.1
Real GDP	1,663.4
GDP % annual growth	1.8
GDP per capita	47,761

Current Account Balance (billions of US $)

2012:	-67.0
2011:	-52.8

Current Account Balance in % of GDP

2012:	-3.7
2011:	-3.0

Total Number of Fatal Road Accidents (most recent data)

2010:	2,209

Capital	Ottawa
Population	34,568,211
Population Density	9.8
Surface Area sq km/miles	9,093,507 / 3,511,022

Road Infrastructure

Paved Road Surface %	39.9
Total Road Length in km/miles	1,042,300 / 647,655
Road Traffic (Vehicles per mile)	33.6

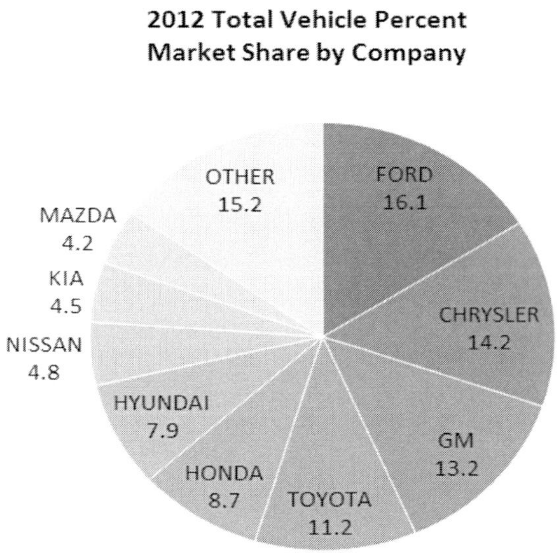

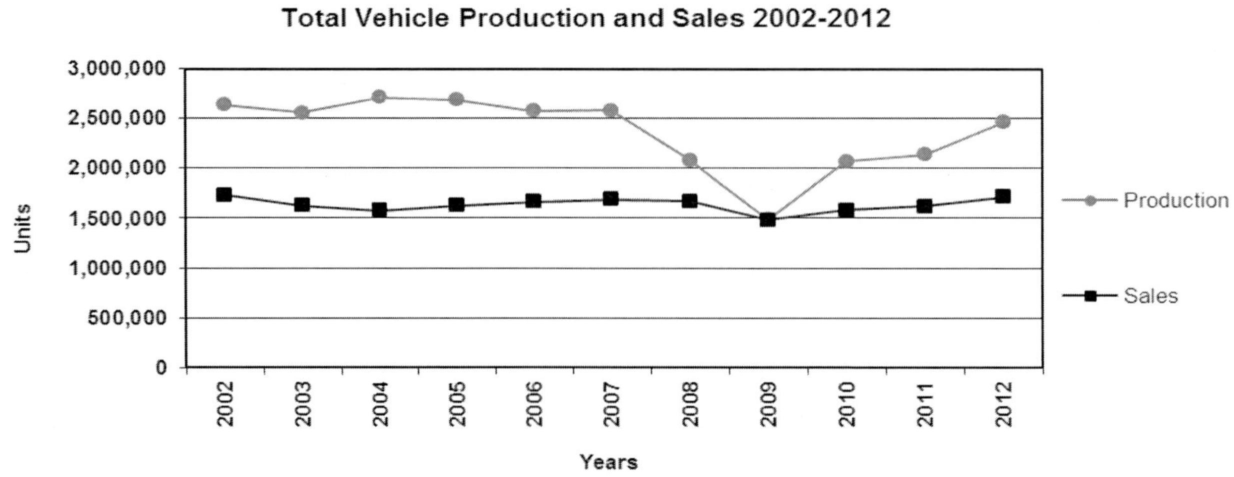

CANADA

PRODUCTION

Year	Cars	Commercial Vehicles	Total
2012	1,040,298	1,423,066	2,463,364
2011	990,482	1,144,639	2,135,121
2010	967,077	1,101,112	2,068,189
2009	822,267	668,215	1,490,482
2008	1,195,436	886,805	2,082,241
2007	1,342,133	1,236,657	2,578,790
2006	1,427,582	1,143,784	2,571,366
2005	1,406,777	1,281,115	2,687,892
2004	1,400,129	1,311,407	2,711,536
2003	1,340,175	1,212,687	2,552,862
2002	1,369,042	1,264,259	2,633,301
2001	1,274,853	1,259,998	2,534,851
2000	1,550,500	1,412,597	2,963,097
1999	1,626,316	1,430,667	3,056,983
1998	1,481,141	1,089,180	2,570,321
1997	1,372,588	1,197,765	2,570,353
1996	1,279,767	1,117,444	2,397,211
1995	1,336,725	1,071,274	2,407,999
1994	1,215,830	1,105,709	2,321,539
1993	1,352,834	900,305	2,253,139
1992	1,027,881	945,096	1,972,977
1991	1,059,608	827,814	1,887,422
1990	1,097,670	849,701	1,947,371
1989	1,017,028	947,103	1,964,131
1988	1,027,693	948,818	1,976,511
1987	809,818	825,641	1,635,459
1986	1,061,738	791,209	1,852,947
1985	1,077,932	854,699	1,932,631
1984	1,180,085	649,299	1,829,384
1983	970,933	554,327	1,525,260
1982	807,645	468,395	1,276,040

CANADA

PRODUCTION BY MANUFACTURER

	2012	% Chg.	2011
Cars			
Chrysler	*92,422*	*67.8*	*55,072*
Dodge	*146,615*	*8.1*	*135,591*
Lancia	*1,156*	*-70.9*	*3,968*
Chrysler Total	240,193	23.4	194,631
Ford	–	–	*66,096*
Lincoln	–	–	*10,192*
Mercury	–	–	*165*
Ford Total	–	–	76,453
Buick	*21,394*	*-9.8*	*23,715*
Cadillac	*28,475*	–	–
Chevrolet	*269,687*	*-8.3*	*294,225*
GM Total	319,556	0.5	317,940
Acura	–	–	*1,170*
Honda	*226,390*	*26.7*	*178,732*
Honda Total	226,390	25.8	179,902
Toyota	254,159	14.7	221,556
Total Cars	**1,040,298**	**5.0**	**990,482**
Light Trucks			
Chrysler	*121,444*	*9.9*	*110,483*
Dodge	*197,890*	*16.9*	*169,234*
Lancia	*6,502*	*42.7*	*4,555*
Ram	*11,227*	*297.1*	*2,827*
Volkswagen	*8,662*	*-39.6*	*14,351*
Chrysler Total	345,725	14.7	301,450
Ford	*215,728*	*6.6*	*202,448*
Lincoln	*40,196*	*-3.1*	*41,495*
Ford Total	255,924	4.9	243,943
Chevrolet	*244,788*	*4.9*	*233,244*
GMC	*118,714*	*7.2*	*110,700*
GM Total	363,502	5.7	343,944
Acura	*68,915*	*27.6*	*54,009*
Honda	*114,644*	–	*140*
Honda Total	183,559	239.0	54,149
Lexus	*86,696*	*32.9*	*65,242*
Toyota	*178,360*	*41.5*	*126,030*
Toyota Total	265,056	38.6	191,272
Total Light Trucks	**1,413,766**	**24.6**	**1,134,758**
Medium/Heavy Trucks			
International	–	–	1
Kenworth	*4,538*	*-18.1*	*5,543*
Peterbilt	*4,762*	*9.8*	*4,337*
Kenworth Total	9,300	-5.9	9,880
Total Med./Hvy. Trks.	**9,300**	**-5.9**	**9,881**
Total Comm. Vehs.	**1,423,066**	**24.3**	**1,144,639**
Total Vehicles	**2,463,364**	**15.4**	**2,135,121**

CANADA

SALES

Year	Cars	Commercial Vehicles	Total
2012	748,530	967,648	1,716,178
2011	681,956	938,265	1,620,221
2010	694,349	889,039	1,583,388
2009	729,023	753,209	1,482,232
2008	872,720	800,802	1,673,522
2007	841,585	848,760	1,690,345
2006	858,826	807,182	1,666,008
2005	847,436	782,706	1,630,142
2004	821,550	753,253	1,574,803
2003	864,989	760,061	1,625,050
2002	934,057	797,766	1,731,823
2001	868,188	729,687	1,597,875
2000	849,132	736,951	1,586,083
1999	806,440	733,939	1,540,379
1998	740,816	685,776	1,426,592
1997	738,555	684,439	1,422,994
1996	660,804	541,590	1,202,394
1995	670,981	493,617	1,164,598
1994	750,328	506,654	1,256,982
1993	740,868	449,714	1,190,582
1992	800,327	424,051	1,224,378
1991	875,788	408,861	1,284,649
1990	887,981	426,230	1,314,211
1989	990,819	487,727	1,478,546
1988	1,053,037	508,435	1,561,472
1987	1,064,814	463,174	1,527,988
1986	1,087,579	424,509	1,512,088
1985	1,133,103	395,921	1,529,024
1984	971,210	312,292	1,283,502
1983	843,318	237,770	1,081,088
1982	713,481	207,421	920,902

CANADA

SALES BY COMPANY

	2012	% Chg.	2011
Cars			
Audi	12,195	12.9	10,800
BMW	*16,636*	*-1.1*	*16,827*
Mini	*4,574*	*27.3*	*3,592*
BMW Total	21,210	3.9	20,419
Chrysler	*19,885*	*89.0*	*10,519*
Dodge	*14,150*	*-12.2*	*16,119*
Fiat	*8,474*	*57.2*	*5,392*
Chrysler Total	42,509	32.7	32,030
Mercedes-Benz	*16,817*	*-2.3*	*17,207*
Smart	*2,387*	*29.0*	*1,851*
Daimler Total	19,204	0.8	19,058
Ford	*66,755*	*-1.1*	*67,476*
Lincoln	*1,352*	*-44.6*	*2,440*
Ford Total	68,107	-2.6	69,916
Buick	*9,785*	*23.0*	*7,956*
Cadillac	*2,622*	*-20.3*	*3,288*
Chevrolet	*59,560*	*-10.2*	*66,313*
Pontiac	–	–	*42*
GM Total	71,967	-7.3	77,599
Acura	*7,076*	*5.0*	*6,739*
Honda	*80,034*	*20.3*	*66,503*
Honda Total	87,110	18.9	73,242
Hyundai	96,939	8.6	89,236
Jaguar	644	-17.0	776
Kia	48,357	39.5	34,671
Mazda	50,112	-1.2	50,697
Mitsubishi	7,956	9.2	7,285
Infiniti	*3,323*	*-17.2*	*4,011*
Nissan	*39,466*	*-8.5*	*43,137*
Nissan Total	42,789	-9.2	47,148
Porsche	1,375	24.0	1,109
Saab	21	-87.4	167
Subaru	14,403	33.6	10,779
Suzuki	3,651	-1.9	3,721
Lexus	*7,336*	*20.7*	*6,076*
Scion	*5,783*	*22.5*	*4,720*
Toyota	*93,514*	*25.5*	*74,490*
Toyota Total	106,633	25.0	85,286
Volkswagen	50,883	13.7	44,765
Volvo	2,465	-24.2	3,252
Total Cars	**748,530**	**9.8**	**681,956**
Light Trucks			
Audi	7,805	28.6	6,067
BMW	*14,766*	*14.1*	*12,946*
Mini	*1,731*	*10.7*	*1,563*
BMW Total	16,497	13.7	14,509
Chrysler	*3,991*	*-12.0*	*4,536*
Dodge	*82,929*	*-2.9*	*85,441*
Jeep	*42,957*	*0.1*	*42,909*
Ram	*69,838*	*7.9*	*64,708*
Chrysler Total	199,715	1.1	197,594
Mercedes-Benz	*16,299*	*17.6*	*13,856*
Mitsubishi Fuso	*12*	*-20.0*	*15*
Daimler Total	16,311	17.6	13,871
Ford	*199,467*	*1.4*	*196,624*
Lincoln	*4,788*	*-16.3*	*5,722*
Ford Total	204,255	0.9	202,346
Buick	*3,283*	*-11.2*	*3,696*
Cadillac	*4,022*	*-6.2*	*4,289*
Chevrolet	*79,430*	*-5.7*	*84,227*
GMC	*68,123*	*-6.7*	*72,989*
Hummer	–	–	*2*
Pontiac	–	–	*13*
GM Total	154,858	-6.3	165,216
Acura	*10,078*	*18.1*	*8,533*
Honda	*51,524*	*24.6*	*41,346*
Honda Total	61,602	23.5	49,879
Hyundai	39,344	-1.6	40,004
Isuzu	36	-12.2	41
Kia	29,443	-3.3	30,452

CANADA

SALES BY COMPANY

	2012	% Chg.	2011
Land Rover	4,268	32.2	3,228
Mazda	21,526	16.4	18,489
Mitsubishi	11,715	-11.4	13,226
Infiniti	*4,670*	*59.7*	*2,925*
Nissan	*34,735*	*0.4*	*34,594*
Nissan Total	39,405	5.0	37,519
Porsche	1,628	47.3	1,105
Subaru	16,600	2.4	16,205
Suzuki	1,807	-5.1	1,904
Lexus	*7,765*	*6.5*	*7,288*
Toyota	*77,660*	*11.4*	*69,686*
Toyota Total	85,425	11.0	76,974
Volkswagen	8,249	5.2	7,839
Volvo	3,113	-12.6	3,563
Total Light Trucks	**923,602**	**2.6**	**900,031**

Medium/Heavy Trucks

	2012	% Chg.	2011
Chrysler	1,621	18.5	1,368
Freightliner	*8,731*	*14.3*	*7,638*
Mitsubishi Fuso	*412*	*16.1*	*355*
Western Star	*2,456*	*46.1*	*1,681*
Daimler Total	11,599	19.9	9,674
Ford	3,706	-0.3	3,716
GM (GMC)	–	–	15
Hino	1,726	32.3	1,305
International	7,735	-4.6	8,106
Isuzu	492	2.1	482
Kenworth	*6,456*	*11.8*	*5,774*
Peterbilt	*4,667*	*41.7*	*3,293*
PACCAR Total	11,123	22.7	9,067
Mack Truck	*2,327*	*42.2*	*1,636*
Volvo Truck	*3,717*	*29.7*	*2,865*
Volvo Truck Total	6,044	34.3	4,501
Total Med./Hvy. Trks.	**44,046**	**15.2**	**38,234**
Total Comm. Vehs.	**967,648**	**3.1**	**938,265**
Total Vehicles	**1,716,178**	**5.9**	**1,620,221**

MEXICO

ECONOMIC SUMMARY

Common Language	Spanish
Currency	Peso
Steering	Left Hand Drive

2012 Vehicles in Operation

Passenger Cars	22,961,571
Commercial Vehicles	9,881,728
Total	32,843,299
Persons per Car	5.1
Persons per Vehicle	3.5

2012 GDP (billions of US $)

Nominal GDP	1,177.1
Real GDP	731.9
GDP % annual growth	3.9
GDP per capita	6,372

Current Account Balance (billions of US $)

2012:	-9.2
2011:	-9.7

Current Account Balance in % of GDP

2012:	-0.8
2011:	-0.8

Total Number of Fatal Road Accidents (most recent data)

2010:	16,714

Capital	Mexico City
Population	116,220,947
Population Density	154.8
Surface Area sq km/miles	1,943,945 / 750,561

Road Infrastructure

Paved Road Surface %	37.2
Total Road Mileage in km	371,936 / 231,110
Road Traffic (Vehicles per mile)	142.1

2012 Total Vehicle Percent Market Share by Company

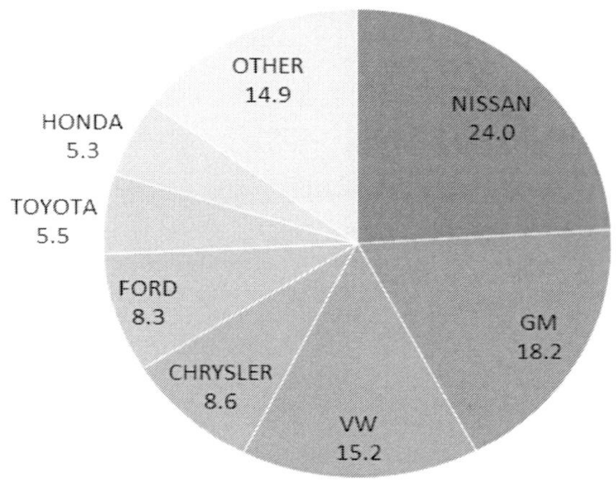

Total Vehicle Production and Sales 2002-2012

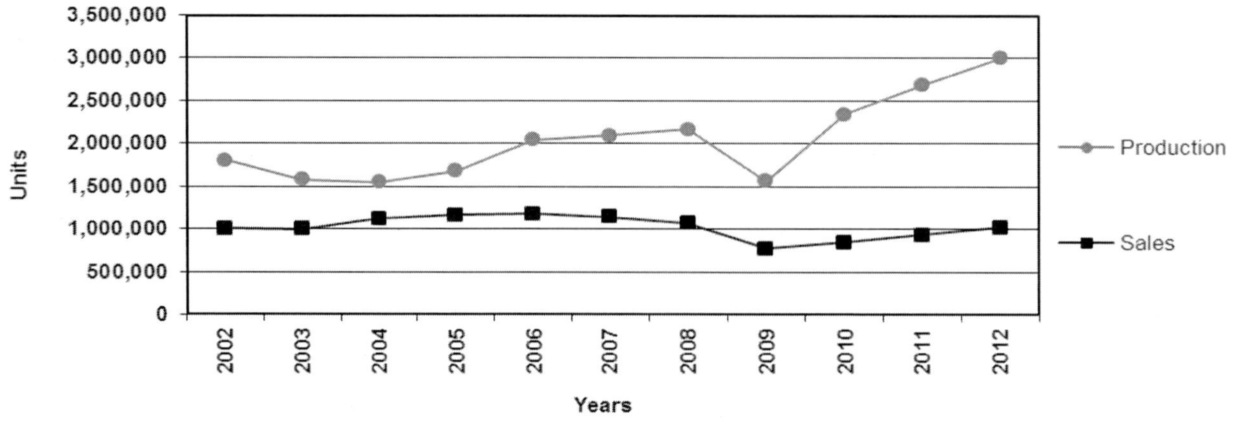

MEXICO

PRODUCTION

Year	Cars	Commercial Vehicles	Total
2012	1,810,007	1,191,807	3,001,814
2011	1,657,080	1,023,970	2,681,050
2010	1,386,148	958,956	2,345,104
2009	942,876	618,176	1,561,052
2008	1,217,458	950,486	2,167,944
2007	1,209,097	886,148	2,095,245
2006	1,097,619	947,899	2,045,518
2005	846,048	838,190	1,684,238
2004	781,827	771,147	1,552,974
2003	774,048	801,399	1,575,447
2002	960,097	844,573	1,804,670
2001	1,000,715	856,399	1,857,114
2000	1,130,488	792,401	1,922,889
1999	994,104	540,056	1,534,160
1998	956,354	503,537	1,459,891
1997	854,818	499,796	1,354,614
1996	797,681	420,637	1,218,318
1995	699,186	235,531	934,717
1994	856,563	265,122	1,121,685
1993	835,090	242,243	1,077,333
1992	776,185	302,442	1,078,627
1991	720,384	266,741	987,125
1990	598,093	220,962	819,055
1989	438,632	201,969	640,601
1988	353,783	158,240	512,023
1987	266,142	128,907	395,049
1986	190,069	149,760	339,829
1985	268,235	188,794	457,029
1984	244,704	112,173	356,877
1983	207,137	77,992	285,129
1982	300,579	170,735	471,314

MEXICO

PRODUCTION BY MANUFACTURER

	2012	% Chg.	2011
Cars			
Chrysler (Fiat)	81,874	43.7	56,972
Ford	*423,423*	*0.8*	*420,081*
Lincoln	*21,002*	*-40.7*	*35,399*
Ford Total	444,425	-2.4	455,480
GM (Chevrolet)	109,997	4.2	105,611
Nissan	569,203	7.6	528,976
Volkswagen	604,508	18.5	510,041
Total Cars	**1,810,007**	**9.2**	**1,657,080**
Light Trucks			
Dodge	*134,246*	*23.6*	*108,601*
Fiat	*44,470*	*34.0*	*33,192*
Ram	*181,836*	*36.9*	*132,818*
Chrysler Total	360,552	31.3	274,611
Cadillac	*92,885*	*5.7*	*87,869*
Chevrolet	*271,987*	*3.2*	*263,448*
GMC	*94,732*	*9.5*	*86,518*
Saab	–	–	*700*
GM Total	459,604	4.8	438,535
Honda	61,813	36.1	45,426
Nissan	114,317	46.4	78,111

	2012	% Chg.	2011
Toyota	55,676	12.3	49,596
Total Light Trucks	**1,051,962**	**18.7**	**886,279**
Medium/Heavy Trucks			
Ford	*7,221*	*3.4*	*6,982*
International	*3,360*	*-37.1*	*5,346*
Blue Diamond Total	10,581	-14.2	12,328
Chrysler	12,300	39.8	8,800
Dina	568	54.8	367
Freightliner	*61,183*	*3.5*	*59,133*
Mercedes-Benz	*3,383*	*-2.5*	*3,468*
Freightliner Total	64,566	3.1	62,601
Hino	432	35.4	319
International	30,964	-14.8	36,363
Kenworth	19,329	21.8	15,870
MAN (VW Truck & Bus)	868	-3.9	903
Volvo Truck	237	69.3	140
Total Med./Hvy. Trks.	**139,845**	**1.6**	**137,691**
Total Comm. Vehs.	**1,191,807**	**16.4**	**1,023,970**
Total Vehicles	**3,001,814**	**12.0**	**2,681,050**

MEXICO

SALES

Year	Cars	Commercial Vehicles	Total
2012	637,355	385,338	1,022,693
2011	586,056	349,037	935,093
2010	499,567	347,314	846,881
2009	434,679	338,706	773,385
2008	580,974	487,762	1,068,736
2007	626,392	517,913	1,144,305
2006	665,678	511,422	1,177,100
2005	709,020	453,138	1,162,158
2004	736,952	382,633	1,119,585
2003	685,586	313,382	998,968
2002	706,081	297,636	1,003,717
2001	667,565	274,866	942,431
2000	599,285	303,087	902,372
1999	465,063	241,317	706,380
1998	430,199	235,185	665,384
1997	303,558	199,113	502,671
1996	200,102	139,951	340,053
1995	117,393	71,198	188,591
1994	413,819	206,987	620,806
1993	398,743	200,278	599,021
1992	445,303	258,125	703,428
1991	392,110	248,467	640,577
1990	352,608	196,196	548,804
1989	274,505	170,678	445,183
1988	210,066	131,245	341,311
1987	154,152	93,575	247,727
1986	160,670	96,945	257,615
1985	242,187	147,638	389,825
1984	217,650	111,571	329,221
1983	192,052	80,488	272,540
1982	286,761	178,472	465,233

MEXICO

SALES BY COMPANY

	2012	% Chg.	2011
Cars			
Audi	7,428	15.0	6,458
BMW	*4,909*	*18.7*	*4,135*
Mini	*3,092*	*26.1*	*2,452*
BMW Total	8,001	21.5	6,587
Chrysler	*1,597*	*1.7*	*1,570*
Dodge	*28,486*	*16.9*	*24,377*
Chrysler Total	30,083	15.9	25,947
Mercedes-Benz	*7,059*	*36.4*	*5,175*
Smart	*1,634*	*11.2*	*1,470*
Daimler Total	8,693	30.8	6,645
Alfa Romeo	*93*	*210.0*	*30*
Fiat	*3,972*	*97.7*	*2,009*
Fiat Total	4,065	99.4	2,039
Ford	*36,424*	*25.0*	*29,141*
Lincoln	*692*	*-3.1*	*714*
Ford Total	37,116	24.3	29,855
Buick	*298*	*33.0*	*224*
Cadillac	*207*	*-3.3*	*214*
Chevrolet	*143,622*	*15.6*	*124,208*
Pontiac	*--*	*--*	*3*
GM Total	144,127	15.6	124,649
Acura	*994*	*15.7*	*859*
Honda	*19,162*	*20.4*	*15,911*
Honda Total	20,156	20.2	16,770
Jaguar	117	116.7	54
Mazda	16,336	-8.8	17,910
Mitsubishi	2,750	-24.7	3,654
Infiniti	*442*	*183.3*	*156*
Nissan	*167,165*	*4.1*	*160,540*
Nissan Total	167,607	4.3	160,696
Peugeot	4,451	-8.5	4,866
Porsche	314	58.6	198
Renault	13,041	-29.4	18,482
Saab	--	--	1
Subaru	151	-54.0	328
Suzuki	8,193	19.8	6,838
Toyota	24,701	8.9	22,679
SEAT	*21,114*	*16.6*	*18,115*
Volkswagen	*117,977*	*4.9*	*112,435*
Volkswagen Total	139,091	6.5	130,550
Volvo	934	9.9	850
Total Cars	**637,355**	**8.8**	**586,056**
Light Trucks			
Audi	2,054	28.4	1,600
BMW	*3,097*	*23.9*	*2,499*
Mini	*729*	*32.3*	*551*
BMW Total	3,826	25.4	3,050
Chrysler	*3,321*	*5.5*	*3,149*
Dodge	*14,491*	*2.8*	*14,101*
Jeep	*22,907*	*8.2*	*21,173*
Ram	*17,221*	*-2.7*	*17,702*
Chrysler Total	57,940	3.2	56,125
Daimler (Mercedes-Benz)	1,907	-28.0	2,648
Fiat	1,066	86.7	571
Ford	*44,018*	*-20.2*	*55,166*
Lincoln	*1,362*	*4.6*	*1,302*
Ford Total	45,380	-19.6	56,468
Buick	*824*	*-6.3*	*879*
Cadillac	*1,049*	*-12.6*	*1,200*
Chevrolet	*35,798*	*-4.2*	*37,366*
GMC	*4,585*	*4.1*	*4,404*
Hummer	*--*	*--*	*4*
GM Total	42,256	-3.6	43,853
Acura	*1,170*	*26.6*	*924*
Honda	*33,189*	*79.3*	*18,515*
Honda Total	34,359	76.8	19,439
Isuzu	1,706	4.7	1,629
Land Rover	927	51.2	613
Mazda	9,088	-23.9	11,950
Mitsubishi	6,003	-12.3	6,842

MEXICO

SALES BY COMPANY

	2012	% Chg.	2011		2012	% Chg.	2011
Infiniti	*294*	*292.0*	*75*	Dina	349	164.4	132
Nissan	*77,797*	*21.6*	*63,969*	Ford	1,886	12.2	1,681
Nissan Total	78,091	21.9	64,044	Hino	1,080	20.9	893
Peugeot	753	-19.0	930	International	8,029	8.4	7,406
Porsche	351	2.9	341	Isuzu	1,737	0.6	1,727
Renault	11,989	157.8	4,650	*MAN*	*40*	*29.0*	*31*
Subaru	132	-49.0	259	*Volkswagen Truck & Bus*	*755*	*-2.8*	*777*
Suzuki	2,540	8.5	2,341	MAN Total	795	-1.6	808
Toyota	31,577	21.9	25,910	PACCAR (Kenworth)	11,314	33.6	8,470
Volkswagen	15,987	11.1	14,396	*Mack Truck*	*134*	*97.1*	*68*
Volvo	579	19.6	484	*Volvo Truck*	*1,315*	*58.2*	*831*
Total Light Trucks	**348,511**	**9.5**	**318,143**	Volvo Truck Total	1,449	61.2	899
				Total Med./Hvy. Trks.	**36,827**	**19.2**	**30,894**
Medium/Heavy Trucks							
Freightliner	*6,950*	*25.3*	*5,546*	**Total Comm. Vehs.**	**385,338**	**10.4**	**349,037**
Mercedes-Benz	*3,238*	*-2.8*	*3,332*				
Daimler Total	10,188	14.8	8,878	**Total Vehicles**	**1,022,693**	**9.4**	**935,093**

UNITED STATES

ECONOMIC SUMMARY

Common Language	English
Currency	US Dollar
Steering	Left Hand Drive

2012 Vehicles in Operation

Passenger Cars	120,901,628
Commercial Vehicles	130,595,459
Total	251,497,087
Persons per Car	2.6
Persons per Vehicle	1.3

2012 GDP (billions of US $)

Nominal GDP	15,684.8
Real GDP	13,593.2
GDP % annual growth	2.2
GDP per capita	43,265

Current Account Balance (billions of US $)

2012:	-475.0
2011:	-465.9

Current Account Balance in % of GDP

2012:	-3.0
2011:	-3.1

Total Number of Fatal Road Accidents (most recent data)

2012:	34,080

Capital	Washington D.C.
Population	316,668,567
Population Density	89.7
Surface Area sq km/miles	9,147,593 / 3,531,905

Road Infrastructure

Paved Road Surface %	67.4
Total Road Length in km/miles	6,531,376 / 4,058,409
Road Traffic (Vehicles per mile)	62.0

2012 Total Vehicle Percent Market Share by Company

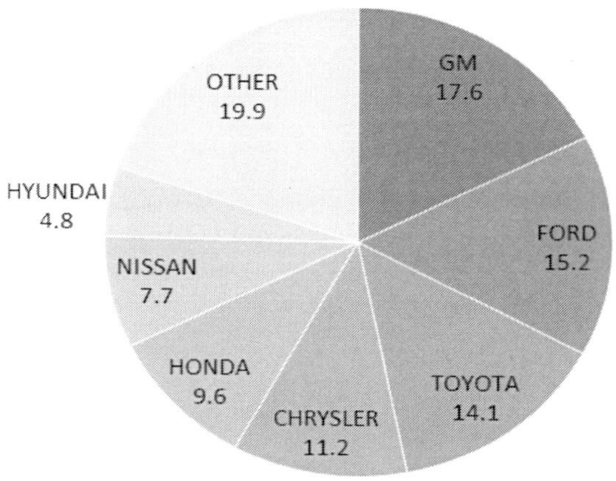

Total Vehicle Production and Sales 2002-2012

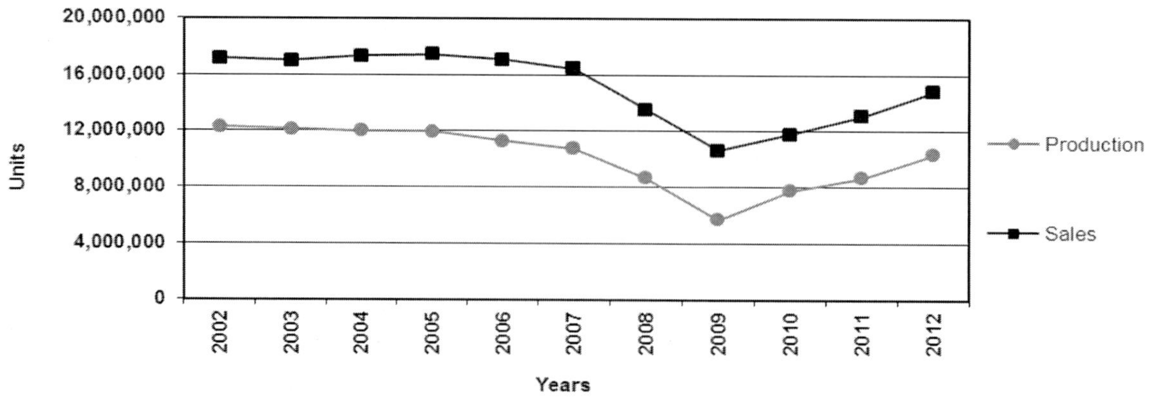

UNITED STATES

PRODUCTION

Year	Cars	Commercial Vehicles	Total
2012	4,105,874	6,226,752	10,332,626
2011	2,976,991	5,684,544	8,661,535
2010	2,731,105	5,011,988	7,743,093
2009	2,195,588	3,513,843	5,709,431
2008	3,731,241	4,940,900	8,672,141
2007	3,867,268	6,885,042	10,752,310
2006	4,311,696	6,948,581	11,260,277
2005	4,265,872	7,680,781	11,946,653
2004	4,165,925	7,794,429	11,960,354
2003	4,453,369	7,633,659	12,087,028
2002	4,957,377	7,322,205	12,279,582
2001	4,808,019	6,616,670	11,424,689
2000	5,470,917	7,302,797	12,773,714
1999	5,577,749	7,447,229	13,024,978
1998	5,492,473	6,510,190	12,002,663
1997	5,878,221	6,252,354	12,130,575
1996	6,035,235	5,794,922	11,830,157
1995	6,325,967	5,669,281	11,995,248
1994	6,601,220	5,638,068	12,239,288
1993	5,982,120	4,873,342	10,855,462
1992	5,666,891	4,024,552	9,691,443
1991	5,439,864	3,349,976	8,789,840
1990	6,077,903	3,689,536	9,767,439
1989	6,821,291	4,035,501	10,856,792
1988	7,137,433	4,084,575	11,222,008
1987	7,099,854	3,812,074	10,911,928
1986	7,829,271	3,490,853	11,320,124
1985	8,186,043	3,452,094	11,638,137
1984	7,773,342	3,165,716	10,939,058
1983	6,782,061	2,443,637	9,225,698
1982	5,073,214	1,912,099	6,985,313

UNITED STATES

PRODUCTION BY MANUFACTURER

	2012	% Chg.	2011
Cars			
Ford	*95,438*	*21.9*	*78,286*
Mazda	*37,566*	*-5.0*	*39,550*
AA/Flat Rock Assy Total	133,004	12.9	117,836
Chrysler	*150,321*	*22.6*	*122,625*
Dodge	*183,186*	*43.3*	*127,792*
Lancia	*1,318*	–	–
Chrysler Total	334,825	33.7	250,417
Ford	*405,212*	*35.4*	*299,178*
Lincoln	*14,695*	*16.3*	*12,636*
Ford Total	419,907	34.7	311,814
Buick	*113,996*	*34.0*	*85,082*
Cadillac	*74,525*	*1.6*	*73,357*
Chevrolet	*645,548*	*15.2*	*560,457*
Opel	*6,631*	*154.5*	*2,605*
GM Total	840,700	16.5	721,501
Acura	*69,665*	*94.8*	*35,763*
Honda	*569,443*	*81.2*	*314,206*
Honda Total	639,108	82.6	349,969
Hyundai	361,348	6.9	338,127
Kia	128,536	265.9	35,132
Mitsubishi	13,716	-48.4	26,583
Nissan	406,782	5.6	385,109
Subaru	*50,403*	*8.2*	*46,568*
Toyota	*93,126*	*17.6*	*79,170*
Subaru Total	143,529	14.1	125,738
Toyota	531,876	98.2	268,314
Volkswagen	152,543	228.4	46,451
Total Cars	**4,105,874**	**37.9**	**2,976,991**
Light Trucks			
BMW	301,519	9.2	276,065
Dodge	*53,854*	*-45.0*	*97,838*
Jeep	*714,651*	*18.5*	*603,156*
Ram	*227,453*	*7.7*	*211,142*
Chrysler Total	995,958	9.2	912,136
Ford	*1,543,758*	*5.1*	*1,468,278*
Lincoln	*9,639*	*-5.4*	*10,193*
Mazda	–	–	*3,977*
Ford Total	1,553,397	4.8	1,482,448
Buick	*62,203*	*-11.4*	*70,244*
Cadillac	*25,748*	*3.6*	*24,854*
Chevrolet	*730,033*	*0.8*	*724,289*
GMC	*325,958*	*-4.7*	*341,966*
GM Total	1,143,942	-1.5	1,161,353
Acura	*37,586*	*94.0*	*19,377*
Honda	*542,632*	*19.4*	*454,304*
Honda Total	580,218	22.5	473,681
Isuzu Truck	1,472	75.0	841
Hyundai	*98,091*	*7.6*	*91,155*
Kia	*131,572*	*-9.9*	*146,017*
Kia Total	229,663	-3.2	237,172
Mercedes-Benz	180,449	21.7	148,253
Mitsubishi	23,303	120.5	10,567
Infiniti	*32,751*	–	–
Nissan	*203,406*	*15.6*	*175,916*
Suzuki	*2,000*	*-8.7*	*2,190*
Nissan Total	238,157	33.7	178,106
Subaru	134,380	16.7	115,148
Toyota	576,350	29.3	445,727
Total Light Trucks	**5,958,808**	**9.5**	**5,441,497**
Medium/Heavy Trucks			
Ford	45,909	7.4	42,765
Freightliner	*49,515*	*23.4*	*40,133*
Western Star	*7,461*	*38.2*	*5,399*
Freightliner Total	56,976	25.1	45,532
Hino	8,199	71.6	4,777
International	41,041	-4.6	42,998
Isuzu Truck	2,597	-3.3	2,685
Kenworth	32,667	6.7	30,625
Mack Truck	24,985	23.1	20,299
Peterbilt	29,355	4.4	28,124
Volvo Truck	26,184	3.9	25,209
Other	31	-6.1	33
Total Med./Hvy. Trks.	**267,944**	**10.2**	**243,047**
Total Comm. Vehs.	**6,226,752**	**9.5**	**5,684,544**
Total Vehicles	**10,332,626**	**19.3**	**8,661,535**

UNITED STATES

SALES

Year	Cars	Commercial Vehicles	Total
2012	7,241,900	7,544,036	14,785,936
2011	6,089,403	6,951,210	13,040,613
2010	5,635,432	6,136,787	11,772,219
2009	5,400,890	5,200,478	10,601,368
2008	6,769,107	6,724,058	13,493,165
2007	7,562,334	8,897,981	16,460,315
2006	7,761,592	9,287,389	17,048,981
2005	7,659,983	9,784,346	17,444,329
2004	7,482,555	9,816,018	17,298,573
2003	7,555,551	9,411,891	16,967,442
2002	8,042,255	9,096,397	17,138,652
2001	8,352,000	9,120,378	17,472,378
2000	8,777,723	9,033,950	17,811,673
1999	8,637,708	8,777,020	17,414,728
1998	8,084,989	7,882,298	15,967,287
1997	8,217,480	7,280,380	15,497,860
1996	8,478,545	6,977,567	15,456,112
1995	8,620,159	6,496,166	15,116,325
1994	8,990,517	6,420,857	15,411,374
1993	8,517,859	5,680,995	14,198,854
1992	8,213,113	4,904,331	13,117,444
1991	8,184,979	4,364,544	12,549,523
1990	9,303,215	4,846,163	14,149,378
1989	9,778,517	5,066,744	14,845,261
1988	10,546,808	5,244,736	15,791,544
1987	10,191,877	5,001,069	15,192,946
1986	11,404,239	4,918,782	16,323,021
1985	10,979,187	4,746,104	15,725,291
1984	10,323,695	4,159,446	14,483,141
1983	9,148,038	3,163,478	12,311,516
1982	7,956,460	2,581,902	10,538,362

UNITED STATES

SALES BY COMPANY

	2012	% Chg.	2011
Cars			
Audi	96,516	15.4	83,655
BMW	*186,146*	*7.4*	*173,375*
Mini	*45,111*	*10.5*	*40,828*
BMW Total	231,257	8.0	214,203
Chrysler	*196,223*	*56.1*	*125,698*
Dodge	*258,100*	*23.6*	*208,892*
Fiat	*43,772*	*121.4*	*19,769*
Chrysler Total	498,095	40.6	354,359
Mercedes-Benz	*177,745*	*13.7*	*156,313*
Smart	*10,009*	*92.2*	*5,208*
Daimler Total	187,754	16.2	161,521
Ford	*719,068*	*6.8*	*673,047*
Lincoln	*41,578*	*-15.5*	*49,206*
Mercury	–	–	*248*
Ford Total	760,646	5.3	722,501
Buick	*123,705*	*3.7*	*119,241*
Cadillac	*69,665*	*-0.5*	*69,981*
Chevrolet	*837,488*	*9.7*	*763,173*
GM Total	1,030,858	8.2	952,395
Acura	*75,067*	*18.6*	*63,268*
Honda	*709,170*	*30.6*	*542,981*
Honda Total	784,237	29.4	606,249
Hyundai	574,881	11.6	514,922
Jaguar	12,011	-2.2	12,276
Kia	384,133	35.6	283,271
Mazda	182,817	15.3	158,513
Mitsubishi	31,448	-27.1	43,131
Infiniti	*68,974*	*-0.1*	*69,064*
Nissan	*608,017*	*5.7*	*575,194*
Nissan Total	676,991	5.1	644,258
Porsche	19,498	21.5	16,045
Saab	1,060	-80.2	5,343
Subaru	133,070	59.2	83,597
Suzuki	18,426	-5.3	19,464
Lexus	*132,741*	*31.2*	*101,181*
Scion	*73,505*	*49.2*	*49,271*
Toyota	*992,638*	*32.9*	*746,861*
Toyota Total	1,198,884	33.6	897,313
Volkswagen	385,366	38.4	278,404
Volvo	33,952	-10.6	37,983
Total Cars	**7,241,900**	**18.9**	**6,089,403**
Light Trucks			
Audi	42,794	26.2	33,906
BMW	*95,314*	*27.9*	*74,532*
Mini	*21,012*	*25.9*	*16,683*
BMW Total	116,326	27.5	91,215
Chrysler	*111,744*	*16.8*	*95,648*
Dodge	*266,889*	*10.2*	*242,148*
Jeep	*474,131*	*13.1*	*419,349*
Ram	*290,621*	*16.2*	*250,083*
Chrysler Total	1,143,385	13.5	1,007,228
Mercedes-Benz	*117,239*	*11.2*	*105,437*
Mitsubishi Fuso	*395*	*78.7*	*221*
Daimler Total	117,634	11.3	105,658
Ford	*1,404,329*	*3.9*	*1,351,893*
Lincoln	*40,572*	*11.3*	*36,437*
Ford Total	1,444,901	4.1	1,388,330
Buick	*56,703*	*-2.9*	*58,392*
Cadillac	*80,117*	*-2.8*	*82,408*
Chevrolet	*1,014,158*	*0.2*	*1,012,629*
GMC	*413,881*	*4.0*	*397,973*
GM Total	1,564,859	0.9	1,551,402
Acura	*81,149*	*35.2*	*60,031*
Honda	*557,399*	*15.9*	*481,005*

UNITED STATES

SALES BY COMPANY

	2012	% Chg.	2011
Honda Total	638,548	18.0	541,036
Hyundai	128,126	-2.0	130,769
International	30	-96.4	845
Isuzu	2,570	39.7	1,840
Kia	173,466	-14.2	202,221
Land Rover	43,664	14.6	38,099
Mazda	94,229	2.5	91,913
Mitsubishi	26,342	-26.6	35,889
Infiniti	*50,903*	*73.2*	*29,397*
Nissan	*413,762*	*12.2*	*368,879*
Nissan Total	464,665	16.7	398,276
Porsche	15,545	19.8	12,978
Saab	221	-17.2	267
Subaru	203,371	10.9	183,392
Suzuki	6,932	-3.1	7,155
Lexus	*111,425*	*14.4*	*97,371*
Toyota	*772,195*	*18.8*	*649,976*
Toyota Total	883,620	18.2	747,347
Volkswagen	52,767	14.7	45,998
Volvo	34,165	16.8	29,257
Total Light Trucks	**7,198,160**	**8.3**	**6,645,021**

Medium/Heavy Trucks

	2012	% Chg.	2011
Chrysler	10,307	36.9	7,527
Freightliner	*97,765*	*18.5*	*82,504*
Mitsubishi Fuso	*2,449*	*16.6*	*2,100*
Sterling	*–*	*–*	*6*
Western Star	*2,623*	*25.5*	*2,090*
Daimler Total	102,837	18.6	86,700
Ford	44,618	17.5	37,975
Chevrolet	*–*	*–*	*10*
GMC	*–*	*–*	*13*
GM Total	–	–	23
Hino	6,305	24.2	5,078
International	68,380	-5.1	72,091
Isuzu	10,208	24.0	8,234
Kenworth	*32,298*	*23.7*	*26,111*
Peterbilt	*31,659*	*14.2*	*27,728*
PACCAR Total	63,957	18.8	53,839
Mack Truck	*17,327*	*34.0*	*12,928*
UD Trucks	*1,004*	*22.6*	*819*
Volvo Truck	*20,921*	*-0.2*	*20,955*
Volvo Truck Total	39,252	13.1	34,702
Other	12	-40.0	20
Total Med./Hvy. Trks.	**345,876**	**13.0**	**306,189**
Total Comm. Vehs.	**7,544,036**	**8.5**	**6,951,210**
Total Vehicles	**14,785,936**	**13.4**	**13,040,613**

Ward's World Motor Vehicle Data 2013

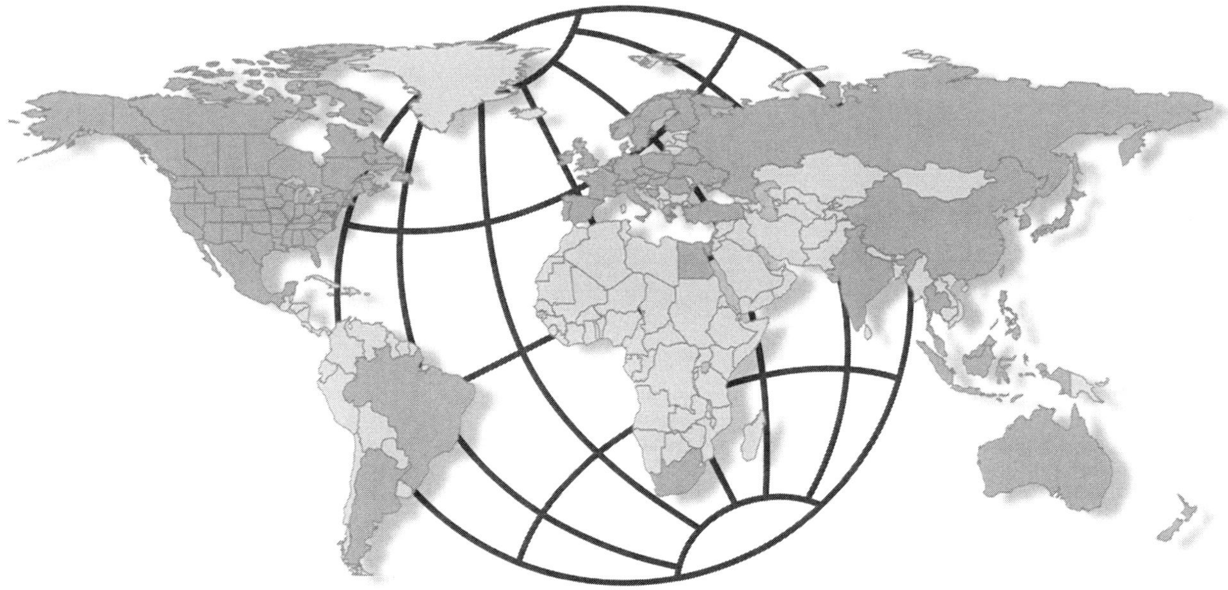

South America Data

ARGENTINA

ECONOMIC SUMMARY

Common Language	Spanish/English
Currency	Argentine Peso
Steering	Left Hand Drive

2012 Vehicles in Operation

Passenger Cars	9,100,000
Commercial Vehicles	3,000,000
Total	12,100,000
Persons per Car	4.7
Persons per Vehicle	3.5

2012 GDP (billions of US $)

Nominal GDP	475.0
Real GDP	95.4
GDP % annual growth	1.9
GDP per capita	2,324

Current Account Balance (billions of US $)

2012:	0.3
2011:	-1.9

Current Account Balance in % of GDP

2012:	0.1
2011:	-0.4

Total Number of Fatal Road Accidents (most recent data)

2010:	5,094

Capital	Buenos Aires
Population	42,610,981
Population Density	40.3
Surface Area sq.km/miles	2,736,690 / 1,056,642

Road Infrastructure

Paved Road Surface %	42.9
Total Road Length in km/miles	231,374 / 143,769
Road Traffic (Vehicles per mile)	84.2

2012 Total Vehicle Percent Market Share by Company

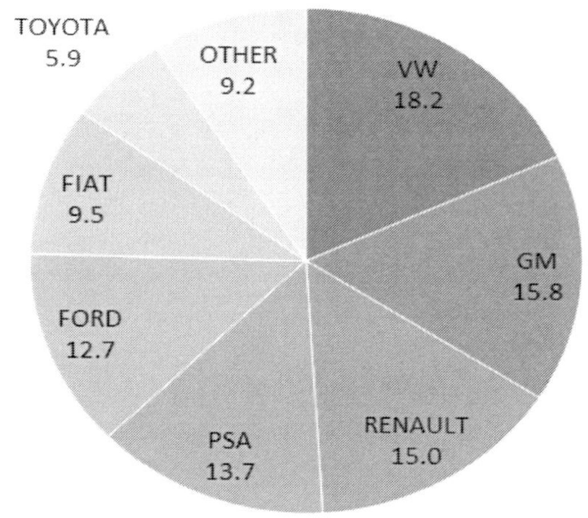

Total Vehicle Production and Sales 2002-2012

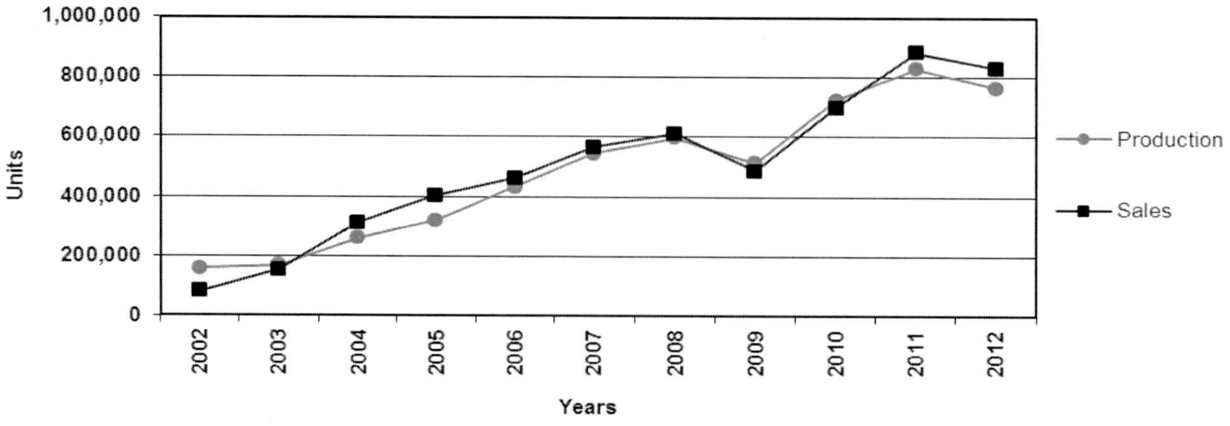

ARGENTINA

PRODUCTION

Year	Cars	Commercial Vehicles	Total
2012	497,376	267,119	764,495
2011	577,233	251,538	828,771
2010	515,884	208,139	724,023
2009	380,067	132,857	512,924
2008	399,577	197,509	597,086
2007	350,735	193,912	544,647
2006	263,120	168,981	432,101
2005	182,761	136,994	319,755
2004	171,400	89,002	260,402
2003	109,364	59,812	169,176
2002	111,340	48,061	159,401
2001	169,580	65,997	235,577
2000	238,706	100,540	339,246
1999	224,733	80,076	304,809
1998	353,074	104,883	457,957
1997	366,466	79,840	446,306
1996	269,439	43,713	313,152
1995	226,656	58,779	285,435
1994	338,355	70,422	408,777
1993	286,964	55,380	342,344
1992	220,502	41,502	262,004
1991	114,113	24,845	138,958
1990	81,107	18,532	99,639
1989	107,597	20,226	127,823
1988	135,776	28,384	164,160
1987	158,743	34,572	193,315
1986	137,889	32,601	170,490
1985	113,788	23,887	137,675
1984	137,206	30,117	167,323
1983	128,962	30,914	159,876
1982	106,886	25,231	132,117

ARGENTINA

PRODUCTION BY MANUFACTURER

	2012	% Chg.	2011
Cars			
Fiat	73,863	-34.9	113,468
Ford	48,492	-16.3	57,934
GM (Chevrolet)	128,501	-5.3	135,678
Honda	8,114	823.1	879
Citroen	*20,782*	*-34.5*	*31,717*
Peugeot	*94,863*	*-2.5*	*97,325*
Peugeot Citroen Total	115,645	-10.4	129,042
Renault	85,159	-7.2	91,759
Volkswagen	37,602	-22.4	48,473
Total Cars	**497,376**	**-15.2**	**577,233**
Light Trucks			
Ford	40,580	-11.7	45,949
GM (Chevrolet)	–	–	750
Mercedes-Benz	13,417	-15.4	15,862
Citroen	*5,910*	*29.0*	*4,582*
Peugeot	*11,979*	*16.6*	*10,274*
Peugeot Citroen Total	17,889	20.4	14,856
Renault	23,746	20.3	19,733
Toyota	93,570	35.7	68,929
Volkswagen	70,302	-8.7	76,965
Total Light Trucks	**259,504**	**6.8**	**243,044**
Heavy Trucks			
Iveco	4,735	-18.6	5,814
Mercedes-Benz	1,241	165.7	467
Total Heavy Trucks	**5,976**	**-4.9**	**6,281**
Buses			
Iveco	34	-8.1	37
Mercedes-Benz	1,605	-26.2	2,176
Total Buses	**1,639**	**-25.9**	**2,213**
Total Comm. Vehs.	**267,119**	**6.2**	**251,538**
Total Vehicles	**764,495**	**-7.8**	**828,771**

ARGENTINA

SALES

Year	Cars	Commercial Vehicles	Total
2012	596,397	233,661	830,058
2011	673,853	209,497	883,350
2010	524,514	173,785	698,299
2009	373,231	113,911	487,142
2008	452,894	158,876	611,770
2007	422,230	142,696	564,926
2006	336,296	124,182	460,478
2005	290,648	112,042	402,690
2004	223,475	88,491	311,966
2003	104,561	50,565	155,126
2002	59,080	23,393	82,473
2001	125,504	51,198	176,702
2000	224,950	81,995	306,945
1999	274,038	106,064	380,102
1998	321,970	133,402	455,372
1997	321,400	104,926	426,326
1996	299,235	76,908	376,143
1995	279,751	48,231	327,982
1994	360,721	53,609	414,330
1993	263,728	47,485	311,213
1992	203,122	40,241	243,363
1991	113,148	24,027	137,175
1990	77,306	17,481	94,787
1989	112,275	20,646	132,921
1988	135,667	26,850	162,517
1987	156,274	34,553	190,827
1986	134,269	31,306	165,575
1985	121,203	24,321	145,524
1984	135,163	29,896	165,059
1983	120,395	30,170	150,565
1982	106,452	25,353	131,805

ARGENTINA

SALES BY COMPANY

	2012	% Chg.	2011
Cars			
Audi	6,572	7.6	6,107
Daimler (Mercedes-Benz)	3,730	7.5	3,469
Fiat	60,096	-17.9	73,221
Ford	59,779	1.5	58,879
GM (Chevrolet)	120,415	-7.1	129,611
Honda	12,604	3.9	12,128
Nissan	–	–	8,889
Citroen	*26,775*	*-10.7*	*29,978*
Peugeot	*70,059*	*3.8*	*67,521*
Peugeot Citroen Total	96,834	-0.7	97,499
Renault	79,202	-15.2	93,376
Suzuki	5	-99.7	1,661
Toyota	15,722	33.7	11,763
SEAT	*3*	*-97.9*	*146*
Volkswagen	*129,885*	*-21.0*	*164,500*
Volkswagen Total	129,888	-21.1	164,646
Other	11,550	-8.4	12,604
Total Cars	**596,397**	**-11.5**	**673,853**
Light Trucks			
Daimler (Mercedes-Benz)	25	-99.3	3,616
Fiat	18,997	0.5	18,906
Ford	42,614	-1.3	43,189
GM (Chevrolet)	11,100	-27.7	15,343
Honda	3,358	-13.1	3,862
Iveco	302	-28.1	420
Nissan	–	–	4,886
Citroen	*5,314*	*9.4*	*4,856*

	2012	% Chg.	2011
Peugeot	*11,738*	*12.5*	*10,435*
Peugeot Citroen Total	17,052	11.5	15,291
Renault	45,087	165.0	17,011
Suzuki	1	-99.5	199
Toyota	33,579	37.4	24,434
Volkswagen	20,919	-4.5	21,900
Other	12,500	7.2	11,660
Total Light Trucks	**205,534**	**13.7**	**180,717**
Heavy Trucks			
Daimler (Mercedes-Benz)	7,740	60.3	4,828
Ford	2,799	-39.5	4,623
Iveco	6,266	-6.5	6,702
MAN (VW Truck & Bus)	1,651	-26.9	2,260
Scania	1,821	-32.6	2,701
Other	1,980	-6.6	2,120
Total Heavy Trucks	**22,257**	**-4.2**	**23,234**
Buses			
Daimler (Mercedes-Benz)	4,699	26.7	3,708
Iveco	1	–	–
MAN (VW Truck & Bus)	103	-3.7	107
Scania	157	-39.8	261
Other	910	-38.1	1,470
Total Buses	**5,870**	**5.8**	**5,546**
Total Comm. Vehs.	**233,661**	**11.5**	**209,497**
Total Vehicles	**830,058**	**-6.0**	**883,350**

BRAZIL

ECONOMIC SUMMARY

Common Language	Portuguese
Currency	Real
Steering	Left Hand Drive

2012 Vehicles in Operation

Passenger Cars	29,566,116
Commercial Vehicles	7,705,144
Total	37,271,260
Persons per Car	6.8
Persons per Vehicle	5.4

2012 GDP (billions of US $)

Nominal GDP	2,396.0
Real GDP	562.3
GDP % annual growth	0.9
GDP per capita	2,835

Current Account Balance (billions of US $)

2012:	-54.2
2011:	-52.5

Current Account Balance in % of GDP

2012:	-2.3
2011:	-2.1

Total Number of Fatal Road Accidents (most recent data)

2009:	37,594

Capital	Brasilia
Population	201,009,622
Population Density	61.5
Surface Area sq km/miles	8,459,417 / 3,266,186

Road Infrastructure

Paved Road Surface %	5.5
Total Road Length in km/miles	1,751,868 / 1,088,560
Road Traffic (Vehicles per mile)	34.2

2012 Total Vehicle Percent Market Share by Company

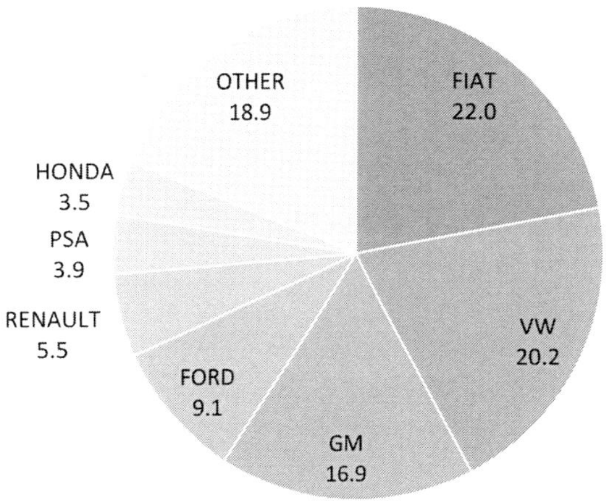

Total Vehicle Production and Sales 2002-2012

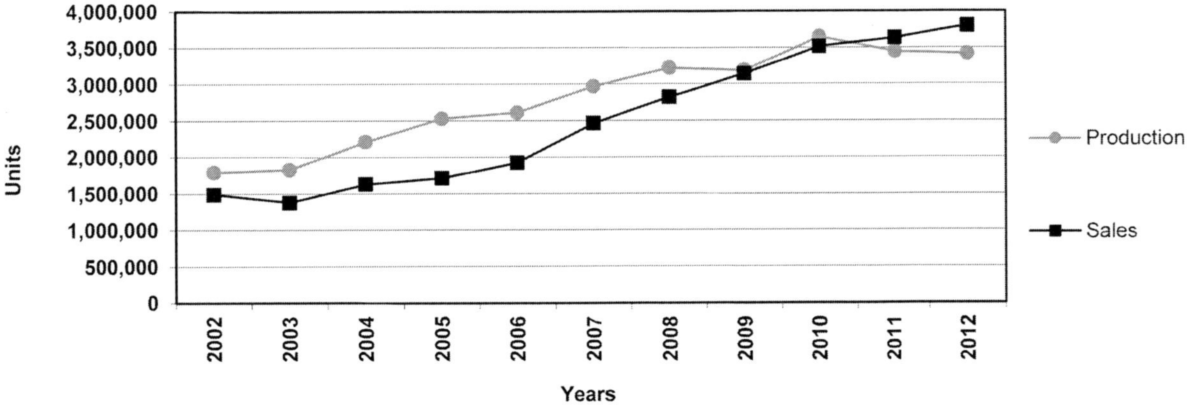

BRAZIL

PRODUCTION

Year	Cars	Trucks	Buses	Total
2012	2,591,344	782,953	41,189	3,415,486
2011	2,505,241	882,433	55,113	3,442,787
2010	2,828,273	774,631	45,869	3,648,773
2009	2,577,739	573,528	34,535	3,185,802
2008	2,561,496	615,273	44,111	3,220,880
2007	2,388,402	543,387	39,087	2,970,876
2006	2,092,003	485,222	33,809	2,611,034
2005	2,009,494	483,540	35,266	2,528,300
2004	1,756,166	425,158	28,738	2,210,062
2003	1,504,998	295,050	26,990	1,827,038
2002	1,520,285	248,419	22,826	1,791,530
2001	1,501,586	292,367	23,163	1,817,116
2000	1,361,721	306,847	22,672	1,691,240
1999	1,107,751	228,143	14,934	1,350,828
1998	1,254,016	310,156	21,458	1,585,630
1997	1,677,858	370,289	21,556	2,069,703
1996	1,458,576	328,409	17,343	1,804,328
1995	1,297,467	309,894	21,647	1,629,008
1994	1,248,773	315,181	17,435	1,581,389
1993	1,100,278	272,263	18,894	1,391,435
1992	815,959	233,616	24,286	1,073,861
1991	705,303	231,904	23,012	960,219
1990	663,084	236,351	15,031	914,466
1989	730,992	267,707	14,553	1,013,252
1988	782,411	267,918	18,427	1,068,756
1987	683,380	223,052	13,639	920,071
1986	815,152	229,962	11,218	1,056,332
1985	759,141	199,182	8,385	966,708
1984	679,386	177,927	7,340	864,653
1983	748,371	141,885	6,206	896,462
1982	672,589	176,895	9,820	859,304

BRAZIL

PRODUCTION BY MANUFACTURER

	2012	% Chg.	2011
Cars			
Fiat	648,520	12.6	576,029
Ford	209,426	1.6	206,160
GM (Chevrolet)	478,683	-7.7	518,487
Honda	135,058	57.9	85,545
Hyundai	26,003	–	–
Nissan	14,364	-19.3	17,797
Citroen	*55,766*	*-23.7*	*73,120*
Peugeot	*37,177*	*-37.9*	*59,828*
Peugeot Citroen Total	92,943	-30.1	132,948
Dacia	*51,670*	*-9.0*	*56,767*
Renault	*120,563*	*-9.0*	*132,461*
Renault Total	172,233	-9.0	189,228
Toyota	66,332	9.7	60,456
Volkswagen	747,782	4.1	718,591
Total Cars	**2,591,344**	**3.4**	**2,505,241**
Light Trucks			
Agrale	783	119.3	357
Fiat	171,521	-7.9	186,152
Ford	61,781	-23.0	80,197
GM (Chevrolet)	136,867	2.0	134,167
Hyundai	25,568	-33.8	38,635
Iveco	4,215	-38.1	6,806
Mahindra & Mahindra	830	39.0	597
Mitsubishi	37,353	-5.3	39,441
Nissan	16,978	14.4	14,843
Citroen	*4,603*	*7.8*	*4,270*
Peugeot	*1,972*	*-80.2*	*9,967*
Peugeot Citroen Total	6,575	-53.8	14,237
Renault	81,192	158.6	31,397
Volkswagen	104,304	-5.1	109,853
Total Light Trucks	**647,967**	**-1.3**	**656,682**
Medium/Heavy Trucks			
Agrale	448	-56.8	1,036
Ford	23,462	-42.0	40,422
International	306	-85.4	2,089
Iveco	10,979	-42.7	19,172
MAN (VW Truck & Bus)	36,845	-46.8	69,275
Mercedes-Benz	31,845	-35.7	49,543
Scania	13,633	-29.9	19,455
Volvo Truck	17,468	-29.4	24,759
Total Med./Hvy. Trks.	**134,986**	**-40.2**	**225,751**
Buses			
Agrale	5,200	-2.5	5,332
International	–	–	100
Iveco	1,963	2.6	1,913
MAN (VW Truck & Bus)	11,072	-19.9	13,831
Mercedes-Benz	18,931	-31.9	27,789
Scania	2,054	-32.5	3,041
Volvo Truck	1,969	-36.6	3,107
Total Buses	**41,189**	**-25.3**	**55,113**
Total Comm. Vehs.	**824,142**	**-12.1**	**937,546**
Total Vehicles	**3,415,486**	**-0.8**	**3,442,787**

BRAZIL

SALES

Year	Cars	Trucks	Buses	Total
2012	2,841,936	930,207	29,716	3,801,859
2011	2,649,032	949,225	34,749	3,633,006
2010	2,651,752	835,051	28,317	3,515,120
2009	2,479,457	639,171	22,598	3,141,226
2008	2,195,499	599,122	26,336	2,820,957
2007	1,977,135	463,619	22,151	2,462,905
2006	1,556,231	350,660	18,422	1,925,313
2005	1,365,449	330,519	16,004	1,711,972
2004	1,295,798	315,699	16,982	1,628,479
2003	1,118,600	241,266	17,413	1,377,279
2002	1,229,146	241,603	16,790	1,487,539
2001	1,295,096	289,608	16,578	1,601,282
2000	1,176,774	296,268	16,439	1,489,481
1999	1,001,078	215,394	10,679	1,227,151
1998	1,195,392	284,192	15,761	1,495,345
1997	1,361,106	265,046	14,091	1,640,243
1996	1,245,972	248,222	12,589	1,506,783
1995	1,106,591	237,902	14,839	1,359,332
1994	975,697	220,813	10,313	1,206,823
1993	850,562	199,515	11,390	1,061,467
1992	577,305	149,314	13,706	740,325
1991	583,072	170,999	16,865	770,936
1990	532,791	169,744	10,091	712,626
1989	566,582	185,558	9,485	761,625
1988	556,744	178,004	12,968	747,716
1987	410,260	159,757	10,068	580,085
1986	672,384	185,856	8,488	866,728
1985	602,069	153,970	7,141	763,180
1984	532,235	138,850	5,997	677,082
1983	608,499	112,658	6,567	727,724
1982	556,229	127,020	8,045	691,294

BRAZIL

SALES BY COMPANY

	2012	% Chg.	2011
Cars			
Aston Martin	–	–	21
Audi	3,949	-18.7	4,856
Bentley	–	–	17
BMW	*5,875*	*-9.9*	*6,524*
Mini	*1,540*	*-47.3*	*2,921*
BMW Total	7,415	-21.5	9,445
Chery	12,164	-31.6	17,787
Chrysler	*699*	*9.4*	*639*
Dodge	*18*	*–*	*18*
Chrysler Total	717	9.1	657
Dacia	33,910	-13.2	39,086
Mercedes-Benz	*5,817*	*-37.7*	*9,338*
Smart	*631*	*-40.4*	*1,058*
Daimler Total	6,448	-38.0	10,396
Effa	220	-68.7	702
Ferrari	46	-27.0	63
Fiat	679,987	13.8	597,366
Ford	255,427	4.7	244,037
Cadillac	*15*	*-53.1*	*32*
Chevrolet	*535,637*	*1.2*	*529,144*
GM Total	535,652	1.2	529,176
Honda	120,043	56.7	76,600
Hyundai	62,967	1.1	62,266
Jaguar	41	-62.7	110
Jianghuai Auto	18,037	-24.0	23,747
Kia	19,651	-58.9	47,774
Lamborghini	–	–	21
Lifan	1,219	-56.8	2,822
Maserati	15	-31.8	22
MG Motors	37	184.6	13
Mitsubishi	6,738	904.2	671
Nissan	87,189	62.9	53,536
Citroen	*60,535*	*-14.5*	*70,779*
Peugeot	*64,367*	*-11.6*	*72,847*
Peugeot Citroen Total	124,902	-13.0	143,626
Porsche	212	-47.7	405
Renault	146,790	6.1	138,334
Rolls-Royce	14	27.3	11

	2012	% Chg.	2011
Subaru	342	-69.4	1,119
Suzuki	1,546	-25.7	2,081
Lexus	*20*	*–*	*–*
Toyota	*63,578*	*19.0*	*53,434*
Toyota Total	63,598	19.0	53,434
Volkswagen	651,237	11.1	586,110
Volvo	564	-77.8	2,540
Other	859	374.6	181
Total Cars	**2,841,936**	**7.3**	**2,649,032**
Light Trucks			
Agrale	48	37.1	35
Audi	1,012	67.8	603
BMW	*2,972*	*-46.5*	*5,550*
Mini	*432*	*–*	*–*
BMW Total	3,404	-38.7	5,550
Brilliance (Jinbei)	1,033	30.4	792
Chana	2,120	-24.4	2,803
Chery	2,052	-47.3	3,895
Dodge	*1,654*	*216.3*	*523*
Jeep	*3,420*	*35.8*	*2,519*
Ram	*1,343*	*–*	*18*
Chrysler Total	6,417	109.7	3,060
Daimler (Mercedes-Benz)	4,738	-25.0	6,316
Fiat	158,173	0.9	156,696
Ford	68,215	-2.5	69,953
Chevrolet	*58,425*	*2.3*	*57,128*
Pontiac	*48,474*	*5.8*	*45,830*
GM Total	106,899	3.8	102,958
Hafei	8,806	-47.3	16,725
Honda	14,895	-8.6	16,289
Huanghai	140	154.5	55
Hyundai	45,384	-13.7	52,595
Iveco	4,615	-7.8	5,006
Kia	21,508	-26.9	29,419
Land Rover	8,180	-0.1	8,185
Mahindra	327	-28.4	457
Mitsubishi	53,834	-1.9	54,851
Infiniti	*–*	*–*	*60*

BRAZIL

SALES BY COMPANY

	2012	% Chg.	2011
Nissan	*17,602*	*27.6*	*13,793*
Nissan Total	17,602	27.1	13,853
Citroen	*14,055*	*-27.0*	*19,248*
Peugeot	*7,683*	*-40.8*	*12,973*
Peugeot Citroen Total	21,738	-32.5	32,221
Porsche	422	-42.8	738
Renault	60,856	260.9	16,863
Shineray	386	–	–
Ssangyong	1,789	-64.1	4,984
Subaru	932	-22.0	1,195
Suzuki	5,569	5.1	5,300
Toyota	50,150	9.6	45,766
Volkswagen	117,101	4.4	112,216
Volvo	2,396	-33.2	3,589
Other	1,744	-51.5	3,596
Total Light Trucks	**792,485**	**2.1**	**776,564**
Medium/Heavy Trucks			
Agrale	533	-36.2	835
Brilliance (Jinbei)	249	–	–
Chana (Jiangling)	477	168.0	178
Daimler (Mercedes-Benz)	34,527	-18.6	42,391
FAW	2	–	–
Fiat	1	–	–
Ford	21,816	-28.1	30,351
Foton	2	–	2
GM (Chevrolet)	2	–	–
Hyundai	133	–	4
International	521	40.4	371
Iveco	10,376	-27.3	14,264
Kamaz	1	–	–
MAN	*456*	–	*12*
VW Truck & Bus	*40,967*	*-19.4*	*50,822*
MAN Total	41,423	-18.5	50,834
PACCAR (DAF)	3	–	–
Scania	11,078	-17.8	13,484
Renault Truck	*5*	–	–
Volvo Truck	*15,877*	*-16.7*	*19,068*
Volvo Truck Total	15,882	-16.7	19,068
Other	696	-20.8	879
Total Med./Hvy. Trks.	**137,722**	**-20.2**	**172,661**
Buses			
Agrale	668	-22.5	862
Brilliance (Jinbei)	5	–	–
Daimler (Mercedes-Benz)	12,856	-12.5	14,686
Fiat	21	–	–
Ford	3	200.0	1
International	58	728.6	7
Iveco	1,743	26.4	1,379
MAN (VW Truck & Bus)	8,051	-27.4	11,090
Peugeot	10	–	–
Scania	1,036	-25.8	1,397
Renault Truck	*39*	–	–
Volvo Truck	*1,689*	*25.3*	*1,348*
Volvo Truck Total	1,728	28.2	1,348
Other	3,537	-11.1	3,979
Total Buses	**29,716**	**-14.5**	**34,749**
Total Comm. Vehs.	**959,923**	**-2.4**	**983,974**
Total Vehicles	**3,801,859**	**4.6**	**3,633,006**

CHILE

ECONOMIC SUMMARY

Common Language	Spanish
Currency	Chilean Peso
Steering	Left Hand Drive

2012 Vehicles in Operation

Passenger Cars	2,383,638
Commercial Vehicles	1,208,390
Total	3,592,028
Persons per Car	7.2
Persons per Vehicle	4.8

2012 GDP (billions of US $)

Nominal GDP	268.2
Real GDP	228.7
GDP % annual growth	5.5
GDP per capita	13,141

Current Account Balance (billions of US $)

2012:	-9.5
2011:	-3.3

Current Account Balance in % of GDP

2012:	-3.5
2011:	-1.3

Total Number of Fatal Road Accidents (most recent data)

2010:	1,595

Capital	Santiago
Population	17,216,945
Population Density	60.0
Surface Area sq km/miles	743,812 / 287,186

Road Infrastructure

Paved Road Surface %	20.8
Total Road Length in km/miles	80,505 / 50,023
Road Traffic (Vehicles per mile)	71.8

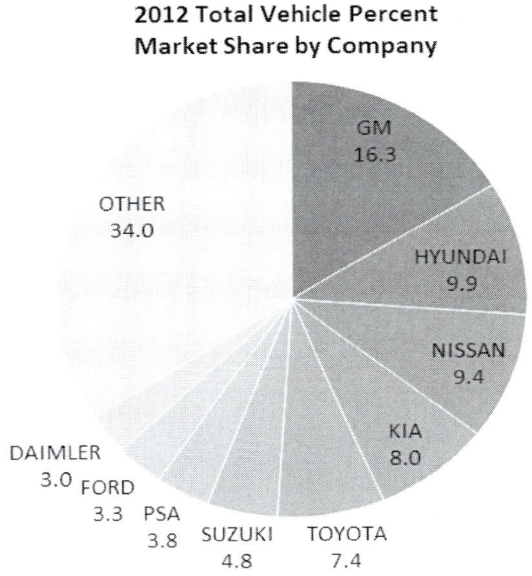

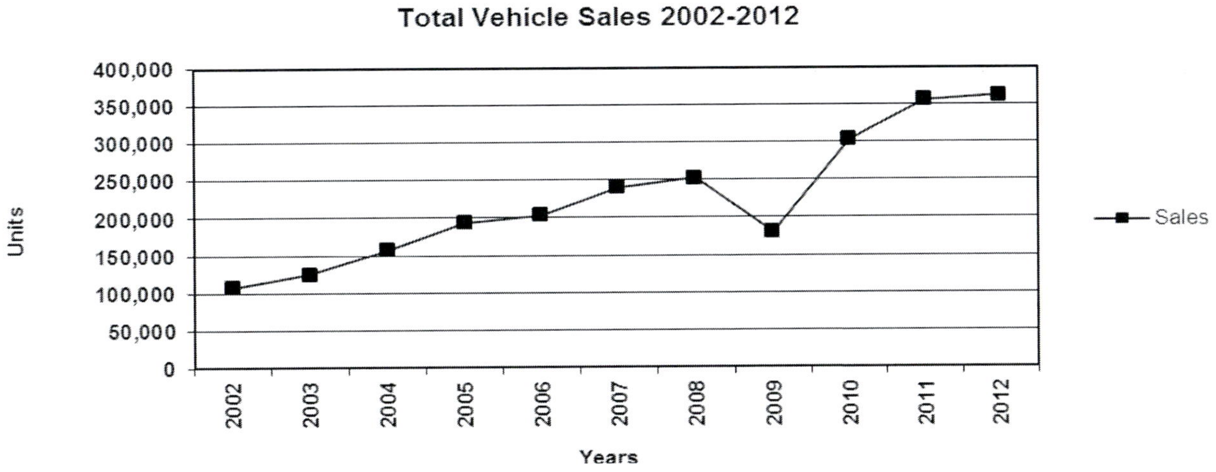

CHILE

SALES

Year	Cars	Commercial Vehicles	Total
2012	257,013	105,318	362,331
2011	245,116	111,067	356,183
2010	158,540	144,820	303,360
2009	98,233	82,463	180,696
2008	140,334	112,356	252,690
2007	139,372	101,105	240,477
2006	114,959	89,073	204,032
2005	112,699	80,694	193,393
2004	111,459	45,676	157,135
2003	88,014	37,075	125,089
2002	71,204	35,896	107,100
2001	64,269	38,710	102,979
2000	71,499	42,221	113,720
1999	68,132	40,846	108,978
1998	89,650	58,744	148,394
1997	115,506	70,238	185,744
1996	105,606	66,023	171,629
1995	95,360	60,432	155,792

CHILE

SALES BY COMPANY

	2012	% Chg.	2011
Cars			
Aston Martin	16	-44.8	29
Audi	2,259	21.3	1,863
BMW	*2,838*	*2.6*	*2,767*
Mini	*444*	*2.1*	*435*
BMW Total	3,282	2.5	3,202
Brilliance	464	592.5	67
BYD	2,360	-2.7	2,425
Chana	*240*	–	*4*
Jiangling	*2*	–	*2*
Chana Total	242	–	6
Chery	5,387	23.2	4,372
Chrysler	7,320	4.4	7,011
Daihatsu	932	24.4	749
Daimler (Mercedes-Benz)	2,523	22.9	2,053
Dongfeng	73	-40.2	122
FAW	4	–	–
Ferrari	21	31.3	16
Alfa Romeo	*279*	*190.6*	*96*
Fiat	*1,977*	*45.9*	*1,355*
Fiat Total	2,256	55.5	1,451
Ford	6,971	34.4	5,185
Foton	57	418.2	11
Chevrolet	*48,642*	*4.0*	*46,783*
GMC	*4*	–	–
Opel	*433*	–	–
GM Total	49,079	4.9	46,783
Great Wall	5,044	11.0	4,546
Hafei	11	-52.2	23
Haima	*836*	*235.7*	*249*
Zhengzhou	*140*	–	–
Haima Total	976	292.0	249
Honda	2,547	13.2	2,250
Huatai	1	–	–
Hyundai	31,691	8.5	29,212
Jaguar	*80*	*1.3*	*79*
Land Rover	*564*	*13.0*	*499*
Jaguar Land Rover Total	644	11.4	578
Jianghuai	3,620	38.5	2,614
Kia	26,038	7.2	24,280
Lifan	1,210	12.2	1,078
Mahindra & Mahindra	1,503	12.0	1,342
Maserati	18	63.6	11
Mazda	7,175	-7.0	7,713
MG Motors	1,460	212.0	468
Mitsubishi	2,017	-34.1	3,061
Infiniti	*33*	–	–
Nissan	*19,617*	*12.3*	*17,476*
Nissan Total	19,650	12.4	17,476
Citroen	*3,175*	*3.7*	*3,061*
Peugeot	*4,430*	*4.9*	*4,223*
Peugeot Citroen Total	7,605	4.4	7,284
Porsche	337	3.7	325
Proton (Lotus)	5	–	–
Renault	3,182	-20.1	3,982
Samsung	5,631	-34.1	8,543
Ssangyong	2,633	1.7	2,588
Subaru	3,926	10.2	3,561
Suzuki	16,977	-19.0	20,967
Lexus	*398*	*-13.3*	*459*
Toyota	*19,288*	*10.7*	*17,424*
Toyota Total	19,686	10.1	17,883
Skoda	*610*	*250.6*	*174*
Volkswagen	*4,234*	*-5.4*	*4,474*
Volkswagen Total	4,844	4.2	4,648
Volvo	1,905	3.9	1,833
Geely	*2,473*	*14.9*	*2,152*
Maple	–	–	*12*
Zhejiang Geely Total	2,473	14.3	2,164
Zotye	822	-11.1	925
ZX Auto	–	–	5
Other	136	-16.0	162
Total Cars	**257,013**	**4.9**	**245,116**
Light Trucks			
BMW	*4*	–	–
Mini	–	–	*1*
BMW Total	4	300.0	1
Brilliance (Jinbei)	283	–	–
Chana	*1,214*	*21.0*	*1,003*
Jiangling	*117*	*-80.8*	*610*
Chana Total	1,331	-17.5	1,613
Chery	3	-25.0	4
Chrysler	854	-55.2	1,908
Daimler (Mercedes-Benz)	952	-46.8	1,790
Dongfeng	1,119	-5.8	1,188
Fiat	2,267	-0.5	2,279
Ford	4,061	-17.8	4,941
Foton	331	-29.1	467
Chevrolet	*8,596*	*-12.8*	*9,857*
GMC	*3*	–	–
GM Total	8,599	-12.8	9,857
GoNow	2	–	–
Great Wall	3,158	58.4	1,994

CHILE

SALES BY COMPANY

	2012	% Chg.	2011
Hafei	1,792	7.0	1,675
Haima (Zhengzhou)	574	–	–
Honda	491	33.8	367
Hyundai	3,432	-41.3	5,851
Iveco	10	-95.7	231
Jaguar Land Rover	11	-42.1	19
Jianghuai	643	-24.1	847
Kia	2,878	-19.9	3,591
Lifan	464	72.5	269
Mahindra & Mahindra	2,874	14.4	2,513
Mazda	678	-61.4	1,757
MG Motors	–	–	1
Mitsubishi	6,582	9.0	6,041
Nissan	14,507	-15.3	17,121
Citroen	*1,953*	*2.2*	*1,911*
Peugeot	*4,263*	*-5.6*	*4,515*
Peugeot Citroen Total	6,216	-3.3	6,426
Porsche	1	–	–
Renault	487	28.8	378
Samsung	1	-50.0	2
Ssangyong	5,421	11.2	4,877
Subaru	–	–	1
Suzuki	412	-29.8	587
Tata	383	-4.5	401
Toyota	7,274	27.0	5,727
Volkswagen	2,489	-4.3	2,601
Volvo	–	–	1
ZX Auto	902	-28.4	1,259
Other	327	-6.8	351
Total Light Trucks	**81,813**	**-8.0**	**88,936**
Medium/Heavy Trucks			
Brilliance (Jinbei)	53	-19.7	66
Chana (Jiangling)	761	74.9	435
Freightliner	*1,960*	*10.5*	*1,773*
Mercedes-Benz	*1,719*	*-19.4*	*2,134*
Mitsubishi Fuso	*610*	*1.2*	*603*
Western Star	*9*	–	–
Daimler Total	4,298	-4.7	4,510
Chrysler (Dodge)	13	–	–
Dongfeng	521	20.0	434
FAW	190	171.4	70
Ford	1,013	-10.6	1,133
Foton	524	12.7	465
GM (Chevrolet)	1,365	-10.2	1,520
Hino	827	1.3	816

	2012	% Chg.	2011
Hyundai	711	-34.7	1,088
International	1,224	20.9	1,012
Iveco	247	-24.5	327
Jianghuai	1,257	-7.9	1,365
Kia	–	–	189
MAN	*297*	*9.2*	*272*
Volkswagen Truck & Bus	*1,090*	*-13.0*	*1,253*
MAN Total	1,387	-9.0	1,525
DAF	*17*	*750.0*	*2*
Kenworth	*123*	*44.7*	*85*
PACCAR Total	140	60.9	87
Scania	653	-3.0	673
Mack Truck	*816*	*-0.7*	*822*
Renault Truck	*133*	*9.0*	*122*
UD Trucks	*4*	*-33.3*	*6*
Volvo Truck	*1,333*	*21.6*	*1,096*
Volvo Truck Total	2,286	11.7	2,046
Yuejin	371	-10.4	414
Other	468	14.1	410
Total Med./Hvy. Trks.	**18,309**	**-1.5**	**18,585**
Buses			
Agrale	2	–	–
Mercedes-Benz	*2,989*	*64.7*	*1,815*
Mitsubishi Fuso	*110*	*-52.2*	*230*
Daimler Total	3,099	51.5	2,045
Dongfeng	33	135.7	14
Ford	–	–	1
Hyundai	66	-50.4	133
International	–	–	7
Iveco	19	-72.9	70
Jianghuai	69	305.9	17
Golden Dragon	*10*	–	–
King Long	*100*	–	–
King Long Total	110	–	–
MAN (VW Truck & Bus)	183	3.4	177
Scania	534	100.0	267
Toyota	1	–	–
Volvo Truck	194	11.5	174
Yuejin	–	–	11
Yutong	49	–	–
Other	837	32.9	630
Total Buses	**5,196**	**46.5**	**3,546**
Total Comm. Vehs.	**105,318**	**-5.2**	**111,067**
Total Vehicles	**362,331**	**1.7**	**356,183**

COLOMBIA

ECONOMIC SUMMARY

Common Language	Spanish
Currency	Colombian Peso
Steering	Left Hand Drive

2012 Vehicles in Operation

Passenger Cars	2,526,000
Commercial Vehicles	974,000
Total	3,500,000
Persons per Car	18.1
Persons per Vehicle	13.1

2012 GDP (billions of US $)

Nominal GDP	366.0
Real GDP	266.1
GDP % annual growth	4.0
GDP per capita	5,710

Current Account Balance (billions of US $)

2012: -12.4
2011: -10.0

Current Account Balance in % of GDP

2012: -3.4
2011: -3.0

Total Number of Fatal Road Accidents (most recent data)

2010: 5,502

Capital	Bogota
Population	45,745,783
Population Density	114.1
Surface Area sq km/miles	1,038,700 / 401,043

Road Infrastructure

Paved Road Surface %	15.0
Total Road Length in km/miles	141,374 / 87,846
Road Traffic (Vehicles per mile)	39.8

2012 Total Vehicle Percent Market Share by Company

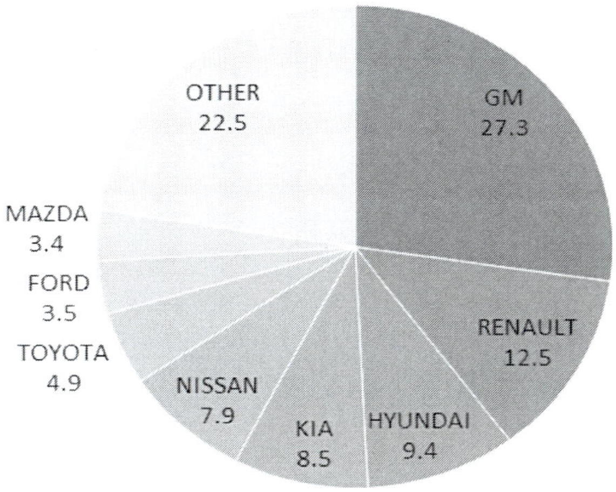

Total Vehicle Sales 2003-2012

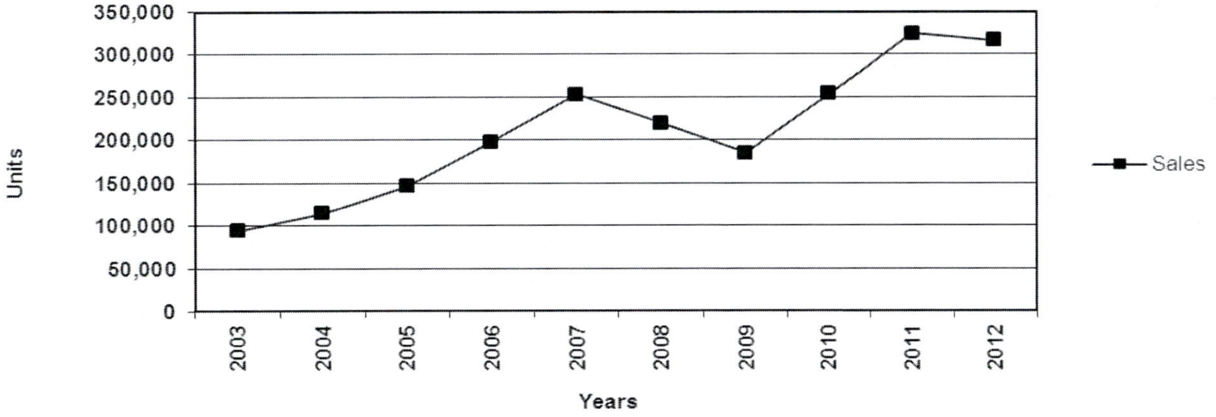

COLOMBIA

SALES

Year	Light Vehicles	Med./Hvy. Trucks	Total
2012	281,257	34,711	315,968
2011	296,302	28,268	324,570
2010	237,969	15,900	253,869
2009	171,731	13,398	185,129
2008	199,035	20,463	219,498
2007	222,777	30,257	253,034
2006	175,056	23,128	198,184
2005	132,598	13,969	146,567
2004	103,876	10,692	114,568
2003	84,018	9,875	93,893

COLOMBIA

SALES BY COMPANY

	2012	% Chg.	2011
Light Vehicles			
Audi	–	–	1,335
BMW	–	–	*2,500*
Mini	–	–	*455*
BMW Total	–	–	2,955
BYD	301	-24.8	400
Chana	225	-18.5	276
Chery	2,238	23.4	1,813
Dodge	*3,195*	*9.6*	*2,915*
Jeep	*983*	*10.4*	*890*
Chrysler Total	4,178	9.8	3,805
Daihatsu	448	-29.2	633
Daimler (Mercedes-Benz)	612	17.5	521
Dongfeng	2,064	33.5	1,546
Alfa Romeo	*61*	*-41.9*	*105*
Fiat	*1,785*	*57.8*	*1,131*
Fiat Total	1,846	49.4	1,236
Ford	11,212	5.1	10,669
GM (Chevrolet)	81,067	-19.5	100,743
Great Wall	1,112	10.9	1,003
Hafei	659	-3.4	682
Honda	2,670	55.6	1,716
Hyundai	28,225	-2.5	28,941
Jaguar	–	–	*4*
Land Rover	–	–	*162*
Jaguar Land Rover Total	–	–	166
Kia	26,931	0.7	26,736
Lifan	127	13.4	112
Mazda	10,747	-19.5	13,345
Mitsubishi	1,001	-37.1	1,591
Nissan	25,110	4.4	24,051
Citroen	*693*	*29.5*	*535*
Peugeot	*722*	*-22.8*	*935*
Peugeot Citroen Total	1,415	-3.7	1,470
Porsche	–	–	126
Renault	39,411	-10.3	43,936
Ssangyong	2,326	4.3	2,230
Subaru	401	-26.4	545
Suzuki	1,255	–	–
Toyota	15,470	14.3	13,534
SEAT	*423*	*0.2*	*422*
Skoda	*264*	*-45.8*	*487*
Volkswagen	*7,955*	*29.2*	*6,157*
Volkswagen Total	8,642	22.3	7,066
Volvo	54	10.2	49
Zhejiang Geely	743	41.3	526
Zotye	2,244	23.6	1,815
Other	8,523	–	730
Total Light Vehicles	**281,257**	**-5.1**	**296,302**
Medium/Heavy Trucks			
Brilliance (Jinbei)	148	-27.1	203
Chana (Jiangling)	232	-11.5	262
Freightliner	–	–	*1,346*
Mercedes-Benz	*2,269*	*17.5*	*1,931*
Mitsubishi Fuso	*250*	*-20.1*	*313*
Daimler Total	2,519	-29.8	3,590
Dongfeng	1	-88.9	9
GM (Chevrolet)	5,175	1.9	5,080
Hino	1,627	-8.0	1,768
Hyundai	1,485	118.1	681
MAN (VW Truck & Bus)	2,375	53.2	1,550
Scania	56	–	–
Mack Truck	*182*	*-22.9*	*236*
Renault Truck	*4,380*	*50.8*	*2,905*
Volvo Truck	*751*	*7.9*	*696*
Volvo Truck Total	5,313	38.5	3,837
Other	15,780	39.8	11,288
Total Med./Hvy. Trucks	**34,711**	**22.8**	**28,268**
Total Vehicles	**315,968**	**-2.7**	**324,570**

URUGUAY

ECONOMIC SUMMARY

Common Language	Spanish
Currency	Uruguayan Peso
Steering	Left Hand Drive

2012 Vehicles in Operation

Passenger Cars	656,500
Commercial Vehicles	89,900
Total	746,400
Persons per Car	5.1
Persons per Vehicle	4.5

2012 GDP (billions of US $)

Nominal GDP	49.4
Real GDP	32.1
GDP % annual growth	3.8
GDP per capita	9,500

Current Account Balance (billions of US $)

2012: 1.7
2011: -1.3

Current Account Balance in % of GDP

2012: -3.4
2011: -2.8

Total Number of Fatal Road Accidents (most recent data)

2010: 556

Capital	Montevideo
Population	3,324,460
Population Density	49.2
Surface Area sq km/miles	175,015 / 67,573

Road Infrastructure

Paved Road Surface %	10.0
Total Road Length in km/miles	77,732 / 48,300
Road Traffic (Vehicles per mile)	15.5

2012 Total Vehicle Percent Market Share by Company

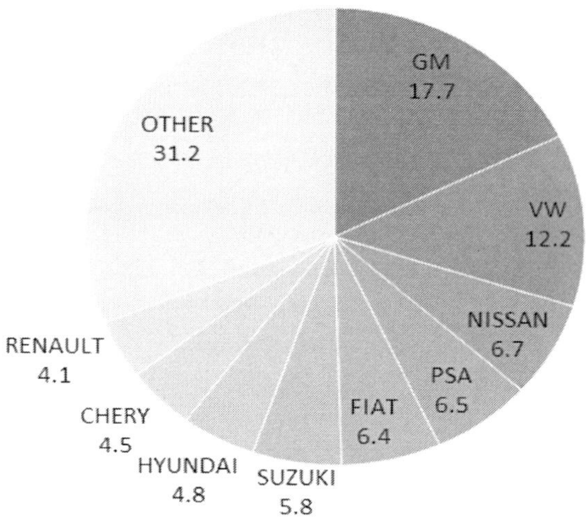

Total Vehicle Sales 2002-2012

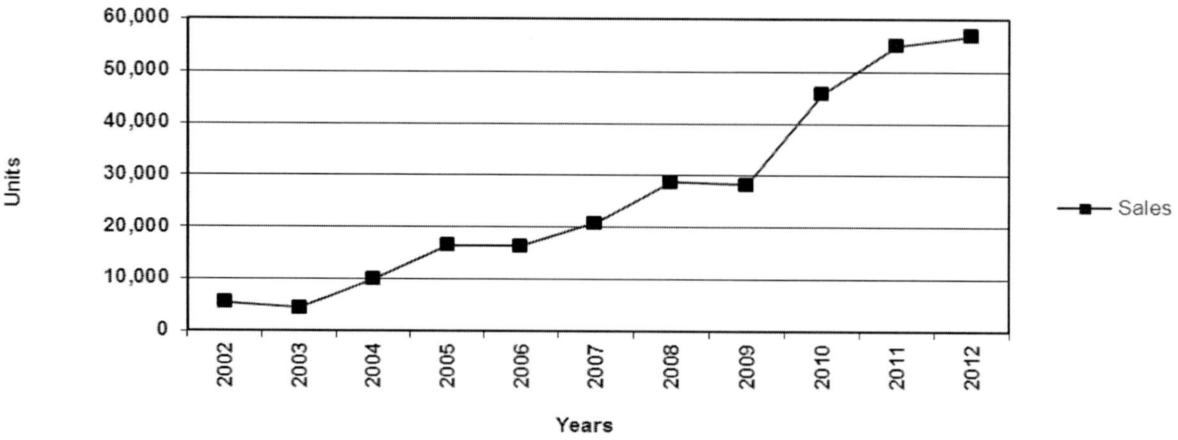

URUGUAY

SALES

Year	Light Vehicles	Med./Hvy. Trucks	Total
2012	53,233	3,604	56,837
2011	51,443	3,561	55,004
2010	42,675	3,046	45,721
2009	25,974	2,293	28,267
2008	25,942	2,834	28,776
2007	19,072	1,719	20,791
2006	15,155	1,211	16,366
2005	15,405	1,065	16,470
2004	9,392	528	9,920
2003	4,365	100	4,465
2002	5,206	273	5,479
2001	14,223	919	15,142
2000	17,389	1,138	18,527
1999	18,221	1,410	19,631
1998	36,909	2,238	39,147
1997	33,959	2,134	36,093
1996	27,053	1,751	28,804
1995	29,972	512	30,484

URUGUAY

SALES BY COMPANY

	2012	% Chg.	2011
Light Vehicles			
Audi	259	0.4	258
BMW	*250*	*-10.7*	*280*
Mini	*42*	*-25.0*	*56*
BMW Total	292	-13.1	336
Brilliance	8	-61.9	21
BYD	898	-12.7	1,029
Chana	*542*	*2.5*	*529*
Jiangling	*–*	*–*	*10*
Chana Total	542	0.6	539
Changhe	157	48.1	106
Chery	2,556	-24.7	3,394
Chrysler	*1*	*–*	*–*
Dodge	*190*	*90.0*	*100*
Jeep	*31*	*6.9*	*29*
Ram	*77*	*48.1*	*52*
Chrysler Total	299	65.2	181
Daimler (Mercedes-Benz)	513	6.4	482
Dongfeng	772	9.5	705
Effa	385	-39.8	640
FAW	1,551	105.4	755
Alfa Romeo	*7*	*16.7*	*6*
Fiat	*3,654*	*7.8*	*3,390*
Fiat Total	3,661	7.8	3,396
Ford	1,855	18.0	1,572
Foton	73	32.7	55
Chevrolet	*9,979*	*-23.4*	*13,033*
Wuling	*54*	*-16.9*	*65*
GM Total	10,033	-23.4	13,098
GoNow	192	50.0	128
Great Wall	855	-10.9	960
Hafei	91	51.7	60
Haima	277	403.6	55
Honda	527	93.8	272
Hyundai	2,568	57.5	1,630
Jaguar	*2*	*100.0*	*1*
Land Rover	*5*	*150.0*	*2*
Jaguar Land Rover Total	7	133.3	3
Jianghuai	581	183.4	205
Kia	627	64.1	382
Lifan	302	-41.8	519
Mahindra & Mahindra	123	434.8	23
Mazda	155	-37.5	248
Mitsubishi	161	-46.7	302
Nissan	3,828	25.6	3,048
Citroen	*1,035*	*27.9*	*809*
Peugeot	*2,648*	*-6.0*	*2,818*
Peugeot Citroen Total	3,683	1.5	3,627
Porsche	14	-36.4	22
Renault	2,354	11.9	2,104
Shineray	30	15.4	26
Ssangyong	3	–	3
Subaru	29	-58.0	69
Suzuki	3,290	33.1	2,471
Lexus	*1*	*–*	*–*
Toyota	*1,506*	*9.4*	*1,376*
Toyota Total	1,507	9.5	1,376
Volkswagen	6,938	11.7	6,212
Volvo	27	28.6	21
Geely	*932*	*33.0*	*701*
Maple	*–*	*–*	*1*
Zhejiang Geely Total	932	32.8	702
Zhongneng (Anchi)	3	–	–
Zotye	107	256.7	30
ZX Auto	73	-70.7	249
Other	95	-26.4	129
Total Light Vehicles	**53,233**	**3.5**	**51,443**
Medium/Heavy Trucks			
Chana	*17*	*-39.3*	*28*
Jiangling	*648*	*23.2*	*526*
Chana Total	665	20.0	554
China National	47	-6.0	50
Freightliner	*37*	*270.0*	*10*
Mercedes-Benz	*199*	*-47.9*	*382*
Daimler Total	236	-39.8	392
Aeolus	*93*	*-21.8*	*119*
Dongfeng	*2*	*-75.0*	*8*
Dongfeng Total	95	-25.2	127
Effa	–	–	7
FAW	6	-53.8	13
Ford	49	-48.4	95
Foton	311	-18.4	381
Hualing	6	–	–
Hyundai	171	16.3	147
International	4	300.0	1
Iveco	122	-32.2	180
Jianghuai	443	42.4	311
Kia	240	252.9	68
MAN	*11*	*–*	*–*
Volkswagen Truck & Bus	*672*	*-14.8*	*789*
MAN Total	683	-13.4	789
Scania	226	-2.2	231
Renault Truck	*28*	*-26.3*	*38*
Volvo Truck	*130*	*14.0*	*114*
Volvo Truck Total	158	3.9	152
Yuejin	24	–	2
Yutong	112	–	–
Zhongtong	2	–	–
Other	4	-93.4	61
Total Med./Hvy. Trks.	**3,604**	**1.2**	**3,561**
Total Vehicles	**56,837**	**3.3**	**55,004**

VENEZUELA

ECONOMIC SUMMARY

Common Language	Spanish
Currency	Bolivar Fuerte
Steering	Left Hand Drive

2012 Vehicles in Operation

Passenger Cars	2,531,529
Commercial Vehicles	807,867
Total	3,339,396
Persons per Car	11.2
Persons per Vehicle	8.5

2012 GDP (billions of US $)

Nominal GDP	382.4
Real GDP	14.3
GDP % annual growth	5.5
GDP per capita	484

Current Account Balance (billions of US $)

2012:	11.0
2011:	24.4

Current Account Balance in % of GDP

2012:	2.9
2011:	7.7

Total Number of Fatal Road Accidents (most recent data)

2008:	7,714

Capital	Caracas
Population	28,459,085
Population Density	83.6
Surface Area sq km/miles	882,050 / 340,560

Road Infrastructure

Paved Road Surface %	33.6
Total Road Length in km/miles	96,155 / 59,748
Road Traffic (Vehicles per mile)	55.9

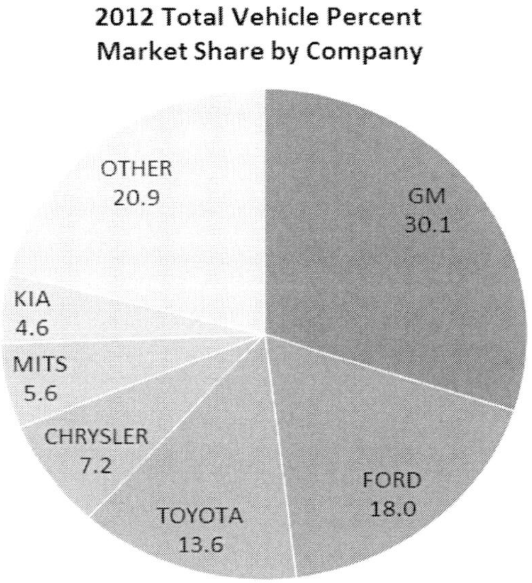

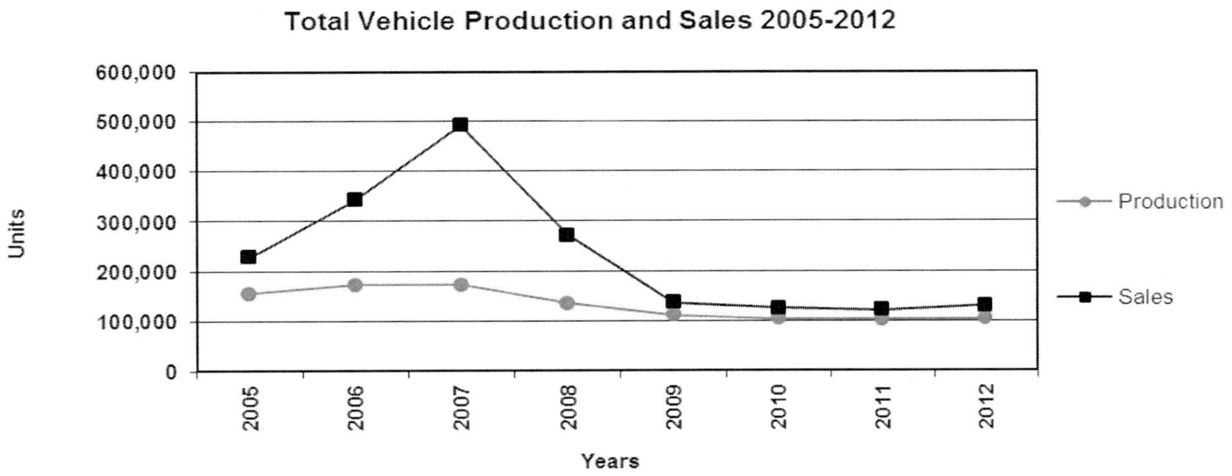

VENEZUELA

PRODUCTION

Year	Cars	Trucks	Buses	Total
2012	53,797	50,286	–	104,083
2011	58,032	44,377	–	102,409
2010	60,810	43,547	–	104,357
2009	58,579	52,778	197	111,554
2008	78,322	56,429	291	135,042
2007	113,443	58,625	350	172,418
2006	102,250	69,283	182	171,715
2005	95,096	59,712	153	154,961
2004	69,114	41,829	110	111,053
2003	29,862	16,899	123	46,884
2002	54,098	28,522	185	82,805
2001	94,489	42,121	157	136,767
2000	58,599	30,172	661	89,432
1999	41,145	32,661	374	74,180
1998	95,903	32,565	400	128,868
1997	85,538	70,826	416	156,780
1996	35,358	35,868	544	71,770
1995	44,081	51,678	644	96,403

VENEZUELA

PRODUCTION BY MANUFACTURER

	2012	% Chg.	2011
Cars			
Chrysler (Dodge)	952	-49.6	1,889
Ford	7,623	-27.9	10,578
GM (Chevrolet)	28,310	2.4	27,650
Hyundai	3,872	-52.0	8,071
Mitsubishi	6,541	51.0	4,331
Toyota	6,499	17.9	5,513
Total Cars	**53,797**	**-7.3**	**58,032**
Light Trucks			
Chrysler (Jeep)	8,320	9.1	7,625
Ford	14,252	32.3	10,774
GM (Chevrolet)	7,581	4.0	7,290
Iveco	825	16.0	711
Mitsubishi	811	42.0	571
Mitsubishi Fuso	1,094	14.2	958
Toyota	6,330	19.0	5,318
Total Light Trucks	**39,213**	**17.9**	**33,247**
Medium/Heavy Trucks			
Ford	1,341	-37.9	2,158
Chevrolet	*3,148*	*-31.7*	*4,610*
Isuzu	*4,410*	*79.7*	*2,454*
GM Total	7,558	7.0	7,064
Iveco	1,374	13.6	1,209
Mack Truck	800	15.8	691
Mercedes-Benz	–	–	8
Total Med./Hvy. Trks.	**11,073**	**-0.5**	**11,130**
Total Comm. Vehs.	**50,286**	**13.3**	**44,377**
Total Vehicles	**104,083**	**1.6**	**102,409**

VENEZUELA

SALES

Year	Cars	Trucks	Buses	Total
2012	60,776	69,222	555	130,553
2011	65,339	54,993	357	120,689
2010	68,210	56,499	493	125,202
2009	67,047	68,605	865	136,517
2008	135,882	134,420	1,320	271,622
2007	267,932	222,079	1,888	491,899
2006	184,108	158,064	1,179	343,351
2005	128,601	98,723	1,054	228,378

VENEZUELA

SALES BY COMPANY

	2012	% Chg.	2011
Cars			
Audi	1	–	–
BMW	4	100.0	2
Chrysler (Dodge)	1,064	-41.1	1,807
Dacia	250	–	–
Dongfeng	693	–	–
Fiat	1,050	-34.3	1,599
Ford	7,691	-29.5	10,915
GM (Chevrolet)	28,528	-3.7	29,633
Hyundai	4,308	-45.0	7,836
Kia	2,603	4.6	2,489
Mitsubishi	6,493	39.8	4,645
Nissan	–	–	12
Peugeot Citroen	1,562	19.1	1,312
Renault	395	-7.1	425
Toyota	6,134	35.9	4,514
Other	–	–	150
Total Cars	**60,776**	**-7.0**	**65,339**
Light Trucks			
Audi	2	–	–
BMW	14	-17.6	17
Jeep	*8,358*	*9.7*	*7,618*
Ram	*1*	*-99.6*	*249*
Chrysler Total	8,359	6.3	7,867
Mercedes-Benz	*18*	*-94.3*	*315*
Mitsubishi Fuso	*1,061*	*-4.2*	*1,108*
Daimler Total	1,079	-24.2	1,423
Dongfeng	1,902	–	–
Ford	7,556	3.8	7,276
GM (Chevrolet)	7,607	4.5	7,276
Haima	352	–	–
Hyundai	2	-77.8	9
Iveco	1,379	8.8	1,267
Kia	3,414	10.6	3,088
Mazda	3,102	227.9	946
Mitsubishi	815	41.2	577
Nissan	11	37.5	8
Citroen	*320*	–	–
Peugeot	*440*	*-3.5*	*456*
Peugeot Citroen Total	760	66.7	456
Renault	448	-33.2	671
Suzuki	5	–	5
Toyota	11,578	76.0	6,580
Volkswagen	1,563	446.5	286
Total Light Trucks	**49,948**	**32.3**	**37,752**
Medium/Heavy Trucks			
Mercedes-Benz	*1*	*-99.0*	*98*
Mitsubishi Fuso	*2*	*-88.9*	*18*
Daimler Total	3	-97.4	116
Dongfeng	531	103.4	261
Ford	8,200	5.3	7,787
GM (Chevrolet)	3,142	-32.2	4,636
Hyundai	291	–	–
International	–	–	18
Isuzu	3,973	69.0	2,351
Iveco	1,557	13.6	1,371
MAN (VW Truck & Bus)	–	–	20
Scania	261	–	7
Volvo Truck (Mack)	1,316	95.3	674
Total Med./Hvy. Trks.	**19,274**	**11.8**	**17,241**
Buses			
Daimler (Mercedes-Benz)	39	-83.0	230
Isuzu	453	256.7	127
MAN (VW Truck & Bus)	1	–	–
Scania	32	–	–
Toyota	30	–	–
Total Buses	**555**	**55.5**	**357**
Total Comm. Vehs.	**69,777**	**26.1**	**55,350**
Total Vehicles	**130,553**	**8.2**	**120,689**

Ward's World Motor Vehicle Data 2013

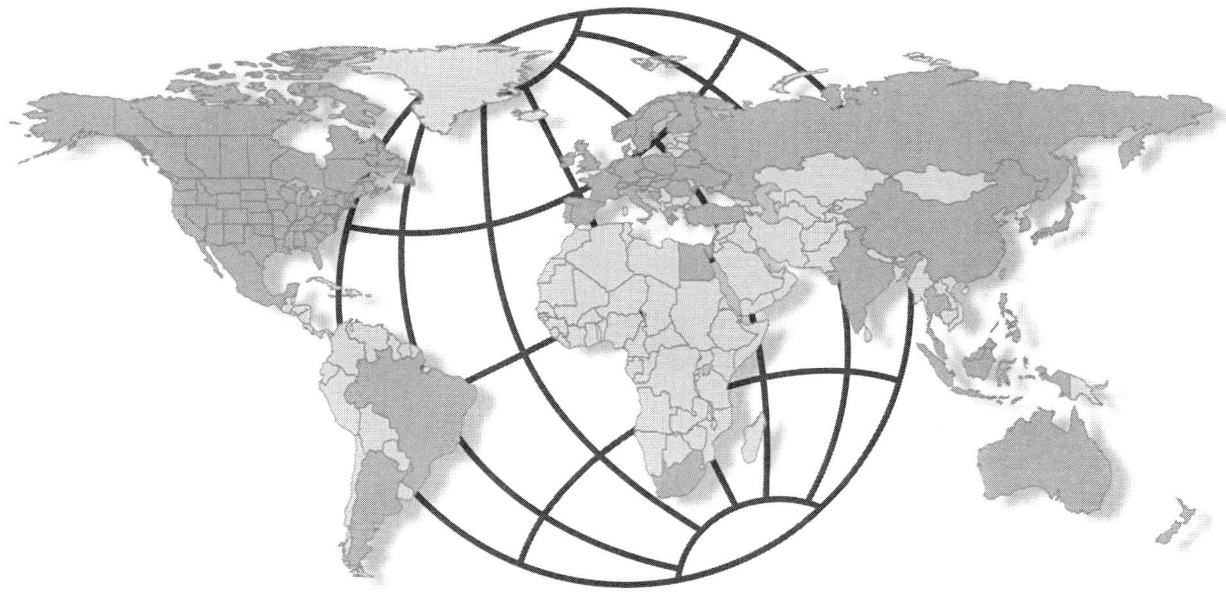

Other Data

EGYPT

ECONOMIC SUMMARY

Common Language	Arabic
Currency	Egyptian Pound
Steering	Left Hand Drive

2012 Vehicles in Operation

Passenger Cars	3,231,513
Commercial Vehicles	1,136,388
Total	4,367,901
Persons per Car	26.4
Persons per Vehicle	19.5

2012 GDP (billions of US $)

Nominal GDP	256.7
Real GDP	95.1
GDP % annual growth	2.2
GDP per capita	1,152

Current Account Balance (billions of US $)

2012:	-7.9
2011:	-6.1

Current Account Balance in % of GDP

2012:	-3.1
2011:	-2.6

Total Number of Fatal Road Accidents (most recent data)

2011:	6,807

Capital	Cairo
Population	85,294,388
Population Density	221.9
Surface Area sq km/miles	995,450 / 384,345

Road Infrastructure

Paved Road Surface %	73.0
Total Road Length in km/miles	65,050 / 40,420
Road Traffic (Vehicles per mile)	108.1

2012 Total Vehicle Percent Market Share by Company

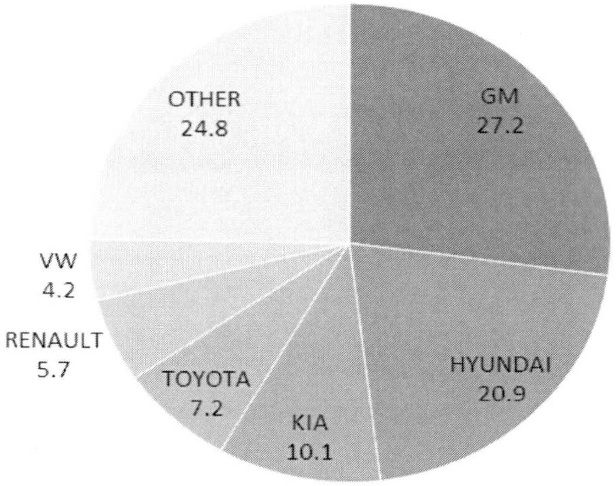

Total Vehicle Production and Sales 2002-2012

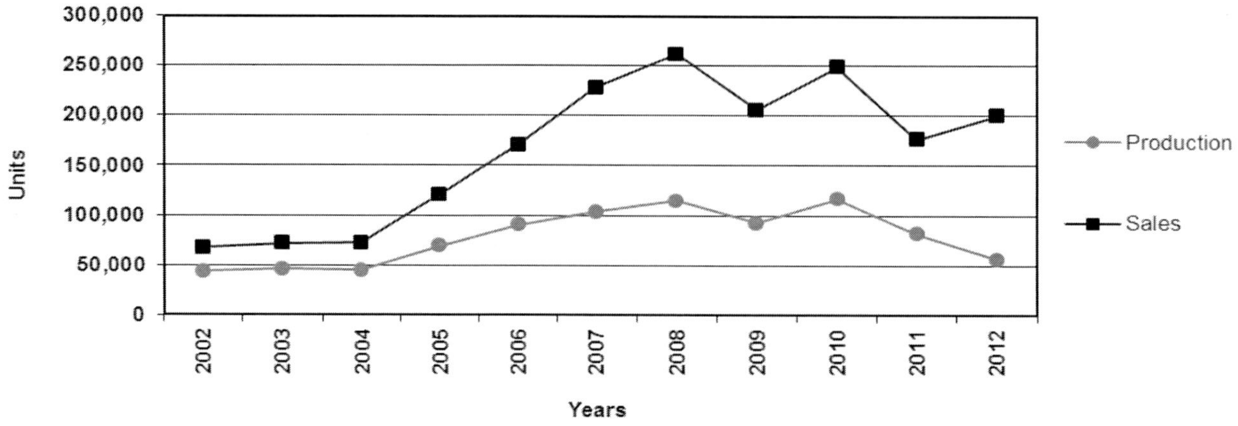

EGYPT

PRODUCTION

Year	Cars	Commercial Vehicles	Total
2012	36,880	19,600	56,480
2011	53,072	28,659	81,731
2010	76,412	40,271	116,683
2009	60,249	32,090	92,339
2008	72,485	42,297	114,782
2007	67,149	36,403	103,552
2006	59,462	31,016	90,478
2005	48,197	21,037	69,234
2004	32,390	12,601	44,991
2003	32,975	13,447	46,422
2002	28,150	15,805	43,955
2001	37,422	17,781	55,203
2000	41,236	21,406	62,642
1999	45,416	30,632	76,048
1998	39,135	33,358	72,493
1997	31,142	28,003	59,145
1996	30,910	23,535	54,445
1995	31,428	23,890	55,318

EGYPT

PRODUCTION BY MANUFACTURER

	2012	% Chg.	2011
Cars			
AutoVaz (Lada)	1,214	-27.6	1,676
BMW	1,290	-22.4	1,662
Chrysler	604	2.0	592
GB Auto	1,011	–	–
Daewoo	*11,650*	*-31.6*	*17,031*
GM/Opel	*4,545*	*-40.7*	*7,669*
GM Total	16,195	-34.4	24,700
Hyundai	9,286	-35.0	14,281
Kia	142	3.6	137
Mercedes-Benz	716	-23.6	937
Nasco	5,944	-27.2	8,162
Peugeot	377	-28.9	530
Suzuki	101	-74.4	395
Total Cars	**36,880**	**-30.5**	**53,072**
Trucks			
GM	11,633	-29.4	16,478
MAN	48	11.6	43

	2012	% Chg.	2011
Mercedes-Benz	487	-7.4	526
Mitsubishi	184	-47.0	347
Nissan	2,212	-42.6	3,855
Scania	17	-29.2	24
Suzuki	966	-29.7	1,374
Volvo Truck	33	3.1	32
Total Trucks	**15,580**	**-31.3**	**22,679**
Buses			
GM	1,230	-35.3	1,902
Hyundai	99	-16.1	118
MAN	71	2.9	69
Mercedes-Benz	1,303	-33.8	1,967
Mitsubishi	603	-10.8	676
Suzuki	714	-42.8	1,248
Total Buses	**4,020**	**-32.8**	**5,980**
Total Comm. Vehs.	**19,600**	**-31.6**	**28,659**
Total Vehicles	**56,480**	**-30.9**	**81,731**

NOTE: Individual manufacturer figures estimated from vehicle type totals.

EGYPT

SALES

Year	Cars	Trucks	Buses	Total
2012	144,123	36,113	20,016	200,252
2011	133,165	29,963	13,029	176,157
2010	192,848	39,306	16,763	248,917
2009	158,926	33,446	13,149	205,521
2008	198,800	45,685	17,469	261,954
2007	179,168	33,749	14,901	227,818
2006	132,373	27,207	11,034	170,614
2005	94,322	19,530	7,318	121,170
2004	55,471	11,383	5,561	72,415
2003	53,375	13,537	4,924	71,836
2002	46,748	15,074	5,523	67,345
2001	57,360	17,044	4,613	79,017
2000	64,567	20,678	4,792	90,037

EGYPT

SALES BY COMPANY

	2012	% Chg.	2011
Cars			
Audi	581	81.6	320
BMW	*2,607*	*17.8*	*2,213*
Mini	*122*	*18.4*	*103*
BMW Total	2,729	17.8	2,316
Brilliance	674	10.3	611
Chery	6,297	-32.4	9,310
Chrysler	*75*	*56.3*	*48*
Dodge	*184*	*32.4*	*139*
Jeep	*1,723*	*-1.5*	*1,749*
Chrysler Total	1,982	2.4	1,936
Dacia	551	-80.4	2,807
Daihatsu	1,670	14.4	1,460
Daimler (Mercedes-Benz)	2	-99.9	2,609
Fiat	502	-59.7	1,246
Ford	863	54.1	560
Chevrolet	*25,010*	*2.5*	*24,398*
Opel	*1,154*	*380.8*	*240*
GM Total	26,164	6.2	24,638
Great Wall	192	-15.8	228
Honda	691	123.6	309
Hyundai	41,238	-3.9	42,924
Jaguar	*47*	*67.9*	*28*
Land Rover	*155*	*192.5*	*53*
Jaguar Land Rover Total	202	149.4	81
Jianghuai Auto Co.	298	60.2	186
Jianglin (Landwind)	39	21.9	32
Kia	20,305	77.7	11,426
Mahindra & Mahindra	185	30.3	142
Mazda	9	-98.4	553
Mitsubishi	5,843	110.5	2,776
Nissan	400	-78.1	1,828
Peugeot	1,647	54.8	1,064
Proton	1,544	65.3	934
Renault	11,344	26.1	8,999
Subaru	270	164.7	102
Suzuki	2,007	-10.8	2,250
Toyota	6,640	95.3	3,400
SEAT	–	–	*487*
Skoda	*4,793*	*32.3*	*3,622*
Volkswagen	*3,030*	*10.0*	*2,754*
Volkswagen Total	7,823	14.0	6,863
Volvo	854	95.9	436
Englon	*53*	*-64.9*	*151*
Geely	*404*	*-14.0*	*470*
Maple	*120*	*-4.0*	*125*
Zhejiang Geely Total	577	-22.7	746
Zotye	–	–	73
Total Cars	**144,123**	**8.2**	**133,165**

	2012	% Chg	2011
Trucks			
Changhe	289	189.0	100
Daihatsu	528	16.6	453
Daimler (Mercedes-Benz)	588	20.0	490
Fiat	33	-41.1	56
Ford	–	–	70
GM (Chevrolet)	25,559	27.9	19,982
Great Wall	–	–	5
Hyundai	465	32.1	352
Iveco	106	-19.7	132
Jianghuai Auto Co.	672	21.3	554
Jiangling	528	-37.9	850
Kia	–	–	3
Mahindra & Mahindra	57	-53.7	123
Mitsubishi	978	132.9	420
Nissan	–	–	1,919
Peugeot	81	189.3	28
Renault	–	–	39
Suzuki	1,124	50.3	748
Toyota	4,533	29.5	3,500
Volkswagen	463	432.2	87
Volvo Truck	–	–	52
Other	109	–	–
Total Trucks	**36,113**	**20.5**	**29,963**
Buses			
Brilliance	631	53.2	412
Changhe	578	–	50
Daimler (Mercedes-Benz)	355	-39.2	584
Ford	–	–	4
GM (Chevrolet)	2,655	42.1	1,868
Hafei	90	718.2	11
Hyundai	220	-9.8	244
Golden Dragon	*3,976*	–	–
King Long	*3,676*	*15.1*	*3,195*
King Long Total	7,652	139.5	3,195
Mitsubishi	372	-7.5	402
Nissan	–	–	397
Suzuki	4,150	34.3	3,090
Toyota	3,230	20.8	2,673
Volkswagen	45	-37.5	72
Volvo Truck	–	–	2
Other	38	52.0	25
Total Buses	**20,016**	**53.6**	**13,029**
Total Comm. Vehs.	**56,129**	**30.6**	**42,992**
Total Vehicles	**200,252**	**13.7**	**176,157**

ISRAEL

ECONOMIC SUMMARY

Common Language	Hebrew/Arabic
Currency	New Shekel
Steering	Left Hand Drive

2012 Vehicles in Operation

Passenger Cars	2,246,053
Commercial Vehicles	391,198
Total	2,637,251
Persons per Car	3.4
Persons per Vehicle	2.9

2012 GDP (billions of US $)

Nominal GDP	240.9
Real GDP	214.5
GDP % annual growth	3.1
GDP per capita	27,868

Current Account Balance (billions of US $)

2012:	-0.2
2011:	3.4

Current Account Balance in % of GDP

2012:	-0.1
2011:	1.4

Total Number of Fatal Road Accidents (most recent data)

2012:	263

Capital	Jerusalem
Population	7,707,042
Population Density	981.9
Surface Area sq km/miles	20,330 / 7,849

Road Infrastructure

Paved Road Surface %	100.0
Total Road Length in km/miles	18,470 / 11,477
Road Traffic (Vehicles per mile)	229.8

2012 Total Vehicle Percent Market Share by Company

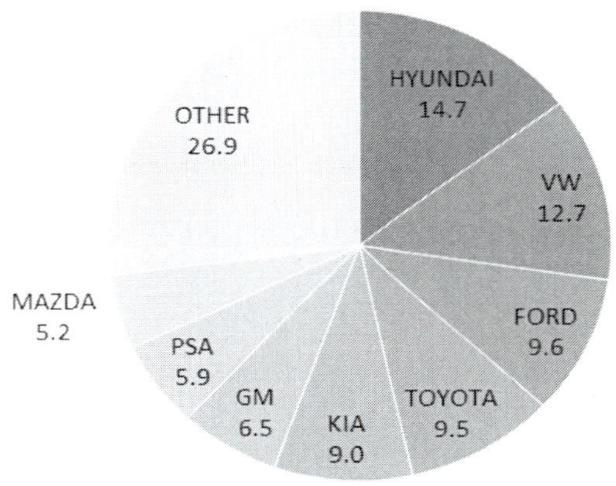

Total Vehicle Sales 2002-2012

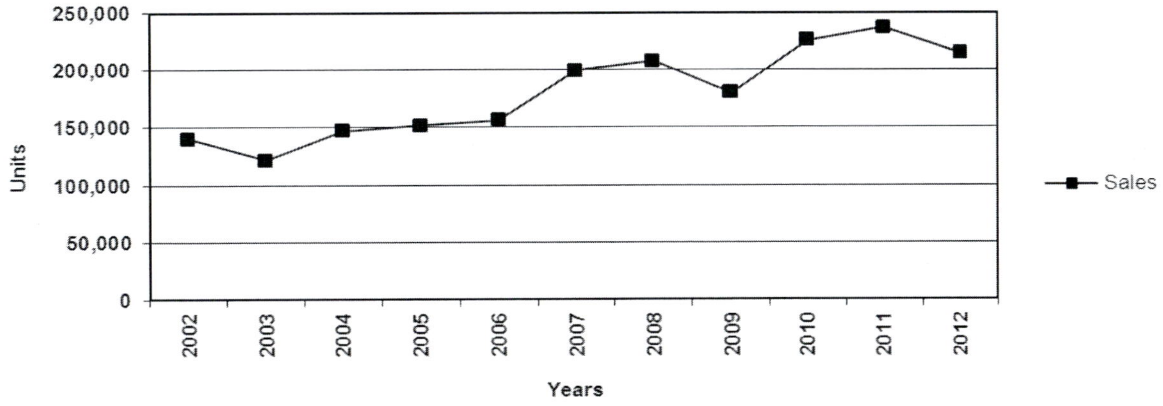

ISRAEL

SALES

Year	Cars	Commercial Vehicles	Total
2012	169,380	45,164	214,544
2011	188,904	48,021	236,925
2010	181,526	44,295	225,821
2009	145,364	34,846	180,210
2008	156,698	50,852	207,550
2007	157,221	42,050	199,271
2006	122,590	33,935	156,525
2005	119,667	31,776	151,443
2004	120,292	26,534	146,826
2003	94,084	27,738	121,822
2002	109,536	30,726	140,262
2001	130,094	37,810	167,904
2000	111,921	30,280	142,201
1999	90,935	24,792	115,727
1998	101,954	27,270	129,224
1997	102,263	26,431	128,694
1996	112,820	26,618	139,438
1995	106,264	28,723	134,987

ISRAEL

SALES BY COMPANY

	2012	% Chg.	2011
Cars			
Audi	1,195	1.3	1,180
BMW	*1,278*	*60.6*	*796*
Mini	*–*	*–*	*135*
BMW Total	1,278	37.3	931
Chrysler	338	463.3	60
Daihatsu	–	–	1,892
Mercedes-Benz	*574*	*-4.2*	*599*
Smart	*181*	*56.0*	*116*
Daimler Total	755	5.6	715
Alfa Romeo	*486*	*-25.0*	*648*
Fiat	*1,785*	*-32.1*	*2,627*
Lancia	*155*	*9.9*	*141*
Fiat Total	2,426	-29.0	3,416
Ford	18,042	101.2	8,966
Buick	*558*	*-44.2*	*1,000*
Cadillac	*205*	*-18.0*	*250*
Chevrolet	*6,181*	*-50.6*	*12,519*
Opel	*4,105*	*0.7*	*4,078*
GM Total	11,049	-38.1	17,847
Great Wall	2	-33.3	3
Honda	5,742	-14.6	6,722
Hyundai	29,199	-0.2	29,251
Jaguar	74	-24.5	98
Kia	15,779	27.4	12,381
Mazda	9,002	-57.4	21,145
Mitsubishi	1,150	-72.4	4,168
Nissan	7,167	-4.6	7,516
Citroen	*4,096*	*-23.1*	*5,326*
Peugeot	*4,988*	*-9.5*	*5,510*
Peugeot Citroen Total	9,084	-16.2	10,836
Porsche	17	-19.0	21
Renault	5,444	-20.7	6,864
Shanghai AIC	168	-48.6	327
Subaru	897	-64.4	2,519
Suzuki	9,282	-20.6	11,697
Lexus	*370*	*13.8*	*325*
Toyota	*18,320*	*2.1*	*17,941*
Toyota Total	18,690	2.3	18,266
SEAT	*5,887*	*33.0*	*4,425*
Skoda	*8,691*	*16.5*	*7,463*
Volkswagen	*7,506*	*-20.7*	*9,464*
Volkswagen Total	22,084	3.4	21,352
Volvo	516	-29.4	731
Total Cars	**169,380**	**-10.3**	**188,904**
Light Trucks			
Audi	746	29.5	576
BMW	401	31.5	305
Chrysler	1,187	14.8	1,034
Daihatsu	–	–	578
Daimler (Mercedes-Benz)	942	-4.2	983
Fiat	1,802	27.8	1,410
Ford	2,330	-1.6	2,369
Chevrolet	*1,734*	*19.4*	*1,452*
Opel	*26*	*–*	*–*
GM Total	1,760	21.2	1,452
Great Wall	203	395.1	41
Honda	370	-32.6	549
Hyundai	2,326	-38.9	3,809
Isuzu	1,243	-14.9	1,460
Iveco	5	-89.1	46
Kia	3,447	6.9	3,225
Land Rover	207	5.1	197
Mazda	2,090	–	–
Mitsubishi	1,402	-49.6	2,782

ISRAEL

SALES BY COMPANY

	2012	% Chg.	2011
Infiniti	*259*	*-14.2*	*302*
Nissan	*2,645*	*5.2*	*2,514*
Nissan Total	2,904	3.1	2,816
Citroen	*2,023*	*-9.6*	*2,239*
Peugeot	*1,400*	*-46.0*	*2,594*
Peugeot Citroen Total	3,423	-29.2	4,833
Piaggio	94	-17.5	114
Porsche	29	-32.6	43
Renault	1,201	-37.5	1,921
Ssangyong	207	-11.2	233
Subaru	849	22.7	692
Suzuki	258	-43.7	458
Lexus	*110*	*3.8*	*106*
Toyota	*1,680*	*-5.5*	*1,778*
Toyota Total	1,790	-5.0	1,884
Skoda	*2,585*	*47.0*	*1,758*
Volkswagen	*2,532*	*6.1*	*2,387*
Volkswagen Total	5,117	23.4	4,145
Volvo	196	-26.6	267
Volvo Truck (Renault Truck)	76	533.3	12
Other	2	–	–
Total Light Trucks	**36,607**	**-4.3**	**38,234**
Medium/Heavy Trucks			
BMC	–	–	1
Chrysler	251	16.2	216
Mercedes-Benz	*995*	*2.7*	*969*
Mitsubishi Fuso	*142*	*-23.7*	*186*
Daimler Total	1,137	-1.6	1,155
Fiat	670	15.1	582
Ford	149	-39.9	248
Chevrolet	*1,178*	*-30.7*	*1,701*
GM Total	1,178	-30.7	1,701
Hino	76	–	–

	2012	% Chg.	2011
Hyundai	1	–	–
International	3	–	–
Isuzu	591	5.5	560
Iveco	355	-21.3	451
MAN	*437*	*6.8*	*409*
Volkswagen Truck & Bus	*200*	*19.0*	*168*
MAN Total	637	10.4	577
DAF	*877*	*-2.8*	*902*
Kenworth	*21*	–	–
PACCAR Total	898	-0.4	902
Peugeot	190	-26.9	260
Scania	343	-4.5	359
Renault Truck	*420*	*21.4*	*346*
UD Trucks	*160*	*-14.4*	*187*
Volvo Truck	*842*	*-17.6*	*1,022*
Volvo Truck Total	1,422	-8.6	1,555
Other	7	-92.9	98
Total Med./Hvy. Trks.	**7,908**	**-8.7**	**8,665**
Buses			
BMC	20	-81.0	105
Daimler (Mercedes-Benz)	72	-22.6	93
Iveco	7	-87.7	57
MAN	142	-60.0	355
PACCAR (DAF)	–	–	37
Scania	131	-22.0	168
VDL	103	–	–
Volvo Truck	174	-34.8	267
Other	–	–	40
Total Buses	**649**	**-42.2**	**1,122**
Total Comm. Vehs.	**45,164**	**-5.9**	**48,021**
Total Vehicles	**214,544**	**-9.4**	**236,925**

SOUTH AFRICA

ECONOMIC SUMMARY

Common Language	IsiZulu
Currency	Rand
Steering	Right Hand Drive

2012 Vehicles in Operation

Passenger Cars	6,112,000
Commercial Vehicles	2,832,000
Total	8,944,000
Persons per Car	8.0
Persons per Vehicle	5.4

2012 GDP (billions of US $)

Nominal GDP	384.3
Real GDP	230.3
GDP % annual growth	2.5
GDP per capita	4,497

Current Account Balance (billions of US $)

2012: -24.1
2011: -13.6

Current Account Balance in % of GDP

2012: -6.3
2011: -3.4

Total Number of Fatal Road Accidents (most recent data)

2012: 14,140

Capital	Pretoria
Population	48,601,098
Population Density	103.6
Surface Area sq km/miles	1,214,470 / 468,909

Road Infrastructure

Paved Road Surface %	20.3
Total Road Length in km/miles	362,099 / 224,998
Road Traffic (Vehicles per mile)	39.8

2012 Total Vehicle Percent Market Share by Company

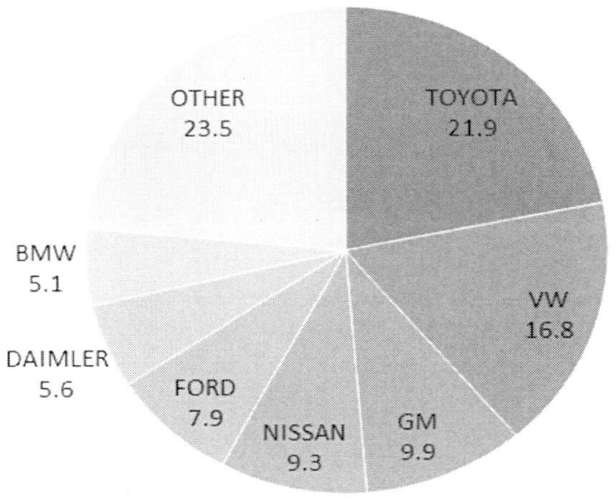

Total Vehicle Production and Sales 2002-2012

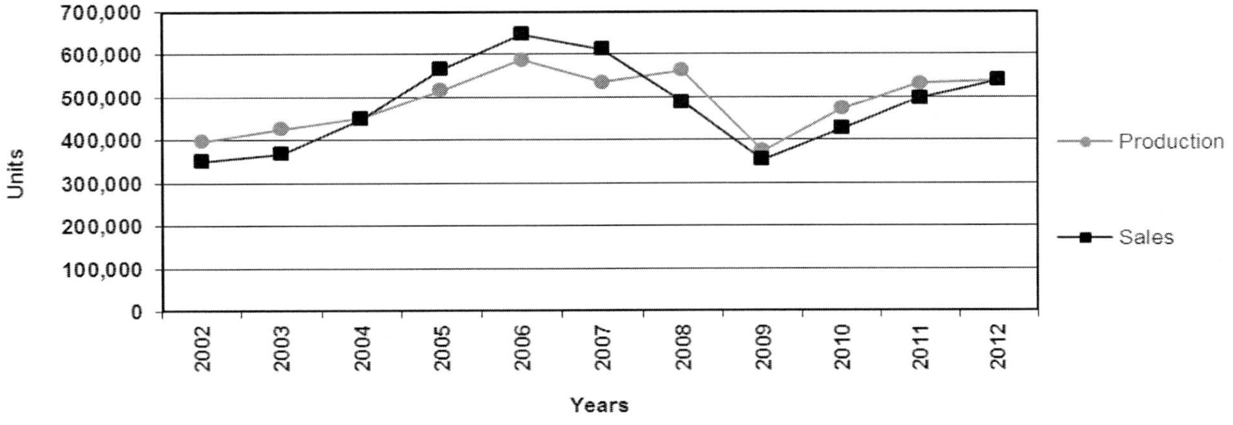

SOUTH AFRICA

PRODUCTION

Year	Cars	Commercial Vehicles	Total
2012	274,873	264,665	539,538
2011	312,265	220,280	532,545
2010	295,394	176,655	472,049
2009	222,981	150,942	373,923
2008	321,124	241,841	562,965
2007	276,018	258,472	534,490
2006	334,482	253,237	587,719
2005	324,613	191,855	516,468
2004	290,061	161,216	451,277
2003	305,064	120,746	425,810
2002	282,774	114,352	397,126
2001	273,495	118,980	392,475
2000	217,661	108,356	326,017
1999	212,291	113,774	326,065
1998	193,212	118,843	312,055
1997	226,242	135,862	362,104
1996	235,359	150,952	386,311
1995	242,488	146,988	389,476
1994	191,979	111,144	303,123
1993	195,032	103,019	298,051
1992	183,662	100,297	283,959
1991	197,750	110,325	308,075
1990	209,636	125,143	334,779
1989	221,554	130,943	352,497
1988	230,629	129,102	359,731
1987	200,906	108,244	309,150
1986	174,576	90,100	264,676
1985	204,322	101,005	305,327
1984	268,751	137,059	405,810
1983	272,822	132,317	405,139
1982	283,433	142,690	426,123

SOUTH AFRICA

PRODUCTION BY MANUFACTURER

	2012	% Chg.	2011
Cars			
BMW	43,161	-18.4	52,908
Ford	15,004	0.8	14,887
GM	22,724	27.8	17,786
Mercedes-Benz	61,301	14.8	53,375
Nissan	2,305	-21.9	2,950
Toyota	26,646	-21.2	33,800
Volkswagen	103,732	-24.0	136,559
Total Cars	**274,873**	**-12.0**	**312,265**
Commercial Vehicles			
DAF	665	31.4	506
Fiat	*1,650*	*32.0*	*1,250*
Iveco	*570*	*34.1*	*425*
Fiat Total	2,220	32.5	1,675
Ford	35,438	156.3	13,829
GM	19,128	7.5	17,786
Isuzu	16,013	12.9	14,180
MAN	2,126	16.5	1,825
Mazda	564	-90.3	5,844
Mercedes-Benz	1,850	8.2	1,710
Mitsubishi	4	-99.7	1,558
Nissan	52,154	28.3	40,651
Scania	640	24.3	515
Toyota	132,783	11.3	119,252
Volvo Truck	1,080	13.8	949
Total Comm. Vehs.	**264,665**	**20.1**	**220,280**
Total Vehicles	**539,538**	**1.3**	**532,545**

SOUTH AFRICA

SALES

Year	Cars	Commercial Vehicles	Total
2012	364,724	174,635	539,359
2011	330,703	166,983	497,686
2010	279,081	146,656	425,737
2009	224,705	129,056	353,761
2008	295,064	193,947	489,011
2007	384,431	228,277	612,708
2006	426,812	219,744	646,556
2005	376,845	188,129	564,974
2004	301,151	148,443	449,594
2003	247,259	121,201	368,460
2002	231,602	118,452	350,054
2001	239,060	127,839	366,899
2000	224,122	116,960	341,082
1999	189,370	106,069	295,439
1998	203,640	110,721	314,361
1997	239,691	127,138	366,829
1996	249,838	143,142	392,980
1995	236,584	140,200	376,784
1994	191,979	111,144	303,123
1993	195,032	103,019	298,051
1992	183,662	100,297	283,959
1991	197,750	110,325	308,075
1990	209,636	125,143	334,779
1989	221,554	130,943	352,497
1988	230,629	129,102	359,731
1987	200,906	108,244	309,150
1986	174,576	90,100	264,676
1985	204,322	101,005	305,327
1984	268,751	137,059	405,810
1983	272,822	132,317	405,139
1982	283,433	142,690	426,123

SOUTH AFRICA

SALES BY COMPANY

	2012	% Chg.	2011
Cars			
Audi	16,743	15.2	14,531
BMW	*24,744*	*5.0*	*23,560*
Mini	*2,794*	*11.4*	*2,509*
BMW Total	27,538	5.6	26,069
Chana	58	-83.3	348
Chrysler	*584*	*-1.0*	*590*
Dodge	*1,765*	*-15.4*	*2,086*
Jeep	*7,237*	*65.1*	*4,383*
Chrysler Total	9,586	35.8	7,059
Dacia	36	-95.2	744
Mercedes-Benz	*22,438*	*-4.6*	*23,509*
Smart	*122*	*-1.6*	*124*
Daimler Total	22,560	-4.5	23,633
Alfa Romeo	*944*	*16.3*	*812*
Fiat	*3,305*	*24.1*	*2,663*
Fiat Total	4,249	22.3	3,475
Ford	25,891	-3.0	26,679
Cadillac	*–*	*–*	*2*
Chevrolet	*31,175*	*4.8*	*29,754*
Hummer	*8*	*60.0*	*5*
Opel	*2,704*	*-38.1*	*4,365*
GM Total	33,887	-0.7	34,126
Honda	9,093	16.2	7,828
Jaguar	*909*	*74.8*	*520*
Land Rover	*6,686*	*30.3*	*5,133*
Jaguar Land Rover Total	7,595	34.4	5,653
Mahindra & Mahindra	1,447	220.8	451
Maserati	65	-9.7	72
Mazda	4,827	0.8	4,790
Mitsubishi	2,619	51.0	1,734
Infiniti	*127*	*–*	*–*
Nissan	*20,627*	*20.9*	*17,065*
Nissan Total	20,754	21.6	17,065
Citroen	*1,945*	*0.2*	*1,941*
Peugeot	*3,393*	*20.6*	*2,813*
Peugeot Citroen Total	5,338	12.3	4,754
Porsche	1,435	26.9	1,131
Renault	10,180	6.0	9,603
Ssangyong	80	–	–
Subaru	861	-31.1	1,250
Suzuki	4,724	-13.5	5,462
Tata	2,795	46.4	1,909
Lexus	*1,371*	*16.4*	*1,178*
Toyota	*65,645*	*25.2*	*52,424*
Toyota Total	67,016	25.0	53,602
Volkswagen	82,363	9.1	75,495
Volvo	2,984	-7.9	3,240
Total Cars	**364,724**	**10.3**	**330,703**
Light Trucks			
Chana	191	-54.5	420
Daimler (Mercedes-Benz)	535	7.6	497
Fiat	448	-12.5	512
Ford	16,545	12.1	14,764
Chevrolet	*19,324*	*7.4*	*17,996*
Opel	*231*	*-24.8*	*307*
GM Total	19,555	6.8	18,303
Isuzu	13,393	1.2	13,228
Land Rover	776	23.6	628
Mahindra & Mahindra	2,294	73.7	1,321
Mazda	1,961	-48.0	3,772
Mitsubishi	958	-38.5	1,558
Nissan	29,173	27.8	22,827
Citroen	*182*	*119.3*	*83*
Peugeot	*228*	*20.6*	*189*
Peugeot Citroen Total	410	50.7	272
Renault	339	133.8	145
Ssangyong	40	–	–
Tata	1,598	27.4	1,254
Toyota	50,629	-3.4	52,423
Volkswagen	8,505	-2.9	8,759
Total Light Trucks	**147,350**	**4.7**	**140,683**
Medium/Heavy Trucks			
Freightliner	*1,312*	*–*	*1,312*
Mercedes-Benz	*4,779*	*4.5*	*4,575*
Mitsubishi Fuso	*944*	*-4.3*	*986*

SOUTH AFRICA

SALES BY COMPANY

	2012	% Chg.	2011
Daimler Total	7,035	2.4	6,873
Fiat	145	68.6	86
Hino	3,295	6.2	3,103
International	576	-32.2	849
Isuzu	3,105	13.2	2,744
Iveco	1,079	28.9	837
MAN	*1,435*	*-13.2*	*1,654*
Volkswagen Truck & Bus	*1,033*	*14.7*	*901*
MAN Total	2,468	-3.4	2,555
PACCAR (DAF)	86	-34.8	132
Citroen	*11*	*266.7*	*3*
Peugeot	*185*	*-17.4*	*224*
Peugeot Citroen Total	196	-13.7	227
Scania	1,089	-6.3	1,162
Tata	1,463	14.6	1,277
Toyota	336	17.5	286
Renault Truck	*305*	*2.3*	*298*
UD Trucks	*2,992*	*-7.5*	*3,234*
Volvo Truck	*1,542*	*4.9*	*1,470*
Volvo Truck Total	4,839	-3.3	5,002
Other	484	167.4	181
Total Med./Hvy. Trks.	**26,196**	**3.5**	**25,314**
Buses			
Daimler (Mercedes-Benz)	209	3.0	203
Isuzu	17	6.3	16
Iveco	–	–	76
MAN	457	4.1	439
Scania	243	85.5	131
Tata	57	35.7	42
VDL	33	37.5	24
Volvo Truck	73	32.7	55
Total Buses	**1,089**	**10.4**	**986**
Total Comm. Vehs.	**174,635**	**4.6**	**166,983**
Total Vehicles	**539,359**	**8.4**	**497,686**

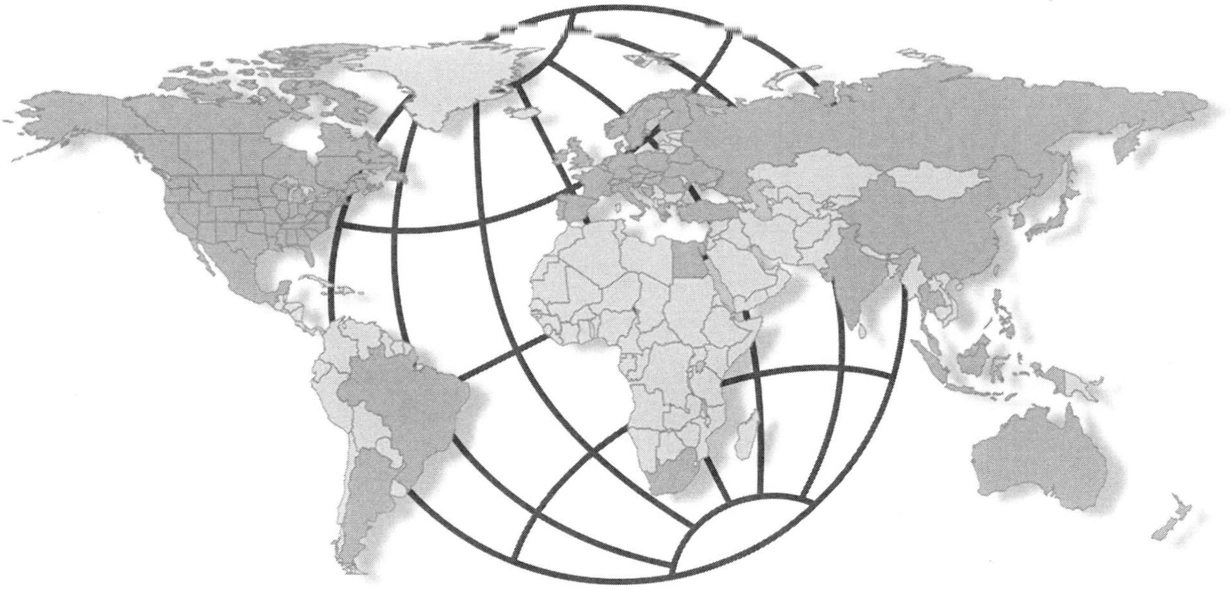

World Summary

WORLD

PRODUCTION BY VEHICLE TYPE

Year	Cars	Commercial Vehicles	Total
2012	55,247,580	28,333,054	83,580,634
2011	52,231,386	26,563,447	78,794,833
2010	51,229,399	25,275,673	76,505,072
2009	42,525,214	18,013,355	60,538,569
2008	48,870,515	20,839,008	69,709,523
2007	50,490,507	21,892,714	72,383,221
2006	46,577,235	21,983,975	68,561,210
2005	44,173,067	21,735,769	65,908,836
2004	42,488,957	21,246,548	63,735,505
2003	40,656,604	19,825,232	60,481,836
2002	40,864,405	17,842,377	58,706,782
2001	40,150,342	16,803,139	56,953,481
2000	41,229,160	17,716,514	58,945,674
1999	39,793,444	16,517,766	56,311,210
1998	37,925,000	15,062,000	52,987,000
1997	38,453,000	15,981,000	54,434,000
1996	37,196,868	14,135,097	51,331,965
1995	36,070,056	13,912,763	49,982,819
1994	35,638,599	13,861,569	49,500,168
1993	34,197,046	12,588,334	46,785,380
1992	35,487,740	12,600,310	48,088,050
1991	35,080,275	11,848,142	46,928,417
1990	36,273,082	12,280,887	48,553,969
1989	35,698,786	13,549,024	49,247,810
1988	34,396,619	13,962,867	48,359,486
1987	33,114,256	12,927,384	46,041,640
1986	32,937,324	12,472,954	45,410,278
1985	32,353,081	12,555,926	44,909,007
1984	30,533,439	11,615,852	42,149,291
1983	30,009,006	9,823,128	39,832,134
1982	26,657,855	9,513,196	36,171,051

CAR PRODUCTION BY MANUFACTURER GROUP AND COUNTRY, 2012

Manufacturer	Australia	China	India	Indo-nesia	Japan	Malaysia	Pakistan	S. Korea	Taiwan	Thailand	Total Asia Pacific	Austria	Bela[illegible]s	Belgium	Czech Republic
AutoVaz	–	–	–	–	–	–	–	–	–	–	–	–	–	–	–
AvtoFramos	–	–	–	–	–	–	–	–	–	–	–	–	–	–	–
Avtotor	–	–	–	–	–	–	–	–	–	–	–	–	–	–	–
Beijing AIC	–	777,032	–	–	–	–	–	–	–	–	777,032	–	–	–	–
BMW	–	–	6,140	1,682	–	4,918	–	–	–	6,065	18,805	103,545	–	–	–
Brilliance China	–	247,213	–	–	–	–	–	–	–	–	247,213	–	–	–	–
BYD	–	366,201	–	–	–	–	–	–	–	–	366,201	–	–	–	–
Chana	–	891,053	–	–	–	–	–	–	–	–	891,053	–	–	–	–
Chery	–	408,265	–	–	–	–	–	–	–	–	408,265	–	–	–	–
China Motor	–	101,879	–	–	–	–	–	–	21,289	–	123,168	–	–	–	–
Chrysler	–	–	–	–	–	–	–	–	–	–	–	–	–	–	–
Daimler	–	–	6,113	2,922	–	5,163	–	–	–	5,223	19,421	9,850	–	–	–
Dongfeng	–	1,620,190	–	–	–	–	–	–	–	–	1,620,190	–	–	–	–
FAW	–	2,064,381	–	–	–	–	–	–	–	–	2,064,381	–	–	–	–
Fiat	–	–	9,992	–	–	–	–	–	–	–	9,992	–	–	–	–
Ford	15,035	–	115,297	–	–	–	–	–	22,927	134,404	287,663	–	–	126,791	–
GM	73,490	–	70,320	–	–	–	–	568,776	–	41,712	754,298	–	–	–	–
Great Wall Zhongxing	–	181,870	–	–	–	–	–	–	–	–	181,870	–	–	–	–
Guangzhou AC	–	457,068	–	–	–	–	–	–	–	–	457,068	–	–	–	–
Honda	–	28,730	75,201	55,828	996,832	26,177	14,835	–	18,532	222,263	1,438,398	–	–	–	–
Hyundai Group	–	–	637,960	3,540	–	11,425	–	2,194,476	12,191	–	2,859,592	–	–	–	303,035
Isuzu	–	–	–	5,100	–	–	–	–	–	–	5,100	–	–	–	–
Jianghuai Auto Co	–	137,300	–	–	–	–	–	–	–	–	137,300	–	–	–	–
Lifan Motor Co	–	157,894	–	–	–	–	–	–	–	–	157,894	–	–	–	–
Mahindra & Mahindra	–	–	16,468	–	–	–	–	–	–	–	16,468	–	–	–	–
Mazda	–	–	–	–	830,294	2,599	–	–	–	–	832,893	–	–	–	–
Mitsubishi	–	–	–	6,327	448,598	–	–	–	–	84,134	539,059	–	–	–	–
Nissan	–	–	152,707	65,900	1,035,726	15,070	–	–	32,506	168,721	1,470,630	–	–	–	–
Perodua	–	–	–	–	–	152,995	–	–	–	–	152,995	–	–	–	–
Peugeot Citroen	–	–	–	–	–	2,036	–	–	–	–	2,036	9,828	–	–	–
Proton	–	–	–	–	–	130,701	–	–	–	–	130,701	–	–	–	–
Renault	–	–	–	–	–	–	–	100,013	–	–	100,013	–	–	–	–
Shanghai AIC	–	2,633,226	–	–	–	–	–	–	–	–	2,633,226	–	–	–	–
Subaru	–	–	–	–	551,812	–	–	–	–	–	551,812	–	–	–	–
Suzuki	–	–	996,450	61,655	886,781	5,750	62,397	–	–	17,684	2,030,717	–	–	–	–
Tata	–	–	229,974	–	–	–	–	–	–	–	229,974	–	–	–	–
Toyota	101,424	–	98,868	539,036	3,804,176	38,904	40,466	–	157,084	277,417	5,057,375	–	–	–	214,915
Volkswagen	–	–	102,064	784	–	1,662	–	–	–	–	104,510	–	–	123,111	653,813
Volvo Truck	–	–	–	–	–	–	–	–	–	–	–	–	–	–	–
Zhejiang Geely	–	462,525	–	727	–	2,513	–	–	–	–	465,765	–	–	257,302	–
Other	–	170,135	2,707	–	–	–	–	4,577	9,429	–	186,848	379	[illegible]70	–	11
TOTAL CAR 2012	**189,949**	**10,704,962**	**2,520,261**	**743,501**	**8,554,219**	**399,913**	**117,698**	**2,867,842**	**273,958**	**957,623**	**27,329,926**	**123,602**	**[illegible]70**	**507,204**	**1,171,774**
TOTAL CAR 2011	**189,503**	**10,052,854**	**2,479,131**	**561,863**	**7,158,525**	**384,935**	**124,267**	**2,901,707**	**280,641**	**537,987**	**24,671,413**	**130,343**	**[illegible]30**	**562,191**	**1,191,968**
Vol. % Chg.	**0.2**	**6.5**	**1.7**	**32.3**	**19.5**	**3.9**	**-5.3**	**-1.2**	**-2.4**	**78.0**	**10.8**	**-5.2**	**[illegible]7.4**	**-9.8**	**-1.7**

Note: Total World represents countries for which data is presented.
Source: WardsAuto InfoBank.

WORLD

CAR PRODUCTION BY MANUFACTURER GROUP AND COUNTRY, 2012

Manufacturer	France	Germany	Hungary	Italy	Nether-lands	Poland	Portugal	Romania	Russia	Serbia	Slovakia	Slovenia	Spain	Sweden	Turkey	Ukraine
AutoVaz	–	–	–	–	–	–	–	–	607,369	–	–	–	–	–	–	–
AvtoFramos	–	–	–	–	–	–	–	–	168,099	–	–	–	–	–	–	–
Avtotor	–	–	–	–	–	–	–	–	249,635	–	–	–	–	–	–	–
Beijing AIC	–	–	–	–	–	–	–	–	–	–	–	–	–	–	–	–
BMW	–	1,043,732	–	–	–	–	–	–	–	–	–	–	–	–	–	–
Brilliance China	–	–	–	–	–	–	–	–	–	–	–	–	–	–	–	–
BYD	–	–	–	–	–	–	–	–	–	–	–	–	–	–	–	–
Chana	–	–	–	–	–	–	–	–	–	–	–	–	–	–	–	–
Chery	–	–	–	–	–	–	–	–	–	–	–	–	–	–	–	–
China Motor	–	–	–	–	–	–	–	–	–	–	–	–	–	–	–	–
Chrysler	–	–	–	–	–	–	–	–	–	–	–	–	–	–	–	–
Daimler	104,154	963,633	41,035	–	–	–	–	–	–	–	–	–	19,290	–	–	–
Dongfeng	–	–	–	–	–	–	–	–	–	–	–	–	–	–	–	–
FAW	–	–	–	–	–	–	–	–	–	–	–	–	–	–	–	–
Fiat	–	–	–	394,620	–	330,954	–	–	–	25,281	–	–	–	–	81,695	–
Ford	–	685,358	–	–	–	–	–	30,591	107,724	–	–	–	149,745	–	–	–
GM	–	387,525	–	–	–	125,469	–	–	169,623	–	–	–	264,858	–	–	–
Great Wall Zhongxing	–	–	–	–	–	–	–	–	23,904	–	–	–	–	–	–	–
Guangzhou AC	–	–	–	–	–	–	–	–	–	–	–	–	–	–	–	–
Honda	–	–	–	–	–	–	–	–	–	–	–	–	–	–	24,378	–
Hyundai Group	–	–	–	–	–	–	–	–	58,400	–	292,000	–	–	–	90,857	–
Isuzu	–	–	–	–	–	–	–	–	–	–	–	–	–	–	–	–
Jianghuai Auto Co	–	–	–	–	–	–	–	–	–	–	–	–	–	–	–	–
Lifan Motor Co	–	–	–	–	–	–	–	–	–	–	–	–	–	–	–	–
Mahindra & Mahindra	–	–	–	–	–	–	–	–	–	–	–	–	–	–	–	–
Mazda	–	–	–	–	–	–	–	–	–	–	–	–	–	–	–	–
Mitsubishi	–	–	–	–	–	–	–	–	–	–	–	–	–	–	–	–
Nissan	–	–	–	–	–	–	–	–	126,058	–	–	–	55,469	–	–	–
Perodua	–	–	–	–	–	–	–	–	–	–	–	–	–	–	–	–
Peugeot Citroen	1,017,436	–	–	–	–	–	3,185	–	40,258	–	215,000	–	176,500	–	–	–
Proton	–	–	–	–	–	–	–	–	–	–	–	–	–	–	–	–
Renault	359,532	–	–	–	–	–	–	295,965	–	–	–	126,836	264,698	–	292,066	–
Shanghai AIC	–	–	–	–	–	–	–	–	–	–	–	–	–	–	–	–
Subaru	–	–	–	–	–	–	–	–	–	–	–	–	–	–	–	–
Suzuki	–	–	140,852	–	–	–	–	–	–	–	–	–	–	–	–	–
Tata	–	–	–	–	–	–	–	–	–	–	–	–	–	–	–	–
Toyota	201,692	–	–	–	–	–	–	–	98,229	–	–	–	–	–	88,300	–
Volkswagen	–	2,158,444	33,553	2,197	–	77,495	112,550	–	179,554	–	337,110	–	664,513	–	–	14,556
Volvo Truck	–	–	–	–	–	–	–	–	–	–	–	–	–	–	–	–
Zhejiang Geely	–	–	–	–	–	–	–	–	–	–	–	–	–	162,814	–	–
Other	–	149,767	–	–	–	–	–	–	139,935	–	–	–	–	–	–	55,131
TOTAL CAR 2012	**1,682,814**	**5,388,459**	**215,440**	**396,817**	**–**	**533,918**	**115,735**	**326,556**	**1,968,788**	**25,281**	**844,110**	**126,836**	**1,595,073**	**162,814**	**577,296**	**69,687**
TOTAL CAR 2011	**1,931,030**	**5,871,918**	**211,218**	**485,606**	**40,772**	**722,285**	**141,779**	**310,243**	**1,738,162**	**10,227**	**573,660**	**168,955**	**1,867,657**	**188,969**	**639,734**	**97,585**
Vol. % Chg.	**-12.9**	**-8.2**	**2.0**	**-18.3**	**–**	**-26.1**	**-18.4**	**5.3**	**13.3**	**147.2**	**47.1**	**-24.9**	**-14.6**	**-13.8**	**-9.8**	**-28.6**

Note: Total World represents countries for which data is presented.

Source: WardsAuto InfoBank.

CAR PRODUCTION BY MANUFACTURER GROUP AND COUNTRY, 2012

Manufacturer	U.K.	Uzbek-istan	Total Europe	Canada	Mexico	U.S.	Total N. America	Argen-tina	Brazil	Vene-zuela	Total S. America	Egypt	South Africa	2012	2011	Vol. % Chg.
AutoVaz	–	–	607,369	–	–	–	–	–	–	–	–	1,214	–	608,583	570,134	6.7
AvtoFramos	–	–	168,099	–	–	–	–	–	–	–	–	–	–	168,099	142,304	18.1
Avtotor	–	–	249,635	–	–	–	–	–	–	–	–	–	–	249,635	222,081	12.4
Beijing AIC	–	–	–	–	–	–	–	–	–	–	–	–	–	777,032	671,327	15.7
BMW	210,220	–	1,357,497	–	–	–	–	–	–	–	–	1,290	43,161	1,420,753	1,373,239	3.5
Brilliance China	–	–	–	–	–	–	–	–	–	–	–	–	–	247,213	221,327	11.7
BYD	–	–	–	–	–	–	–	–	–	–	–	–	–	366,201	386,777	-5.3
Chana	–	–	–	–	–	–	–	–	–	–	–	–	–	891,053	832,315	7.1
Chery	–	–	–	–	–	–	–	–	–	–	–	–	–	408,265	459,392	-11.1
China Motor	–	–	–	–	–	–	–	–	–	–	–	–	–	123,168	126,613	-2.7
Chrysler	–	–	–	240,193	81,874	334,825	656,892	–	–	952	952	604	–	658,448	504,501	30.5
Daimler	–	–	1,137,962	–	–	–	–	–	–	–	–	716	61,301	1,219,400	1,207,335	1.0
Dongfeng	–	–	–	–	–	–	–	–	–	–	–	–	–	1,620,190	1,511,571	7.2
FAW	–	–	–	–	–	–	–	–	–	–	–	–	–	2,064,381	1,940,406	6.4
Fiat	–	–	832,550	–	–	–	–	73,863	648,520	–	722,383	–	–	1,564,925	1,763,019	-11.2
Ford	–	–	1,100,209	–	444,425	552,911	997,336	48,492	209,426	7,623	265,541	–	15,004	2,665,753	2,732,926	-2.5
GM	90,187	210,921	1,248,583	319,556	109,997	840,700	1,270,253	128,501	478,683	28,310	635,494	16,195	22,724	3,947,547	4,127,597	-4.4
Great Wall Zhongxing	–	–	23,904	–	–	–	–	–	–	–	–	–	–	205,774	174,231	18.1
Guangzhou AC	–	–	–	–	–	–	–	–	–	–	–	–	–	457,068	514,966	-11.2
Honda	165,630	–	190,008	226,390	–	639,108	865,498	8,114	135,058	–	143,172	–	–	2,637,076	1,690,546	56.0
Hyundai Group	–	–	744,292	–	–	489,884	489,884	–	26,003	3,872	29,875	9,428	–	4,133,071	3,746,558	10.3
Isuzu	–	–	–	–	–	–	–	–	–	–	–	–	–	5,100	5,310	-4.0
Jianghuai Auto Co	–	–	–	–	–	–	–	–	–	–	–	–	–	37,300	146,405	-6.2
Lifan Motor Co	–	–	–	–	–	–	–	–	–	–	–	–	–	57,894	100,701	56.8
Mahindra & Mahindra	–	–	–	–	–	–	–	–	–	–	–	–	–	16,468	16,305	1.0
Mazda	–	–	–	–	–	–	–	–	–	–	–	–	–	32,893	799,640	4.2
Mitsubishi	–	–	–	–	–	13,716	13,716	–	–	6,541	6,541	–	–	59,316	617,607	-9.4
Nissan	510,572	–	692,099	–	569,203	406,782	975,985	–	14,364	–	14,364	–	2,305	3,155,383	2,890,889	9.1
Perodua	–	–	–	–	–	–	–	–	–	–	–	–	–	52,995	153,683	-0.4
Peugeot Citroen	–	–	1,462,207	–	–	–	–	115,645	92,943	–	208,588	377	–	1,673,208	1,953,986	-14.4
Proton	1,043	–	1,043	–	–	–	–	–	–	–	–	–	–	31,744	144,043	-8.5
Renault	–	–	1,339,097	–	–	–	–	85,159	172,233	–	257,392	–	–	1,696,502	2,036,010	-16.7
Shanghai AIC	–	–	–	–	–	–	–	–	–	–	–	–	–	2,633,226	2,391,942	10.1
Subaru	–	–	–	–	–	143,529	143,529	–	–	–	–	–	–	695,341	492,256	41.3
Suzuki	–	–	140,852	–	–	–	–	–	–	–	–	101	–	2,171,670	2,061,398	5.3
Tata	361,383	–	361,383	–	–	–	–	–	–	–	–	–	–	591,357	565,323	4.6
Toyota	109,429	–	712,565	254,159	–	531,876	786,035	–	66,332	6,499	72,831	–	26,646	6,655,452	5,227,961	27.3
Volkswagen	9,108	–	4,366,004	–	604,508	152,543	757,051	37,602	747,782	–	785,384	–	103,732	6,316,681	6,213,548	-1.6
Volvo Truck	–	–	–	–	–	–	–	–	–	–	–	–	–	–	–	–
Zhejiang Geely	–	–	420,116	–	–	–	–	–	–	–	–	–	–	885,881	880,634	0.6
Other	6,238	–	351,731	–	–	–	–	–	–	–	–	6,955	–	545,534	614,580	-11.2
TOTAL CAR 2012	**1,463,810**	**210,921**	**17,507,205**	**1,040,298**	**1,810,007**	**4,105,874**	**6,956,179**	**497,376**	**2,591,344**	**53,797**	**3,142,517**	**36,880**	**274,873**	**55,247,580**	**52,231,386**	**5.8**
TOTAL CAR 2011	**1,343,794**	**201,251**	**18,429,577**	**990,482**	**1,657,080**	**2,976,991**	**5,624,553**	**577,233**	**2,505,241**	**58,032**	**3,140,506**	**53,072**	**312,265**	**52,231,386**		
Vol. % Chg.	**8.9**	**4.8**	**-5.0**	**5.0**	**9.2**	**37.9**	**23.7**	**-13.8**	**3.4**	**-7.3**	**0.1**	**-30.5**	**-12.0**	**5.8**		

Note: Total World represents countries for which data is presented.

Source: WardsAuto InfoBank.

COMMERCIAL VEHICLE PRODUCTION BY MANUFACTURER GROUP AND COUNTRY, 2012

Manufacturer	Australia	China	India	Indo-nesia	Japan	Malaysia	Pakistan	S. Korea	Taiwan	Thailand	Total Asia Pacific	Austria	Belarus	Belgium	Czech Republic
AutoVaz	–	–	–	–	–	–	–	–	–	–	–	–	–	–	–
AvtoFramos	–	–	–	–	–	–	–	–	–	–	–	–	–	–	–
Avtotor	–	–	–	–	–	–	–	–	–	–	–	–	–	–	–
Beijing AIC	–	895,942	–	–	–	377	–	–	–	–	896,319	–	–	–	–
BMW	–	–	2,030	–	–	310	–	–	–	–	2,340	–	–	–	–
Brilliance China	–	392,609	–	–	–	453	–	–	–	–	393,062	–	–	–	–
BYD	–	89,505	–	–	–	–	–	–	–	–	89,505	–	–	–	–
Chana	–	901,766	–	–	–	–	–	–	–	–	901,766	–	–	–	–
Chery	–	155,686	–	–	–	568	–	–	–	–	156,254	–	–	–	–
China Motor	–	12,655	–	–	–	–	–	–	32,418	–	45,073	–	–	–	–
Chrysler	–	–	–	–	–	–	–	–	–	–	–	–	–	–	–
Daimler	–	22,500	–	8,225	87,086	2,779	–	–	–	6,415	127,005	–	–	–	–
Dongfeng	–	1,373,990	–	–	–	–	–	–	–	–	1,373,990	–	–	–	–
FAW	–	643,393	–	–	–	–	–	–	–	–	643,393	–	–	–	–
Fiat	–	–	–	–	–	–	–	–	–	–	–	–	–	–	2,760
Ford	22,667	–	1,609	–	–	–	–	–	8,272	134,272	166,820	–	–	–	–
GM	8,608	–	20,892	–	–	–	–	216,981	–	82,897	329,378	–	–	–	–
Great Wall Zhongxing	–	442,556	–	–	–	–	–	–	–	–	442,556	–	–	–	–
Guangzhou AC	–	180,855	–	–	–	–	–	–	–	–	180,855	–	–	–	–
Honda	–	–	–	–	32,481	3,241	–	–	–	–	35,722	–	–	–	–
Hyundai Group	–	–	805	1,335	–	5,446	–	1,296,470	3,165	–	1,307,221	–	–	–	–
Isuzu	–	–	–	26,545	241,247	9,895	450	–	–	287,509	565,646	–	–	–	–
Jianghuai Auto Co	–	339,056	–	–	–	–	–	–	–	–	339,056	–	–	–	–
Lifan Motor Co	–	114,763	–	–	–	–	–	–	–	–	114,763	–	–	–	–
Mahindra & Mahindra	–	–	470,896	–	–	–	–	–	–	–	470,896	–	–	–	–
Mazda	–	–	13,697	–	15,256	–	–	–	–	–	28,953	–	–	–	–
Mitsubishi	–	–	–	107,763	66,570	–	–	–	–	272,488	446,821	–	–	–	–
Nissan	–	–	25,608	–	112,539	25,590	–	–	–	82,756	246,493	–	–	–	–
Perodua	–	–	–	–	–	41,990	–	–	–	–	41,990	–	–	–	–
Peugeot Citroen	–	–	–	–	–	959	–	–	–	–	959	–	–	–	–
Proton	–	–	–	–	–	30,711	–	–	–	–	30,711	–	–	–	–
Renault	–	–	–	–	–	–	–	53,878	–	–	53,878	–	–	–	–
Shanghai AIC	–	1,988,688	–	–	–	–	–	–	–	–	1,988,688	–	–	–	–
Subaru	–	–	–	–	18,361	–	–	–	–	–	18,361	–	–	–	–
Suzuki	–	–	186,590	71,784	175,082	–	33,973	–	–	–	467,429	–	–	–	–
Tata	–	–	605,102	–	–	–	–	5,281	–	–	610,383	–	–	–	–
Toyota	–	–	91,453	103,133	615,982	42,547	6,248	–	16,535	628,316	1,504,214	–	–	–	–
Volkswagen	–	–	826	52	–	–	–	–	–	–	878	–	–	–	2,493
Volvo Truck	–	–	1,265	3,219	21,855	1,644	223	–	102	1,110	29,418	–	–	31,500	–
Zhejiang Geely	–	32,123	–	–	–	–	–	–	–	–	32,123	–	–	–	–
Other	–	980,759	207,935	–	2,033	3,197	945	117,286	4,588	331	1,317,074	19,487	27,982	1,000	1,911
TOTAL CV 2012	**31,275**	**8,566,846**	**1,628,708**	**322,056**	**1,388,492**	**169,707**	**41,839**	**1,689,896**	**65,080**	**1,496,094**	**15,399,993**	**19,487**	**27,982**	**32,500**	**7,164**
TOTAL CV 2011	**29,873**	**8,366,022**	**1,461,229**	**276,085**	**1,240,129**	**148,760**	**38,271**	**1,755,387**	**62,655**	**919,808**	**14,298,219**	**22,162**	**24,343**	**34,470**	**7,866**
Vol. % Chg.	**4.7**	**2.4**	**11.5**	**16.7**	**12.0**	**14.1**	**9.3**	**-3.7**	**3.9**	**62.7**	**7.7**	**-12.1**	**14.9**	**-5.7**	**-8.9**

Note: Total World represents countries for which data is presented.

Source: WardsAuto InfoBank.

COMMERCIAL VEHICLE PRODUCTION BY MANUFACTURER GROUP AND COUNTRY 2012

Manufacturer	France	Germany	Hungary	Italy	Nether-lands	Poland	Portugal	Romania	Russia	Serbia	Slovakia	Slovenia	Spain	Sweden	Turkey	Ukraine
AutoVaz	–	–	–	–	–	–	–	–	3,755	–	–	–	–	–	–	–
AvtoFramos	–	–	–	–	–	–	–	–	–	–	–	–	–	–	–	–
Avtotor	–	–	–	–	–	–	–	–	378	–	–	–	–	–	–	–
Beijing AIC	–	–	–	–	–	–	–	–	2,073	–	–	–	–	–	–	–
BMW	–	–	–	–	–	–	–	–	–	–	–	–	–	–	–	–
Brilliance China	–	–	–	–	–	–	–	–	–	–	–	–	–	–	–	–
BYD	–	–	–	–	–	–	–	–	–	–	–	–	–	–	–	–
Chana	–	–	–	–	–	–	–	–	–	–	–	–	–	–	–	–
Chery	–	–	–	–	–	–	–	–	–	–	–	–	–	–	–	–
China Motor	–	–	–	–	–	–	–	–	–	–	–	–	–	–	–	–
Chrysler	–	–	–	–	–	–	–	–	–	–	–	–	–	–	–	–
Daimler	588	243,423	–	–	–	–	4,117	–	4,685	–	–	–	57,150	–	20,002	–
Dongfeng	–	–	–	–	–	–	–	–	–	–	–	–	–	–	–	–
FAW	–	–	–	–	–	–	–	–	–	–	–	–	–	–	–	–
Fiat	17,794	–	–	181,276	–	17,418	–	–	1,027	83	–	–	37,124	–	174,414	–
Ford	–	8,646	–	–	–	–	–	–	–	–	–	–	–	–	272,000	–
GM	–	–	–	–	–	–	–	–	–	–	–	–	–	–	–	–
Great Wall Zhongxing	–	–	–	–	–	–	–	–	–	–	–	–	–	–	–	–
Guangzhou AC	–	–	–	–	–	–	–	–	–	–	–	–	–	–	–	–
Honda	–	–	–	–	–	–	–	–	–	–	–	–	–	–	–	–
Hyundai Group	–	–	–	–	–	–	–	–	243	–	–	–	–	–	–	–
Isuzu	–	–	–	–	–	–	1,573	–	–	–	–	–	–	–	4,741	–
Jianghuai Auto Co	–	–	–	–	–	–	–	–	–	–	–	–	–	–	–	–
Lifan Motor Co	–	–	–	–	–	–	–	–	–	–	–	–	–	–	–	–
Mahindra & Mahindra	–	–	–	–	–	–	–	–	–	–	–	–	–	–	–	–
Mazda	–	–	–	–	–	–	–	–	–	–	–	–	–	–	–	–
Mitsubishi	–	–	–	–	–	–	–	–	–	–	–	–	–	–	–	–
Nissan	–	–	–	–	–	–	–	–	–	–	–	–	87,696	–	–	–
Perodua	–	–	–	–	–	–	–	–	–	–	–	–	–	–	–	–
Peugeot Citroen	96,768	–	–	90,726	–	–	40,755	–	–	–	–	–	197,000	–	16,373	–
Proton	–	–	–	–	–	–	–	–	–	–	–	–	–	–	–	–
Renault	173,039	–	–	–	–	–	–	11,187	–	–	–	4,113	5,060	–	–	–
Shanghai AIC	–	–	–	–	–	–	–	–	–	–	–	–	–	–	–	–
Subaru	–	–	–	–	–	–	–	–	–	–	–	–	–	–	–	–
Suzuki	–	–	–	–	–	–	–	–	–	–	–	–	–	–	–	–
Tata	–	–	–	–	–	–	–	–	–	–	–	–	–	–	–	–
Toyota	–	–	–	–	–	–	1,381	–	–	–	–	–	–	–	–	–
Volkswagen	–	107,232	–	–	–	84,501	–	–	–	–	–	–	–	–	–	–
Volvo Truck	30,500	–	–	–	–	–	–	–	6,252	–	–	–	–	23,900	–	–
Zhejiang Geely	–	–	–	–	–	–	–	–	–	–	–	–	–	–	–	–
Other	9,000	49,711	2,400	2,949	65,754	8,877	–	22	244,970	150	–	–	–	13,600	8,152	6,594
TOTAL CV 2012	**327,689**	**409,012**	**2,400**	**274,951**	**65,754**	**110,796**	**47,826**	**11,209**	**263,383**	**233**	**–**	**4,113**	**384,030**	**37,500**	**495,682**	**6,594**
TOTAL CV 2011	**363,859**	**429,661**	**2,313**	**304,742**	**69,611**	**113,569**	**50,463**	**24,989**	**250,757**	**796**	**–**	**5,164**	**486,025**	**41,460**	**549,397**	**7,068**
Vol. % Chg.	**-9.9**	**-4.8**	**3.8**	**-9.8**	**-5.5**	**-2.4**	**-5.2**	**-55.1**	**5.0**	**-70.7**	**–**	**-20.4**	**-21.0**	**-9.6**	**-9.8**	**-6.7**

Note: Total World represents countries for which data is presented.

Source: WardsAuto InfoBank.

COMMERCIAL VEHICLE PRODUCTION BY MANUFACTURER GROUP AND COUNTRY, 2012

Manufacturer	U.K.	Uzbek-istan	Total Europe	Canada	Mexico	U.S.	Total N. America	Argen-tina	Brazil	Vene-zuela	Total S. America	Egypt	South Africa	2012	2011	Vol. % Chg.
AutoVaz	–	–	3,755	–	–	–	–	–	–	–	–	–	–	3,755	–	–
AvtoFramos	–	–	–	–	–	–	–	–	–	–	–	–	–	–	–	–
Avtotor	–	–	378	–	–	–	–	–	–	–	–	–	–	378	10	–
Beijing AIC	–	–	2,073	–	–	–	–	–	–	–	–	–	–	898,392	844,351	6.4
BMW	261	–	261	–	–	301,519	301,519	–	–	–	–	–	–	304,120	280,502	8.4
Brilliance China	–	–	–	–	–	–	–	–	–	–	–	–	–	393,062	332,814	18.1
BYD	–	–	–	–	–	–	–	–	–	–	–	–	–	89,505	62,971	42.1
Chana	–	–	–	–	–	–	–	–	–	–	–	–	–	90[illegible],766	960,187	-6.1
Chery	–	–	–	–	–	–	–	–	–	–	–	–	–	15[illegible],254	181,205	-13.8
China Motor	–	–	–	–	–	–	–	–	–	–	–	–	–	45,073	54,491	-17.3
Chrysler	–	–	–	345,725	372,852	995,958	1,714,535	–	–	8,320	8,320	–	–	1,722,855	1,504,622	14.5
Daimler	–	–	329,965	–	64,566	237,425	301,991	16,263	50,776	1,094	68,133	1,790	1,850	830,734	834,958	-0.5
Dongfeng	–	–	–	–	–	–	–	–	–	–	–	–	–	1,373,990	1,465,287	-6.2
FAW	–	–	–	–	–	–	–	–	–	–	–	–	–	643,393	698,715	-7.9
Fiat	–	–	431,896	–	–	–	–	4,769	188,678	2,199	195,646	–	2,220	629,762	701,611	-10.2
Ford	28,031	–	308,677	255,924	–	1,599,306	1,855,230	40,580	85,243	15,593	141,416	–	35,438	2,507,581	2,436,867	2.9
GM	59,472	25,280	84,752	363,502	459,604	1,143,942	1,967,048	–	136,867	15,139	152,006	12,863	19,128	2,565,175	2,494,162	2.8
Great Wall Zhongxing	–	–	–	–	–	–	–	–	–	–	–	–	–	442,556	327,682	35.1
Guangzhou AC	–	–	–	–	–	–	–	–	–	–	–	–	–	18[illegible],855	181,949	-0.6
Honda	–	–	–	183,559	61,813	580,218	825,590	–	–	–	–	–	–	86[illegible],312	599,379	43.7
Hyundai Group	–	–	243	–	–	229,663	229,663	–	25,568	–	25,568	99	–	1,56[illegible],794	1,686,175	-7.3
Isuzu	–	–	6,314	–	–	4,069	4,069	–	–	–	–	–	16,013	59[illegible],042	407,512	45.3
Jianghuai Auto Co	–	–	–	–	–	–	–	–	–	–	–	–	–	33[illegible],056	339,451	-0.1
Lifan Motor Co	–	–	–	–	–	–	–	–	–	–	–	–	–	11[illegible],763	105,145	9.1
Mahindra & Mahindra	–	–	–	–	–	–	–	–	830	–	830	–	–	47[illegible],726	370,539	27.3
Mazda	–	–	–	–	–	–	–	–	–	–	–	–	564	2[illegible],517	34,776	-15.1
Mitsubishi	–	–	–	–	–	23,303	23,303	–	37,353	811	38,164	787	4	50[illegible],079	427,018	19.2
Nissan	–	–	87,696	–	114,317	238,157	352,474	–	16,978	–	16,978	2,212	52,154	75[illegible],007	625,327	21.2
Perodua	–	–	–	–	–	–	–	–	–	–	–	–	–	4[illegible],990	38,787	8.3
Peugeot Citroen	–	–	441,622	–	–	–	–	17,889	6,575	–	24,464	–	–	46[illegible],045	531,134	-12.1
Proton	–	–	–	–	–	–	–	–	–	–	–	–	–	3[illegible],711	23,644	29.9
Renault	–	–	193,399	–	–	–	–	23,746	81,192	–	104,938	–	–	35[illegible],215	342,240	2.9
Shanghai AIC	–	–	–	–	–	–	–	–	–	–	–	–	–	1,98[illegible],688	1,756,383	13.2
Subaru	–	–	–	–	–	134,380	134,380	–	–	–	–	–	–	15[illegible],741	167,175	-8.6
Suzuki	–	–	–	–	–	–	–	–	–	–	–	1,680	–	46[illegible],109	390,310	20.2
Tata	6,281	–	6,281	–	–	–	–	–	–	–	–	–	–	61[illegible],664	638,287	-3.4
Toyota	–	–	1,381	265,056	56,108	584,549	905,713	93,570	–	6,330	99,900	–	132,783	2,64[illegible],991	1,954,067	35.3
Volkswagen	–	–	194,226	–	–	–	–	70,302	104,304	–	174,606	–	–	36[illegible],710	385,415	-4.1
Volvo Truck	–	–	92,152	–	237	51,169	51,406	–	19,437	800	20,237	33	1,080	19[illegible],326	206,426	-5.9
Zhejiang Geely	–	–	–	–	–	–	–	–	–	–	–	–	–	3[illegible],123	–	–
Other	17,994	–	480,553	9,300	62,310	103,094	174,704	–	70,341	–	70,341	136	3,431	2,04[illegible],239	2,171,873	-5.8
TOTAL CV 2012	**112,039**	**25,280**	**2,665,624**	**1,423,066**	**1,191,807**	**6,226,752**	**8,841,625**	**267,119**	**824,142**	**50,286**	**1,141,547**	**19,600**	**264,665**	**28,33[illegible],054**	**26,563,447**	**6.7**
TOTAL CV 2011	**120,189**	**20,771**	**2,929,675**	**1,144,639**	**1,023,970**	**5,684,544**	**7,853,153**	**251,538**	**937,546**	**44,377**	**1,233,461**	**28,659**	**220,280**	**26,56[illegible],447**		
Vol. % Chg.	**-6.8**	**21.7**	**-9.0**	**24.3**	**16.4**	**9.5**	**12.6**	**6.2**	**-12.1**	**13.3**	**-7.5**	**-31.6**	**20.1**	**6.7**		

Note: Total World represents countries for which data is presented.

Source: WardsAuto InfoBank.

WORLD

TOTAL VEHICLE PRODUCTION BY MANUFACTURER GROUP AND COUNTRY, 2012

Manufacturer	Australia	China	India	Indo-nesia	Japan	Malaysia	Pakistan	S. Korea	Taiwan	Thailand	Total Asia Pacific	Austria	Belarus	Belgium	Czech Republic
AutoVaz	–	–	–	–	–	–	–	–	–	–	–	–	–	–	–
AvtoFramos	–	–	–	–	–	–	–	–	–	–	–	–	–	–	–
Avtotor	–	–	–	–	–	–	–	–	–	–	–	–	–	–	–
Beijing AIC	–	1,672,974	–	–	–	377	–	–	–	–	1,673,351	–	–	–	–
BMW	–	–	8,170	1,682	–	5,228	–	–	–	6,065	21,145	103,545	–	–	–
Brilliance China	–	639,822	–	–	–	453	–	–	–	–	640,275	–	–	–	–
BYD	–	455,706	–	–	–	–	–	–	–	–	455,706	–	–	–	–
Chana	–	1,792,819	–	–	–	–	–	–	–	–	1,792,819	–	–	–	–
Chery	–	563,951	–	–	–	568	–	–	–	–	564,519	–	–	–	–
China Motor	–	114,534	–	–	–	–	–	–	53,707	–	168,241	–	–	–	–
Chrysler	–	–	–	–	–	–	–	–	–	–	–	–	–	–	–
Daimler	–	22,500	6,113	11,147	87,086	7,942	–	–	–	11,638	146,426	9,850	–	–	–
Dongfeng	–	2,994,180	–	–	–	–	–	–	–	–	2,994,180	–	–	–	–
FAW	–	2,707,774	–	–	–	–	–	–	–	–	2,707,774	–	–	–	–
Fiat	–	–	9,992	–	–	–	–	–	–	–	9,992	–	–	–	2,760
Ford	37,702	–	116,906	–	–	–	–	–	31,199	268,676	454,483	–	–	126,791	–
GM	82,098	–	91,212	–	–	–	–	785,757	–	124,609	1,083,676	–	–	–	–
Great Wall Zhongxing	–	624,426	–	–	–	–	–	–	–	–	624,426	–	–	–	–
Guangzhou AC	–	637,923	–	–	–	–	–	–	–	–	637,923	–	–	–	–
Honda	–	28,730	75,201	55,828	1,029,313	29,418	14,835	–	18,532	222,263	1,474,120	–	–	–	–
Hyundai Group	–	–	638,765	4,875	–	16,871	–	3,490,946	15,356	–	4,166,813	–	–	–	303,035
Isuzu	–	–	–	31,645	241,247	9,895	450	–	–	287,509	570,746	–	–	–	–
Jianghuai Auto Co	–	476,356	–	–	–	–	–	–	–	–	476,356	–	–	–	–
Lifan Motor Co	–	272,657	–	–	–	–	–	–	–	–	272,657	–	–	–	–
Mahindra & Mahindra	–	–	487,364	–	–	–	–	–	–	–	487,364	–	–	–	–
Mazda	–	–	13,697	–	845,550	2,599	–	–	–	–	861,846	–	–	–	–
Mitsubishi	–	–	–	114,090	515,168	–	–	–	–	356,622	985,880	–	–	–	–
Nissan	–	–	178,315	65,900	1,148,265	40,660	–	–	32,506	251,477	1,717,123	–	–	–	–
Perodua	–	–	–	–	–	194,985	–	–	–	–	194,985	–	–	–	–
Peugeot Citroen	–	–	–	–	–	2,995	–	–	–	–	2,995	9,828	–	–	–
Proton	–	–	–	–	–	161,412	–	–	–	–	161,412	–	–	–	–
Renault	–	–	–	–	–	–	–	153,891	–	–	153,891	–	–	–	–
Shanghai AIC	–	4,621,914	–	–	–	–	–	–	–	–	4,621,914	–	–	–	–
Subaru	–	–	–	–	570,173	–	–	–	–	–	570,173	–	–	–	–
Suzuki	–	–	1,183,040	133,439	1,061,863	5,750	96,370	–	–	17,684	2,498,146	–	–	–	–
Tata	–	–	835,076	–	–	–	–	5,281	–	–	840,357	–	–	–	–
Toyota	101,424	–	190,321	642,169	4,420,158	81,451	46,714	–	173,619	905,733	6,561,589	–	–	–	214,915
Volkswagen	–	–	102,890	836	–	1,662	–	–	–	–	105,388	–	–	123,111	656,306
Volvo Truck	–	–	1,265	3,219	21,855	1,644	223	–	102	1,110	29,418	–	–	31,500	–
Zhejiang Geely	–	494,648	–	727	–	2,513	–	–	–	–	497,888	–	–	257,302	–
Other	–	1,150,894	210,642	–	2,033	3,197	945	121,863	14,017	331	1,503,922	19,866	28,352	1,000	1,922
TOTAL VEHICLE 2012	**221,224**	**19,271,808**	**4,148,969**	**1,065,557**	**9,942,711**	**569,620**	**159,537**	**4,557,738**	**339,038**	**2,453,717**	**42,729,919**	**143,089**	**28,352**	**539,704**	**1,178,938**
TOTAL VEHICLE 2011	**219,376**	**18,418,876**	**3,940,360**	**837,948**	**8,398,654**	**533,695**	**162,538**	**4,657,094**	**343,296**	**1,457,795**	**38,969,632**	**152,505**	**24,573**	**596,661**	**1,199,834**
Vol. % Chg.	**0.8**	**4.6**	**5.3**	**27.2**	**18.4**	**6.7**	**-1.8**	**-2.1**	**-1.2**	**68.3**	**9.6**	**-6.2**	**15.0**	**-9.5**	**-1.7**

Note: Total World represents countries for which data is presented.

Source: WardsAuto InfoBank.

WORLD

TOTAL VEHICLE PRODUCTION BY MANUFACTURER GROUP AND COUNTRY, 2012

Manufacturer	France	Germany	Hungary	Italy	Nether-lands	Poland	Portugal	Romania	Russia	Serbia	Slovakia	Slovenia	Spain	Sweden	Turkey	Ukraine
AutoVaz	–	–	–	–	–	–	–	–	611,124	–	–	–	–	–	–	–
AvtoFramos	–	–	–	–	–	–	–	–	168,099	–	–	–	–	–	–	–
Avtotor	–	–	–	–	–	–	–	–	250,013	–	–	–	–	–	–	–
Beijing AIC	–	–	–	–	–	–	–	–	2,073	–	–	–	–	–	–	–
BMW	–	1,043,732	–	–	–	–	–	–	–	–	–	–	–	–	–	–
Brilliance China	–	–	–	–	–	–	–	–	–	–	–	–	–	–	–	–
BYD	–	–	–	–	–	–	–	–	–	–	–	–	–	–	–	–
Chana	–	–	–	–	–	–	–	–	–	–	–	–	–	–	–	–
Chery	–	–	–	–	–	–	–	–	–	–	–	–	–	–	–	–
China Motor	–	–	–	–	–	–	–	–	–	–	–	–	–	–	–	–
Chrysler	–	–	–	–	–	–	–	–	–	–	–	–	–	–	–	–
Daimler	104,742	1,207,056	41,035	–	–	–	4,117	–	4,685	–	–	–	76,440	–	20,002	–
Dongfeng	–	–	–	–	–	–	–	–	–	–	–	–	–	–	–	–
FAW	–	–	–	–	–	–	–	–	–	–	–	–	–	–	–	–
Fiat	17,794	–	–	575,896	–	348,372	–	–	1,027	25,364	–	–	37,124	–	256,109	–
Ford	–	694,004	–	–	–	–	–	30,591	107,724	–	–	–	149,745	–	272,000	–
GM	–	387,525	–	–	–	125,469	–	–	169,623	–	–	–	264,858	–	–	–
Great Wall Zhongxing	–	–	–	–	–	–	–	–	23,904	–	–	–	–	–	–	–
Guangzhou AC	–	–	–	–	–	–	–	–	–	–	–	–	–	–	–	–
Honda	–	–	–	–	–	–	–	–	–	–	–	–	–	–	24,378	–
Hyundai Group	–	–	–	–	–	–	–	–	58,643	–	292,000	–	–	–	90,857	–
Isuzu	–	–	–	–	–	–	1,573	–	–	–	–	–	–	–	4,741	–
Jianghuai Auto Co	–	–	–	–	–	–	–	–	–	–	–	–	–	–	–	–
Lifan Motor Co	–	–	–	–	–	–	–	–	–	–	–	–	–	–	–	–
Mahindra & Mahindra	–	–	–	–	–	–	–	–	–	–	–	–	–	–	–	–
Mazda	–	–	–	–	–	–	–	–	–	–	–	–	–	–	–	–
Mitsubishi	–	–	–	–	–	–	–	–	–	–	–	–	–	–	–	–
Nissan	–	–	–	–	–	–	–	–	126,058	–	–	–	143,165	–	–	–
Perodua	–	–	–	–	–	–	–	–	–	–	–	–	–	–	–	–
Peugeot Citroen	1,114,204	–	–	90,726	–	–	43,940	–	40,258	–	215,000	–	373,500	–	16,373	–
Proton	–	–	–	–	–	–	–	–	–	–	–	–	–	–	–	–
Renault	532,571	–	–	–	–	–	–	307,152	–	–	–	130,949	269,758	–	292,066	–
Shanghai AIC	–	–	–	–	–	–	–	–	–	–	–	–	–	–	–	–
Subaru	–	–	–	–	–	–	–	–	–	–	–	–	–	–	–	–
Suzuki	–	–	140,852	–	–	–	–	–	–	–	–	–	–	–	–	–
Tata	–	–	–	–	–	–	–	–	–	–	–	–	–	–	–	–
Toyota	201,692	–	–	–	–	–	1,381	–	98,229	–	–	–	–	–	88,300	–
Volkswagen	–	2,265,676	33,553	2,197	–	161,996	112,550	–	179,554	–	337,110	–	664,513	–	–	14,556
Volvo Truck	30,500	–	–	–	–	–	–	–	6,252	–	–	–	–	2[illegible]900	–	–
Zhejiang Geely	–	–	–	–	–	–	–	–	–	–	–	–	–	16[illegible]814	–	–
Other	9,000	199,478	2,400	2,949	65,754	8,877	–	22	384,905	150	–	–	–	1[illegible]600	8,152	61,725
TOTAL VEHICLE 2012	**2,010,503**	**5,797,471**	**217,840**	**671,768**	**65,754**	**644,714**	**163,561**	**337,765**	**2,232,171**	**25,514**	**844,110**	**130,949**	**1,979,103**	**20[illegible]314**	**1,072,978**	**76,281**
TOTAL VEHICLE 2011	**2,294,889**	**6,301,579**	**213,531**	**790,348**	**110,383**	**835,854**	**192,242**	**335,232**	**1,988,919**	**11,023**	**573,660**	**174,119**	**2,353,682**	**23[illegible]429**	**1,189,131**	**104,653**
Vol. % Chg.	**-12.4**	**-8.0**	**2.0**	**-15.0**	**-40.4**	**-22.9**	**-14.9**	**0.8**	**12.2**	**131.5**	**47.1**	**-24.8**	**-15.9**	**-13.1**	**-9.8**	**-27.1**

Note: Total World represents countries for which data is presented.
Source: WardsAuto InfoBank.

TOTAL VEHICLE PRODUCTION BY MANUFACTURER GROUP AND COUNTRY, 2012

Manufacturer	U.K.	Uzbek-istan	Total Europe	Canada	Mexico	U.S.	Total N. America	Argen-tina	Brazil	Vene-zuela	Total S. America	Egypt	South Africa	2012	2011	Vol. % Chg.
AutoVaz	–	–	611,124	–	–	–	–	–	–	–	–	1,214	–	612,338	570,134	7.4
AvtoFramos	–	–	168,099	–	–	–	–	–	–	–	–	–	–	168,099	142,304	18.1
Avtotor	–	–	250,013	–	–	–	–	–	–	–	–	–	–	250,013	222,091	12.6
Beijing AIC	–	–	2,073	–	–	–	–	–	–	–	–	–	–	1,675,424	1,515,678	10.5
BMW	210,481	–	1,357,758	–	–	301,519	301,519	–	–	–	–	1,290	43,161	1,724,873	1,653,741	4.3
Brilliance China	–	–	–	–	–	–	–	–	–	–	–	–	–	640,275	554,141	15.5
BYD	–	–	–	–	–	–	–	–	–	–	–	–	–	455,706	449,748	1.3
Chana	–	–	–	–	–	–	–	–	–	–	–	–	–	1,792,819	1,792,502	0.0
Chery	–	–	–	–	–	–	–	–	–	–	–	–	–	564,519	640,597	-11.9
China Motor	–	–	–	–	–	–	–	–	–	–	–	–	–	168,241	181,104	-7.1
Chrysler	–	–	–	585,918	454,726	1,330,783	2,371,427	–	–	9,272	9,272	604	–	2,381,303	2,009,123	18.5
Daimler	–	–	1,467,927	–	64,566	237,425	301,991	16,263	50,776	1,094	68,133	2,506	63,151	2,050,134	2,042,293	0.4
Dongfeng	–	–	–	–	–	–	–	–	–	–	–	–	–	2,994,180	2,976,858	0.6
FAW	–	–	–	–	–	–	–	–	–	–	–	–	–	2,707,774	2,639,121	2.6
Fiat	–	–	1,264,446	–	–	–	–	78,632	837,198	2,199	918,029	–	2,220	2,194,687	2,464,630	-11.0
Ford	28,031	–	1,408,886	255,924	444,425	2,152,217	2,852,566	89,072	294,669	23,216	406,957	–	50,442	5,173,334	5,169,793	0.1
GM	149,659	236,201	1,333,335	683,058	569,601	1,984,642	3,237,301	128,501	615,550	43,449	787,500	29,058	41,852	6,512,722	6,621,759	-1.6
Great Wall Zhongxing	–	–	23,904	–	–	–	–	–	–	–	–	–	–	648,330	501,913	29.2
Guangzhou AC	–	–	–	–	–	–	–	–	–	–	–	–	–	637,923	696,915	-8.5
Honda	165,630	–	190,008	409,949	61,813	1,219,326	1,691,088	8,114	135,058	–	143,172	–	–	3,498,388	2,289,925	52.8
Hyundai Group	–	–	744,535	–	–	719,547	719,547	–	51,571	3,872	55,443	9,527	–	5,695,865	5,432,733	4.8
Isuzu	–	–	6,314	–	–	4,069	4,069	–	–	–	–	–	16,013	597,142	412,822	44.6
Jianghuai Auto Co	–	–	–	–	–	–	–	–	–	–	–	–	–	476,356	485,856	-2.0
Lifan Motor Co	–	–	–	–	–	–	–	–	–	–	–	–	–	272,657	205,846	32.5
Mahindra & Mahindra	–	–	–	–	–	–	–	–	830	–	830	–	–	438,194	386,844	26.2
Mazda	–	–	–	–	–	–	–	–	–	–	–	–	564	862,410	834,416	3.4
Mitsubishi	–	–	–	–	–	37,019	37,019	–	37,353	7,352	44,705	787	4	1,068,395	1,044,625	2.3
Nissan	510,572	–	779,795	–	683,520	644,939	1,328,459	–	31,342	–	31,342	2,212	54,459	3,913,390	3,516,216	11.3
Perodua	–	–	–	–	–	–	–	–	–	–	–	–	–	194,985	192,470	1.3
Peugeot Citroen	–	–	1,903,829	–	–	–	–	133,534	99,518	–	233,052	377	–	2,140,253	2,485,120	-13.9
Proton	1,043	–	1,043	–	–	–	–	–	–	–	–	–	–	162,455	167,687	-3.1
Renault	–	–	1,532,496	–	–	–	–	108,905	253,425	–	362,330	–	–	2,048,717	2,378,250	-13.9
Shanghai AIC	–	–	–	–	–	–	–	–	–	–	–	–	–	4,621,914	4,148,325	11.4
Subaru	–	–	–	–	–	277,909	277,909	–	–	–	–	–	–	848,082	659,431	28.6
Suzuki	–	–	140,852	–	–	–	–	–	–	–	–	1,781	–	2,640,779	2,451,708	7.7
Tata	367,664	–	367,664	–	–	–	–	–	–	–	–	–	–	1,208,021	1,203,610	0.4
Toyota	109,429	–	713,946	519,215	56,108	1,116,425	1,691,748	93,570	66,332	12,829	172,731	–	159,429	9,299,443	7,182,028	29.5
Volkswagen	9,108	–	4,560,230	–	604,508	152,543	757,051	107,904	852,086	–	959,990	–	103,732	6,486,391	6,598,963	-1.7
Volvo Truck	–	–	92,152	–	237	51,169	51,406	–	19,437	800	20,237	33	1,080	194,326	206,426	-5.9
Zhejiang Geely	–	–	420,116	–	–	–	–	–	–	–	–	–	–	918,004	880,634	4.2
Other	24,232	–	832,284	9,300	62,310	103,094	174,704	–	70,341	–	70,341	7,091	3,431	2,591,773	2,786,453	-7.0
TOTAL VEHICLE 2012	**1,575,849**	**236,201**	**20,172,829**	**2,463,364**	**3,001,814**	**10,332,626**	**15,797,804**	**764,495**	**3,415,486**	**104,083**	**4,284,064**	**56,480**	**539,538**	**83,580,634**	**78,794,833**	**6.1**
TOTAL VEHICLE 2011	**1,463,983**	**222,022**	**21,359,252**	**2,135,121**	**2,681,050**	**8,661,535**	**13,477,706**	**828,771**	**3,442,787**	**102,409**	**4,373,967**	**81,731**	**532,545**	**78,794,833**		
Vol. % Chg.	**7.6**	**6.4**	**-5.6**	**15.4**	**12.0**	**19.3**	**17.2**	**-7.8**	**-0.8**	**1.6**	**-2.1**	**-30.9**	**1.3**	**6.1**		

Note: Total World represents countries for which data is presented.
Source: WardsAuto InfoBank.

CAR SALES BY COMPANY GROUP AND COUNTRY, 2012

Company	Australia	China	India	Indonesia	Japan	Malaysia	New Zealand	Pakistan	Philippines	Singapore	S. Korea	Taiwan
Auto Vaz	–	–	–	–	–	–	–	–	–	–	–	–
Beijing AIC	–	20,652	–	–	–	–	–	–	–	–	–	–
BMW	13,156	126,733	3,886	2,188	–	6,027	1,276	–	630	5,373	34,106	9,134
Brilliance	–	118,549	–	–	–	–	–	–	–	1	–	–
BYD	–	368,071	–	–	–	–	–	–	–	–	–	–
Chana	–	227,010	–	–	–	–	–	–	–	2	–	–
Chery	551	405,831	–	–	–	82	165	–	–	5	–	–
China National	–	–	–	–	–	–	–	–	–	–	–	–
Chrysler	3,197	–	–	762	–	–	633	–	–	38	4,123	–
Daimler	15,018	67,735	4,826	4,091	–	5,492	837	–	600	4,325	20,389	10,786
Dongfeng	–	75,084	–	–	–	–	–	–	–	–	–	–
FAW	–	334,875	–	680	–	–	–	–	–	–	–	–
Fiat	1,641	11,288	11,145	–	–	–	304	–	–	238	–	–
Ford	48,238	274,522	86,090	7,174	–	4,052	5,974	–	4,416	354	5,126	11,031
Foton	–	–	–	–	–	–	–	–	–	9	–	–
GAZ	–	–	–	–	–	–	–	–	–	–	–	–
GM	75,425	1,380,223	70,904	5,643	1	952	5,383	–	1,731	537	109,699	667
Great Wall Zhongxing	–	180,592	–	–	–	–	1	–	–	–	–	–
Honda	31,079	396,826	72,840	69,320	712,350	30,872	2,380	13,551	10,882	515	3,944	18,794
Hyundai Group	92,510	1,051,438	390,274	19,397	–	15,391	6,045	–	12,842	1,755	397,736	13,012
Isuzu	–	–	–	11,277	–	–	–	–	–	–	–	–
Jianghuai Auto Co	–	138,320	–	–	–	–	–	–	–	–	–	–
Lifan Motor Co	–	156,345	–	–	–	–	–	–	–	–	–	–
Mahindra & Mahindra	–	–	16,521	–	–	–	–	–	–	–	–	–
Mazda	68,430	312,052	–	8,054	192,463	4,052	3,333	–	45	348	–	13,222
Mitsubishi	15,856	99,972	–	51,830	93,798	1,216	1,500	–	2,742	61	81	20,746
Nissan	15,210	600,375	44,439	67,179	518,509	14,363	989	–	833	759	3,501	40,227
PACCAR	–	–	–	–	–	–	–	–	–	–	–	–
Perodua	–	–	–	–	–	147,752	–	–	–	8	–	–
Peugeot Citroen	5,330	440,028	–	333	–	5,197	1,034	–	–	917	2,662	–
Proton	925	45,421	–	1,017	–	117,004	1	–	–	62	–	–
Renault	2,676	–	11,015	–	–	9	65	–	–	114	54,990	–
Shanghai AIC	–	192,922	–	–	–	–	–	–	–	1	–	–
Subaru	13,014	–	–	204	141,065	53	561	–	1,752	135	627	1,611
Suzuki	19,120	170,037	878,272	51,573	529,212	6,874	4,795	62,558	2,881	144	–	–
Tata Motors	786	–	235,380	331	–	–	113	–	–	767	1,197	–
Toyota	105,981	531,027	81,025	475,211	2,084,322	54,136	11,088	40,660	27,465	4,309	15,771	112,458
Volkswagen	46,361	2,309,633	105,962	1,449	–	12,313	3,564	–	123	5,459	33,656	11,665
Volvo Truck	–	–	–	–	–	–	–	–	–	–	–	–
Zhejiang Geely	1,635	472,257	–	1,040	–	526	97	–	110	1,066	1,768	4,181
Other	716	175,609	2,940	2,032	300,613	1,248	243	–	92	5,989	5,963	14,166
TOTAL CAR 2012	**576,855**	**10,683,427**	**2,015,519**	**780,785**	**4,572,333**	**427,611**	**50,381**	**116,769**	**67,144**	**33,291**	**[illegible]95,339**	**281,700**
TOTAL CAR 2011	**559,314**	**10,040,167**	**1,950,363**	**614,999**	**3,524,789**	**417,990**	**45,062**	**124,508**	**57,576**	**33,732**	**1,[illegible]30,630**	**304,443**
Vol. % Chg.	**3.1**	**6.4**	**3.3**	**27.0**	**29.7**	**2.3**	**11.8**	**-6.2**	**16.6**	**-1.3**	**-3.4**	**-7.5**

Note: Total World represents countries for which data is presented.
Source: WardsAuto InfoBank.

CAR SALES BY COMPANY GROUP AND COUNTRY, 2012

Company	Thailand	Vietnam	Total Asia Pacific	Austria	Belgium	Czech Republic	Denmark	Finland	France	Germany	Greece	Hungary
Auto Vaz	–	–	–	41	26	93	–	–	248	1,575	–	11
Beijing AIC	–	–	20,652	–	–	–	–	–	–	–	–	–
BMW	5,756	–	208,265	17,961	35,891	4,058	3,250	4,467	69,550	284,551	1,996	1,515
Brilliance	–	–	118,550	–	–	–	–	–	–	–	–	–
BYD	–	–	368,071	–	–	–	–	–	–	–	–	–
Chana	–	–	227,012	–	–	–	–	–	–	–	–	–
Chery	35	157	406,826	–	–	–	–	–	–	–	–	–
China National	–	–	–	–	–	–	–	–	–	–	–	–
Chrysler	–	–	8,753	881	643	378	6	251	3,243	7,033	76	69
Daimler	6,274	1,218	141,591	11,353	22,043	3,227	3,829	5,001	53,008	311,059	1,803	1,095
Dongfeng	–	–	75,084	–	–	–	–	–	–	–	–	–
FAW	–	–	335,555	–	–	–	–	–	–	–	–	–
Fiat	–	–	24,616	16,161	19,205	2,898	7,839	995	59,514	84,802	4,954	2,714
Ford	35,016	1,908	483,901	21,624	28,754	12,523	13,123	8,504	92,468	206,128	3,006	5,196
Foton	–	–	9	–	–	–	–	–	–	–	–	–
GAZ	–	–	–	–	–	–	–	–	–	–	–	–
GM	21,486	3,852	1,676,503	23,997	40,821	6,611	13,476	4,951	96,426	243,528	7,996	8,058
Great Wall Zhongxing	–	–	180,593	–	–	–	–	–	–	–	–	–
Honda	163,442	794	1,527,589	2,200	2,439	1,659	2,112	3,140	8,406	26,306	657	1,065
Hyundai Group	684	6,669	2,307,753	31,927	26,475	23,639	19,898	10,751	61,751	155,673	4,133	3,032
Isuzu	–	–	11,277	–	–	–	–	–	–	–	–	–
Jianghuai Auto Co	–	–	138,320	–	–	–	–	–	–	–	–	–
Lifan Motor Co	–	–	156,345	–	–	–	–	–	–	–	8	–
Mahindra & Mahindra	–	–	16,521	–	–	–	–	–	–	–	–	–
Mazda	44,282	–	646,281	8,424	3,174	1,170	2,763	1,345	5,107	40,702	222	486
Mitsubishi	42,427	–	330,229	3,772	3,132	1,056	438	1,291	3,639	23,565	608	169
Nissan	103,504	–	1,409,888	8,635	17,146	3,171	4,470	7,196	70,133	63,084	3,144	2,346
PACCAR	–	–	–	–	–	–	–	–	–	–	–	–
Perodua	–	–	147,760	–	–	–	–	–	–	–	–	–
Peugeot Citroen	331	–	455,832	25,480	77,170	12,273	28,292	6,308	571,870	132,473	6,749	2,724
Proton	1,789	–	166,219	8	37	–	1	–	74	56	–	–
Renault	–	–	68,869	23,150	58,333	14,191	8,695	1,880	424,135	150,740	1,393	5,590
Shanghai AIC	–	–	192,923	–	–	–	–	–	–	–	–	–
Subaru	118	–	159,140	1,638	516	815	51	1,025	971	9,824	41	100
Suzuki	17,951	15	1,743,432	5,880	5,023	2,309	6,420	814	16,026	30,843	2,278	3,300
Tata Motors	36	–	238,610	2,233	4,735	643	82	425	8,667	14,310	15	101
Toyota	224,816	12,679	3,780,948	7,810	17,515	4,054	15,954	13,591	70,815	84,730	5,993	2,322
Volkswagen	650	–	2,530,835	118,618	106,078	74,619	38,174	30,227	263,269	1,155,302	11,900	11,970
Volvo Truck	–	–	–	–	–	–	–	–	–	–	–	–
Zhejiang Geely	1,654	–	484,334	3,126	15,345	1,458	1,646	7,553	13,396	32,750	1,380	1,137
Other	2,209	–	511,820	1,091	2,236	3,164	244	1,534	6,044	23,470	130	60
TOTAL CAR 2012	**672,460**	**27,292**	**21,300,906**	**336,010**	**486,737**	**174,009**	**170,763**	**111,249**	**1,898,760**	**3,082,504**	**58,482**	**53,060**
TOTAL CAR 2011	**361,219**	**41,429**	**19,106,221**	**356,145**	**572,211**	**173,282**	**169,974**	**126,116**	**2,204,229**	**3,173,634**	**97,682**	**45,094**
Vol. % Chg.	**86.2**	**-34.1**	**11.5**	**-5.7**	**-14.9**	**0.4**	**0.5**	**-11.8**	**-13.9**	**-2.9**	**-40.1**	**17.7**

Note: Total World represents countries for which data is presented.
Source: WardsAuto InfoBank.

CAR SALES BY COMPANY GROUP AND COUNTRY, 2012

Company	Ireland	Italy	Lithuania	Luxem-bourg	Nether-lands	Norway	Poland	Portugal	Romania	Russia	Serbia	Slovakia
Auto Vaz	–	69	19	8	3	–	–	–	–	537,625	509	87
Beijing AIC	–	–	–	–	–	–	–	–	–	1,845	–	–
BMW	3,769	60,289	474	5,497	22,676	8,181	5,555	7,422	2,329	40,144	510	2,418
Brilliance	–	–	–	–	–	–	–	–	–	–	–	–
BYD	–	–	–	–	–	–	–	–	–	2,630	–	–
Chana	–	–	–	–	26	–	–	–	–	–	–	–
Chery	–	–	–	–	–	–	–	–	–	19,004	92	–
China National	–	–	–	–	–	–	–	–	–	–	–	–
Chrysler	115	6,555	39	250	395	210	–	33	183	5,104	59	184
Daimler	1,834	63,789	159	3,380	11,733	5,664	5,415	7,189	1,586	41,438	552	1,764
Dongfeng	–	–	–	–	–	–	–	–	–	–	–	–
FAW	–	–	–	–	–	–	–	–	–	1,151	–	–
Fiat	784	408,733	949	1,899	23,548	920	12,932	6,042	1,711	8,623	3,197	1,723
Ford	8,563	98,997	431	2,536	35,861	11,007	16,923	4,838	4,682	130,809	1,763	2,161
Foton	–	–	–	–	–	–	–	–	–	141	–	–
GAZ	–	–	–	–	–	–	–	–	–	90,247	–	–
GM	5,130	110,044	507	2,523	44,257	3,700	30,024	8,444	4,107	376,540	2,251	4,424
Great Wall Zhongxing	–	10	–	–	–	–	–	–	–	14,373	–	–
Honda	1,102	5,314	215	233	2,969	1,731	6,896	1,487	553	21,512	391	1,100
Hyundai Group	8,323	70,478	1,068	2,227	48,949	10,474	30,492	3,036	4,310	361,616	2,679	10,081
Isuzu	–	–	–	–	–	3	–	–	–	156	–	–
Jianghuai Auto Co	–	–	–	–	–	–	–	–	–	–	–	–
Lifan Motor Co	–	–	–	–	–	–	–	–	–	20,544	–	–
Mahindra & Mahindra	–	60	–	–	–	–	–	–	–	–	–	–
Mazda	1,433	4,303	208	124	6,149	2,927	–	376	432	44,443	243	741
Mitsubishi	361	3,553	165	170	3,584	4,654	3,858	868	766	74,294	88	477
Nissan	6,023	49,280	829	1,414	9,417	6,658	13,377	3,921	1,575	162,956	429	1,773
PACCAR	–	–	–	–	–	–	–	–	–	–	–	–
Perodua	8	–	–	–	–	–	–	–	–	–	–	–
Peugeot Citroen	3,387	138,510	944	6,378	65,845	9,792	22,009	13,122	2,707	77,279	295	6,341
Proton	–	–	–	10	1	–	–	1	–	–	–	–
Renault	5,255	85,852	640	5,685	45,979	663	23,157	11,457	24,733	189,852	2,638	5,545
Shanghai AIC	–	–	–	–	–	–	–	–	–	–	–	–
Subaru	143	3,691	315	110	967	3,559	–	3	175	14,296	23	237
Suzuki	630	14,731	81	586	12,519	2,060	5,988	157	1,638	32,684	460	2,431
Tata Motors	698	13,990	102	939	1,683	731	–	531	794	20,549	107	306
Toyota	10,221	57,739	1,231	1,215	37,779	17,396	19,853	3,879	2,378	168,700	804	2,063
Volkswagen	20,136	185,609	3,588	13,462	108,810	35,627	60,718	20,192	16,887	315,937	6,325	24,837
Volvo Truck	–	–	–	–	–	–	–	–	–	–	–	–
Zhejiang Geely	1,154	14,818	166	1,201	16,925	11,550	4,480	1,977	455	37,930	79	456
Other	429	5,675	42	551	2,469	460	11,912	315	178	122,689	16	119
TOTAL CAR 2012	**79,498**	**1,402,089**	**12,172**	**50,398**	**502,544**	**137,967**	**273,589**	**95,290**	**72,179**	**2,935,111**	**23,510**	**69,268**
TOTAL CAR 2011	**89,896**	**1,748,143**	**13,223**	**49,879**	**555,920**	**138,345**	**277,430**	**153,433**	**94,624**	**2,653,408**	**31,780**	**68,203**
Vol. % Chg.	**-11.6**	**-19.8**	**-7.9**	**1.0**	**-9.6**	**-0.3**	**-1.4**	**-37.9**	**-23.7**	**10.6**	**-26.0**	**1.6**

Note: Total World represents countries for which data is presented.
Source: WardsAuto InfoBank.

CAR SALES BY COMPANY GROUP AND COUNTRY, 2012

Company	Slovenia	Spain	Sweden	Switzer-land	Turkey	U.K.	Ukraine	Total Europe	Canada	Mexico	U.S.	Total North America
Auto Vaz	–	37	24	–	–	–	19,718	560,093	–	–	–	–
Beijing AIC	–	–	–	–	–	–	–	1,845	–	–	–	–
BMW	1,217	32,991	17,527	23,561	17,005	178,854	2,674	856,332	21,210	8,001	231,257	260,468
Brilliance	–	–	–	–	–	–	–	–	–	–	–	–
BYD	–	2	–	–	–	–	–	2,632	–	–	–	–
Chana	–	–	–	–	–	–	–	26	–	–	–	–
Chery	–	–	–	–	810	–	3,519	23,425	–	–	–	–
China National	–	–	–	–	–	–	–	–	–	–	–	–
Chrysler	–	1,673	533	3,088	799	5,639	–	37,439	42,509	30,083	498,095	570,687
Daimler	236	23,629	11,454	18,526	12,826	98,049	2,791	724,432	19,204	8,693	187,754	215,651
Dongfeng	–	–	–	–	–	–	–	–	–	–	–	–
FAW	–	–	–	–	–	–	–	1,151	–	–	–	–
Fiat	1,612	21,818	5,244	14,332	45,662	58,761	2,118	819,690	–	4,065	–	4,065
Ford	1,671	51,059	13,852	18,179	47,030	281,917	10,238	1,133,843	68,107	37,116	760,646	865,869
Foton	–	–	–	–	–	–	–	141	–	–	–	–
GAZ	–	–	–	–	–	–	–	90,247	–	–	–	–
GM	4,202	64,759	8,598	19,304	67,635	245,731	16,625	1,464,669	71,967	144,127	,030,858	1,246,952
Great Wall Zhongxing	–	–	–	–	–	–	–	14,383	–	–	–	–
Honda	215	5,917	2,821	5,540	16,516	54,208	3,752	180,456	87,110	20,156	784,237	891,503
Hyundai Group	3,352	47,825	27,761	16,133	55,535	140,914	38,979	1,221,511	145,296	–	959,014	1,104,310
Isuzu	–	–	–	–	–	–	–	159	–	–	–	–
Jianghuai Auto Co	–	–	–	–	–	–	–	–	–	–	–	–
Lifan Motor Co	–	–	–	–	–	–	–	20,552	–	–	–	–
Mahindra & Mahindra	–	3	–	–	–	–	–	63	–	–	–	–
Mazda	216	4,599	2,149	7,152	810	26,183	1,714	167,595	50,112	16,336	182,817	249,265
Mitsubishi	12	4,006	3,525	4,611	1,289	6,549	5,501	156,001	7,956	2,750	31,448	42,154
Nissan	523	35,890	6,462	9,216	18,914	106,365	12,609	626,956	42,789	167,607	676,991	887,387
PACCAR	–	–	–	–	–	–	–	–	–	–	–	–
Perodua	–	–	–	–	–	416	–	424	–	–	–	–
Peugeot Citroen	4,535	108,094	16,167	23,976	29,230	173,142	6,867	1,571,959	–	4,451	–	4,451
Proton	–	2	2	–	760	345	–	1,297	–	–	–	–
Renault	6,124	70,308	10,098	22,393	100,540	40,760	13,567	1,353,353	–	13,041	–	13,041
Shanghai AIC	–	–	–	–	–	782	–	782	–	–	–	–
Subaru	23	995	4,305	8,483	640	2,023	1,142	56,111	14,403	151	133,070	147,624
Suzuki	47	3,717	2,691	7,702	1,780	24,893	1,687	189,375	3,651	8,193	18,426	30,270
Tata Motors	13	7,555	938	4,300	2,358	62,735	1,472	151,012	644	117	12,011	12,772
Toyota	814	37,972	17,314	14,929	31,961	92,967	17,597	759,596	106,633	24,701	1,198,884	1,330,218
Volkswagen	8,979	167,478	73,837	94,407	96,461	400,387	32,486	3,496,320	63,078	146,519	481,882	691,479
Volvo Truck	–	–	–	–	–	–	–	–	–	–	–	–
Zhejiang Geely	76	6,318	52,755	8,059	6,266	31,790	13,144	287,390	2,465	934	33,952	37,351
Other	15	2,912	1,421	4,248	1,453	12,774	29,402	235,053	1,396	314	20,558	22,268
TOTAL CAR 2012	**33,882**	**699,559**	**279,478**	**328,139**	**556,280**	**2,046,184**	**237,602**	**16,206,313**	**748,530**	**637,355**	**7,241,900**	**8,627,785**
TOTAL CAR 2011	**40,816**	**808,036**	**304,984**	**318,958**	**594,307**	**1,942,995**	**207,453**	**17,010,200**	**681,956**	**586,056**	**,089,403**	**7,357,415**
Vol. % Chg.	**-17.0**	**-13.4**	**-8.4**	**2.9**	**-6.4**	**5.3**	**14.5**	**-4.7**	**9.8**	**8.8**	**18.9**	**17.3**

Note: Total World represents countries for which data is presented.
Source: WardsAuto InfoBank.

CAR SALES BY COMPANY GROUP AND COUNTRY, 2012

Company	Argentina	Brazil	Chile	Colombia	Uruguay	Venezuela	Total South America	Egypt	Israel	S. Africa	2012	2011	Vol. % Chg.
Auto Vaz	–	–	–	–	–	–	–	–	–	–	560,093	610,002	-8.2
Beijing AIC	–	–	–	–	–	–	–	–	–	–	22,497	–	–
BMW	–	7,429	3,282	–	292	4	11,007	2,729	1,278	27,538	1,367,617	1,310,164	4.4
Brilliance	–	–	464	–	8	–	472	674	–	–	119,696	134,594	-11.1
BYD	–	–	2,360	301	898	–	3,559	–	–	–	374,262	391,320	-4.4
Chana	–	–	253	884	790	–	1,927	39	–	58	229,062	193,630	18.3
Chery	–	12,164	5,387	2,238	2,556	–	22,345	6,297	–	–	458,893	513,508	-10.6
China National	–	–	–	–	–	–	–	–	–	–	–	–	–
Chrysler	–	717	7,320	4,178	299	1,064	13,578	1,982	338	9,586	642,363	474,838	35.3
Daimler	3,730	6,448	2,523	612	513	–	13,826	2	755	22,560	1,118,817	1,100,110	1.7
Dongfeng	–	–	73	2,064	772	693	3,602	–	–	–	78,686	19,847	296.5
FAW	–	–	980	–	1,828	–	2,808	–	–	–	339,514	465,972	-27.1
Fiat	60,096	680,048	2,295	1,846	3,661	1,050	748,996	502	2,426	4,314	1,604,609	1,713,841	-6.4
Ford	59,779	255,427	6,971	11,212	1,855	7,691	342,935	863	18,042	25,891	2,871,344	2,990,968	-4.0
Foton	–	–	57	–	73	–	130	–	–	–	280	66	324.2
GAZ	–	–	–	–	–	–	–	–	–	–	90,247	89,773	0.5
GM	120,415	535,652	49,079	81,067	10,033	28,528	824,774	26,164	11,049	33,887	5,283,998	5,209,520	1.4
Great Wall Zhongxing	–	–	5,044	1,112	855	–	7,011	192	2	–	202,181	171,853	17.6
Honda	12,604	120,043	2,547	2,670	527	–	138,391	691	5,742	9,093	2,753,465	2,190,142	25.7
Hyundai Group	–	82,618	57,729	55,156	3,195	6,911	205,609	61,543	44,978	–	4,945,704	4,469,142	10.7
Isuzu	–	–	–	–	–	–	–	–	–	–	11,436	10,624	7.6
Jianghuai Auto Co	–	18,037	3,620	–	581	–	22,238	298	–	–	160,356	172,748	-6.9
Lifan Motor Co	–	1,219	1,210	127	302	–	2,858	–	–	–	179,755	121,241	48.3
Mahindra & Mahindra	–	–	1,503	–	123	–	1,626	185	–	1,447	19,342	18,232	8.8
Mazda	–	–	7,175	10,747	155	–	18,077	9	9,002	4,827	1,095,056	951,752	15.1
Mitsubishi	–	6,738	2,017	1,001	161	6,493	16,410	5,843	1,150	2,619	554,106	579,283	-4.3
Nissan	–	87,189	19,650	25,110	3,828	–	135,777	400	7,167	20,754	3,088,329	2,951,181	4.6
PACCAR	–	–	–	–	–	–	–	–	–	–	–	–	–
Perodua	–	–	–	–	–	–	–	–	–	–	148,184	143,169	3.5
Peugeot Citroen	96,834	124,902	7,605	1,415	3,683	1,562	236,001	1,647	9,084	5,338	2,284,312	2,477,703	-7.8
Proton	–	–	5	–	–	–	5	1,544	–	–	169,065	179,734	-5.9
Renault	79,202	180,700	8,813	39,411	2,354	645	311,125	11,895	5,444	10,216	1,773,943	2,056,441	-13.7
Shanghai AIC	–	37	1,460	–	–	–	1,497	–	168	–	195,370	159,507	22.5
Subaru	–	342	3,926	401	29	–	4,698	270	897	861	369,501	281,652	31.2
Suzuki	5	1,546	16,977	1,255	3,290	–	23,073	2,007	9,282	4,724	2,002,[illegible]63	1,922,609	4.1
Tata Motors	–	41	644	–	7	–	692	202	74	10,390	413,752	385,673	7.3
Toyota	15,722	63,598	20,618	15,918	1,507	6,134	123,497	8,310	18,690	67,016	6,088,275	4,962,218	22.7
Volkswagen	136,460	655,186	7,103	8,642	7,197	1	814,589	8,404	23,279	99,106	7,664,012	7,122,007	7.6
Volvo Truck	–	–	–	–	–	–	–	–	–	–	–	–	–
Zhejiang Geely	–	564	4,378	797	959	–	6,698	1,431	516	2,984	820,704	812,622	1.0
Other	11,550	1,291	3,945	13,093	902	–	30,781	–	17	1,515	801,[illegible]54	750,007	6.9
TOTAL CAR 2012	**596,397**	**2,841,936**	**257,013**	**281,257**	**53,233**	**60,776**	**4,090,612**	**144,123**	**169,380**	**364,724**	**50,903,343**	**48,107,693**	**5.8**
TOTAL CAR 2011	**673,853**	**2,649,032**	**245,116**	**296,302**	**51,443**	**65,339**	**3,981,085**	**133,165**	**188,904**	**330,703**	**48,107,693**		
Vol. % Chg.	**-11.5**	**7.3**	**4.9**	**-5.1**	**3.5**	**-7.0**	**2.8**	**8.2**	**-10.3**	**10.3**	**5.8**		

Note: Total World represents countries for which data is presented.
Source: WardsAuto InfoBank.

COMMERCIAL VEHICLE SALES BY COMPANY GROUP AND COUNTRY, 2012

Company	Australia	China	India	Indonesia	Japan	Malaysia	New Zealand	Pakistan	Philippines	Singapore	S. Korea	Taiwan
Auto Vaz	–	–	–	–	–	–	–	–	–	–	–	–
Beijing AIC	–	205,056	–	–	–	–	–	–	–	–	–	–
BMW	7,673	20,641	5,130	–	–	632	793	–	209	–	–	3,508
Brilliance	–	372,056	–	–	–	–	–	–	–	–	–	–
BYD	–	88,247	–	–	–	–	–	–	–	–	–	–
Chana	–	1,067,255	–	–	–	–	–	–	–	–	–	–
Chery	582	157,474	–	–	–	1,554	61	–	–	–	–	–
China National	–	137,802	–	–	–	–	–	–	–	–	–	–
Chrysler	18,204	–	–	–	–	–	775	–	–	–	–	–
Daimler	12,821	57,665	1,280	64,294	34,715	2,593	1,343	–	844	–	–	13,930
Dongfeng	–	906,335	–	–	–	–	–	–	–	–	–	–
FAW	–	404,443	–	–	–	–	–	–	–	–	–	–
Fiat	3,005	17,008	–	–	–	–	129	–	–	–	–	–
Ford	42,170	1,800	1,510	4,784	–	3,056	5,077	–	4,982	–	–	7,428
Foton	111	512,150	–	–	–	–	–	–	26	–	–	–
GAZ	–	–	–	–	–	–	–	–	–	–	–	–
GM	39,781	1,441,497	21,155	–	–	1,074	4,043	–	1,597	–	36,478	287
Great Wall Zhongxing	11,006	444,010	–	–	–	–	995	–	–	–	–	–
Honda	4,733	230,784	294	–	32,668	4,078	623	–	1,771	–	–	–
Hyundai Group	29,849	315,865	1,002	638	–	8,874	4,500	–	16,681	–	452,101	4,051
Isuzu	14,972	1,865	–	21,888	59,805	10,673	885	458	11,759	–	–	–
Jianghuai Auto Co	–	348,440	–	–	–	–	27	–	–	–	–	–
Lifan Motor Co	–	113,958	–	–	–	–	–	–	–	–	–	–
Mahindra & Mahindra	–	–	432,628	–	–	1	–	–	–	–	–	–
Mazda	35,456	24,032	12,169	4,337	25,898	2,280	2,965	–	1,459	–	–	–
Mitsubishi	43,012	7,821	–	33,546	33,928	10,436	3,674	–	31,530	–	–	20,523
Nissan	64,622	178,418	800	–	98,924	20,603	5,370	–	5,571	–	–	674
PACCAR	2,678	–	–	–	–	–	209	–	–	–	–	391
Perodua	–	–	–	–	–	41,385	–	–	–	–	–	–
Peugeot Citroen	1,443	–	–	–	–	917	232	–	–	–	–	–
Proton	149	–	–	–	–	24,117	–	–	–	–	–	–
Renault	1,905	–	24,142	–	–	81	42	–	–	–	4,936	–
Shanghai AIC	–	152,376	–	–	–	–	–	–	114	–	–	–
Subaru	27,175	–	–	–	36,656	–	1,125	–	–	–	–	–
Suzuki	4,937	–	185,327	75,004	142,898	1,213	617	33,205	1,620	–	–	4,008
Tata Motors	7,929	–	602,665	–	–	643	455	–	–	–	7,338	–
Toyota	123,250	220,290	91,216	127,773	307,353	60,371	11,727	6,012	38,411	–	–	18,596
Volkswagen	25,779	299,263	4,450	–	–	2,104	2,436	–	–	–	–	7,957
Volvo Truck	3,473	–	1,274	2,925	9,104	1,445	340	254	270	–	–	141
Zhejiang Geely	3,740	75,251	–	–	–	271	154	–	102	–	–	–
Other	4,722	821,206	192,655	256	15,439	1,741	1,606	883	158	4,341	45,523	2,677
TOTAL CV 2012	**535,177**	**8,623,008**	**1,577,697**	**335,445**	**797,388**	**200,142**	**50,203**	**40,812**	**117,104**	**4,341**	**546,376**	**84,171**
TOTAL CV 2011	**449,123**	**8,465,185**	**1,349,076**	**279,184**	**685,431**	**181,770**	**39,422**	**38,763**	**107,345**	**6,077**	**549,044**	**73,844**
Vol. % Chg.	**19.2**	**1.9**	**16.9**	**20.2**	**16.3**	**10.1**	**27.3**	**5.3**	**9.1**	**-28.6**	**-0.5**	**14.0**

Note: Total World represents countries for which data is presented.
Source: WardsAuto InfoBank.

COMMERCIAL VEHICLE SALES BY COMPANY GROUP AND COUNTRY, 2012

Company	Thailand	Vietnam	Total Asia Pacific	Austria	Belgium	Czech Republic	Denmark	Finland	France	Germany	Greece	Hungary
Auto Vaz	–	–	–	95	14	–	–	–	166	–	–	–
Beijing AIC	–	–	205,056	–	–	–	–	–	–	–	–	–
BMW	500	–	39,086	–	14	–	79	8	117	99	–	–
Brilliance	–	460	372,516	–	–	–	–	–	–	–	–	–
BYD	–	–	88,247	–	–	–	–	–	–	–	–	–
Chana	–	–	1,067,255	–	–	–	–	–	–	–	–	–
Chery	–	–	159,671	–	–	–	–	–	–	–	–	–
China National	–	–	137,802	–	–	–	–	–	–	–	–	–
Chrysler	–	–	18,979	126	111	2	1	6	696	925	–	–
Daimler	2,098	1,165	192,748	3,531	7,081	2,508	3,246	2,341	27,515	82,096	43	1,101
Dongfeng	2,158	–	908,493	–	–	–	–	–	–	–	–	–
FAW	–	–	404,443	–	–	–	–	–	–	–	–	–
Fiat	–	–	20,142	4,196	5,421	3,037	3,176	610	52,687	30,309	445	2,080
Ford	20,885	2,882	94,574	3,903	3,824	1,767	2,921	2,498	18,479	24,754	342	1,745
Foton	–	–	512,287	–	–	–	–	–	–	–	–	–
GAZ	–	–	–	–	–	–	–	–	1	–	–	–
GM	53,971	1,761	1,601,644	2,057	3,587	155	2,172	577	7,291	10,885	239	970
Great Wall Zhongxing	–	–	456,011	–	–	–	–	–	–	–	–	–
Honda	7,766	1,010	283,727	–	–	–	17	4	23	–	–	–
Hyundai Group	5,596	12,196	851,353	151	356	61	274	26	553	540	1	939
Isuzu	213,380	1,493	337,178	216	485	44	1	138	1,876	768	235	–
Jianghuai Auto Co	–	–	348,467	–	–	–	–	–	–	–	–	–
Lifan Motor Co	–	–	113,958	–	–	–	–	–	–	–	–	–
Mahindra & Mahindra	–	–	432,629	2	–	–	–	–	1	–	13	–
Mazda	28,446	–	137,042	50	12	1	45	1	160	28	37	135
Mitsubishi	87,033	1,141	272,644	471	650	79	54	86	1,736	2,266	160	314
Nissan	20,346	313	395,641	842	1,861	117	1,108	435	9,076	4,933	702	400
PACCAR	–	–	3,278	1,001	1,670	1,348	253	102	5,549	6,635	–	414
Perodua	–	–	41,385	–	–	–	–	–	–	–	–	–
Peugeot Citroen	–	–	2,592	3,367	14,704	2,151	3,518	747	129,019	21,362	327	1,566
Proton	–	–	24,266	–	–	–	–	–	–	–	–	–
Renault	–	–	31,106	3,761	9,981	1,501	1,829	597	124,958	18,731	96	1,465
Shanghai AIC	–	–	152,490	–	–	–	–	–	–	–	–	–
Subaru	–	–	64,956	–	12	–	13	6	14	–	–	–
Suzuki	6,734	3,394	458,957	199	77	1	140	–	324	–	–	–
Tata Motors	4,863	–	623,893	308	556	23	70	52	1,478	1,755	–	9
Toyota	308,329	12,880	1,326,208	1,144	1,354	232	1,171	1,539	4,507	2,129	648	596
Volkswagen	249	–	342,238	8,692	6,848	2,348	5,331	3,194	16,241	61,118	353	1,796
Volvo Truck	959	–	20,185	873	2,139	1,546	1,179	1,063	20,811	8,040	–	879
Zhejiang Geely	200	–	79,718	2	57	–	112	20	159	–	–	–
Other	362	14,785	1,106,354	3,832	2,970	2,865	1,673	1,204	9,534	34,125	395	925
TOTAL CV 2012	**763,875**	**53,480**	**13,729,219**	**38,819**	**63,784**	**19,786**	**28,383**	**15,254**	**432,971**	**311,498**	**4,036**	**15,334**
TOTAL CV 2011	**433,725**	**69,687**	**12,727,676**	**40,510**	**72,027**	**22,068**	**28,471**	**18,302**	**482,824**	**334,822**	**6,976**	**15,932**
Vol. % Chg.	**76.1**	**-23.3**	**7.9**	**-4.2**	**-11.4**	**-10.3**	**-0.3**	**-16.7**	**-10.3**	**-7.0**	**-42.1**	**-3.8**

Note: Total World represents countries for which data is presented.
Source: WardsAuto InfoBank.

COMMERCIAL VEHICLE SALES BY COMPANY GROUP AND COUNTRY, 2012

Company	Ireland	Italy	Lithuania	Luxem-bourg	Nether-lands	Norway	Poland	Portugal	Romania	Russia	Serbia	Slovakia
Auto Vaz	–	–	–	–	–	–	–	–	–	–	–	–
Beijing AIC	–	–	–	–	–	–	–	–	–	–	–	–
BMW	–	13	–	–	12	–	–	–	–	–	–	–
Brilliance	–	–	–	–	–	–	–	–	–	–	–	–
BYD	–	–	–	–	1	–	–	–	–	–	–	–
Chana	–	–	–	–	–	–	–	–	–	–	–	–
Chery	–	–	–	–	–	–	–	–	–	–	–	–
China National	–	–	–	–	–	–	–	–	–	–	–	–
Chrysler	2	6	2	13	474	257	–	150	–	–	–	–
Daimler	435	6,965	77	755	9,625	3,316	6,210	1,179	1,716	4,660	–	805
Dongfeng	–	–	–	–	5	–	–	–	–	–	–	1
FAW	–	–	–	–	–	–	–	–	–	–	–	–
Fiat	453	59,905	200	579	4,869	1,588	12,613	1,966	1,958	1,030	–	2,108
Ford	2,278	6,049	125	300	4,607	2,794	2,912	751	1,660	7,670	–	228
Foton	–	–	–	–	–	–	–	–	–	–	–	–
GAZ	–	–	23	–	3	–	–	–	–	32,380	–	–
GM	491	3,976	27	181	5,849	1,067	1,016	1,080	248	–	–	205
Great Wall Zhongxing	–	970	–	–	–	–	–	–	–	–	–	–
Honda	–	–	–	–	–	–	–	–	–	–	–	–
Hyundai Group	195	95	10	13	216	359	255	58	–	920	–	3
Isuzu	81	1,587	17	61	37	669	–	183	38	1,560	–	3
Jianghuai Auto Co	–	–	–	–	–	–	–	–	–	–	–	–
Lifan Motor Co	–	–	–	–	–	–	–	–	–	–	–	–
Mahindra & Mahindra	–	219	–	–	–	–	–	–	–	–	–	5
Mazda	–	93	–	–	1	7	–	54	4	–	–	–
Mitsubishi	138	756	35	16	78	803	443	649	254	–	–	20
Nissan	521	4,843	62	87	816	1,003	903	314	273	–	–	144
PACCAR	117	1,228	–	167	3,762	122	3,545	82	420	–	–	415
Perodua	–	–	–	–	–	–	–	–	–	–	–	–
Peugeot Citroen	1,062	17,026	175	705	8,928	5,250	7,267	4,817	1,686	–	–	1,524
Proton	–	–	–	–	–	–	–	–	–	–	–	–
Renault	1,266	10,497	218	676	7,264	521	5,298	2,630	3,480	–	–	652
Shanghai AIC	–	–	–	–	–	–	–	–	–	–	–	–
Subaru	14	28	–	–	–	–	–	–	–	–	–	–
Suzuki	–	62	–	3	3	–	–	12	–	–	–	–
Tata Motors	442	1,799	1	18	180	223	–	41	–	–	–	5
Toyota	1,015	658	155	24	517	2,813	827	1,161	238	–	–	89
Volkswagen	2,325	6,640	471	412	15,443	11,180	4,366	1,711	1,571	–	–	530
Volvo Truck	346	2,469	–	238	2,546	1,967	3,542	779	680	6,250	–	1,084
Zhejiang Geely	–	100	–	–	38	–	–	–	–	–	–	–
Other	1,049	5,263	2,904	403	4,075	5,003	7,017	504	1,100	90,090	–	1,100
TOTAL CV 2012	**12,230**	**131,247**	**4,502**	**4,651**	**69,349**	**38,942**	**56,214**	**18,121**	**15,326**	**144,560**	**–**	**8,921**
TOTAL CV 2011	**12,552**	**193,078**	**4,690**	**5,136**	**71,945**	**41,968**	**61,336**	**37,883**	**16,044**	**133,800**	**–**	**9,701**
Vol. % Chg.	**-2.6**	**-32.0**	**-4.0**	**-9.4**	**-3.6**	**-7.2**	**-8.4**	**-52.2**	**-4.5**	**8.0**	**–**	**-8.0**

Note: Total World represents countries for which data is presented.
Source: WardsAuto InfoBank.

COMMERCIAL VEHICLE SALES BY COMPANY GROUP AND COUNTRY, 2012

Company	Slovenia	Spain	Sweden	Switzer-land	Turkey	U.K.	Ukraine	Total Europe	Canada	Mexico	U.S.	Total North America
Auto Vaz	–	–	–	–	–	–	–	275	–	–	–	–
Beijing AIC	–	–	–	–	–	–	–	–	–	–	–	–
BMW	275	–	–	–	–	86	–	703	16,497	3,826	116,326	136,649
Brilliance	–	–	–	–	–	–	–	–	–	–	–	–
BYD	–	–	–	–	–	–	–	1	–	–	–	–
Chana	–	–	–	–	–	–	–	–	–	–	–	–
Chery	–	–	–	–	–	–	–	–	–	–	–	–
China National	–	–	–	–	–	–	–	–	–	–	–	–
Chrysler	37	1	251	–	–	–	–	3,060	201,336	57,940	1,153,692	1,412,968
Daimler	732	6,494	4,113	5,357	23,971	29,885	905	236,662	27,910	12,095	220,471	260,476
Dongfeng	–	–	–	–	921	–	–	927	–	–	–	–
FAW	–	–	–	–	–	–	130	130	–	–	–	–
Fiat	685	11,789	2,601	5,491	73,792	15,534	3,241	302,363	–	1,066	–	1,066
Ford	1,152	8,529	3,544	2,812	65,170	65,251	1,476	237,541	207,961	47,266	1,189,519	1,744,746
Foton	–	–	–	–	–	–	–	–	–	–	–	–
GAZ	–	2	–	–	–	–	3,563	35,972	–	–	–	–
GM	1,214	3,112	1,522	2,156	247	26,597	198	77,119	154,858	42,256	1,564,859	1,761,973
Great Wall Zhongxing	–	–	–	–	–	476	–	1,446	–	–	–	–
Honda	85	–	–	–	–	–	–	129	61,602	34,359	638,548	734,509
Hyundai Group	1,651	2	90	159	3,299	379	496	11,101	68,787	–	301,592	370,379
Isuzu	–	223	355	–	5,294	3,904	517	18,292	528	3,443	12,778	16,749
Jianghuai Auto Co	–	–	–	–	–	–	–	–	–	–	–	–
Lifan Motor Co	–	–	–	–	–	–	–	–	–	–	–	–
Mahindra & Mahindra	–	19	–	–	–	–	–	259	–	–	–	–
Mazda	265	1	38	59	203	–	–	1,194	21,526	9,088	94,229	124,843
Mitsubishi	100	313	769	291	3,853	4,853	–	19,187	11,715	6,003	26,342	44,060
Nissan	994	5,033	2,843	1,440	1,247	10,142	–	50,139	39,405	78,091	464,665	582,161
PACCAR	–	1,189	129	161	–	11,153	150	39,612	11,123	11,314	63,957	86,394
Perodua	–	–	–	–	–	–	–	–	–	–	–	–
Peugeot Citroen	4,937	25,088	6,264	4,076	21,085	40,014	2,333	328,998	–	753	–	753
Proton	–	–	–	–	–	–	–	–	–	–	–	–
Renault	3,778	12,458	4,177	4,987	17,629	14,925	–	253,375	–	11,989	–	11,989
Shanghai AIC	–	–	–	–	–	–	–	–	–	–	–	–
Subaru	52	–	–	–	–	–	–	139	16,600	132	203,371	220,103
Suzuki	157	–	–	–	–	–	–	978	1,807	2,540	6,932	11,279
Tata Motors	121	38	18	267	360	5,917	226	13,907	4,268	927	43,664	48,859
Toyota	273	1,666	1,673	1,369	2,615	7,945	–	36,358	87,151	32,657	889,925	1,009,733
Volkswagen	4,475	7,350	12,381	5,538	26,048	31,207	–	237,569	16,054	18,041	95,561	129,656
Volvo Truck	4	4,626	2,551	761	–	7,968	2,420	74,761	6,044	1,449	39,252	46,745
Zhejiang Geely	47	–	–	–	–	–	–	535	3,113	579	34,165	37,857
Other	1,333	3,469	3,223	3,210	15,606	12,918	10,347	226,137	9,363	9,524	84,188	103,075
TOTAL CV 2012	**22,367**	**91,402**	**46,542**	**38,134**	**261,340**	**289,154**	**26,002**	**2,208,869**	**967,648**	**385,338**	**7,544,036**	**8,897,022**
TOTAL CV 2011	**25,588**	**123,353**	**54,082**	**36,298**	**319,060**	**306,488**	**23,437**	**2,498,371**	**938,265**	**349,037**	**6,951,210**	**8,238,512**
Vol. % Chg.	**-12.6**	**-25.9**	**-13.9**	**5.1**	**-18.1**	**-5.7**	**10.9**	**-11.6**	**3.1**	**10.4**	**8.5**	**8.0**

Note: Total World represents countries for which data is presented.
Source: WardsAuto InfoBank.

COMMERCIAL VEHICLE SALES BY COMPANY GROUP AND COUNTRY, 2012

Company	Argentina	Brazil	Chile	Colombia	Uruguay	Venezuela	Total South America	Egypt	Israel	S. Africa	2012	2011	Vol. % Chg.
Auto Vaz	–	–	–	–	–	–	–	–	–	–	275	289	-4.8
Beijing AIC	–	–	–	–	–	–	–	–	–	–	2[illegible]5,056	170,578	20.2
BMW	–	3,404	4	–	–	14	3,422	–	401	–	1[illegible]0,261	131,816	36.8
Brilliance	–	1,287	336	148	–	–	1,771	631	–	–	3[illegible]4,918	339,313	10.5
BYD	–	–	–	–	–	–	–	–	–	–	[illegible]8,248	63,356	39.3
Chana	–	11,403	3,884	232	665	–	16,184	1,485	–	191	1,0[illegible]5,115	993,756	9.2
Chery	–	2,052	3	–	–	–	2,055	–	–	–	1[illegible]1,726	189,392	-14.6
China National	–	–	–	–	47	–	47	–	–	–	1[illegible]7,849	159,162	-13.4
Chrysler	–	6,417	867	–	–	8,359	15,643	–	1,438	–	1,4[illegible]2,088	1,295,831	12.1
Daimler	12,464	52,121	8,349	2,519	236	1,121	76,810	943	2,151	7,779	7[illegible]7,569	744,130	4.5
Dongfeng	–	–	1,673	1	95	2,433	4,202	–	–	–	9[illegible]3,622	1,008,607	-9.4
FAW	–	2	764	–	6	352	1,124	–	–	–	4[illegible]5,697	497,629	-18.5
Fiat	25,566	174,929	2,543	–	122	2,936	206,096	139	2,839	1,672	5[illegible]4,317	622,180	-14.1
Ford	45,413	90,034	5,074	–	49	15,756	156,326	–	2,479	16,545	2,2[illegible]2,211	2,240,341	0.5
Foton	–	2	855	–	311	–	1,168	–	–	–	5[illegible]3,455	541,026	-5.1
GAZ	–	–	–	–	–	–	–	–	–	–	[illegible]5,972	35,935	0.1
GM	11,100	106,901	9,964	5,175	–	10,749	143,889	28,214	2,938	19,555	3,6[illegible]5,332	3,436,174	5.8
Great Wall Zhongxing	–	–	3,158	–	–	–	3,158	–	203	–	4[illegible]0,818	342,412	34.6
Honda	3,358	14,895	491	–	–	–	18,744	–	370	–	1,0[illegible]7,479	869,804	19.3
Hyundai Group	–	67,025	7,087	1,485	411	3,707	79,715	685	5,774	–	1,3[illegible]9,007	1,351,634	-2.4
Isuzu	–	–	–	–	–	4,426	4,426	–	1,834	16,515	3[illegible]4,994	283,610	39.3
Jianghuai Auto Co	–	–	1,969	–	443	–	2,412	672	–	–	3[illegible]1,551	351,929	-0.1
Lifan Motor Co	–	–	464	–	–	–	464	–	–	–	1[illegible]4,422	103,620	10.4
Mahindra & Mahindra	–	327	2,874	–	–	–	3,201	57	–	2,294	4[illegible]8,440	350,108	25.2
Mazda	–	–	678	–	–	3,102	3,780	–	2,090	1,961	2[illegible]0,910	236,495	14.6
Mitsubishi	–	53,834	6,582	–	–	815	61,231	1,350	1,402	958	4[illegible]0,832	402,405	-0.4
Nissan	–	17,602	14,507	–	–	11	32,120	–	2,904	29,173	1,0[illegible]2,138	1,000,340	9.2
PACCAR	–	3	140	–	–	–	143	–	898	86	1[illegible]0,411	117,299	11.2
Perodua	–	–	–	–	–	–	–	–	–	–	[illegible]1,385	37,402	10.6
Peugeot Citroen	17,052	21,748	6,216	–	–	760	45,776	81	3,613	606	3[illegible]2,419	444,662	-14.0
Proton	–	–	–	–	–	–	–	–	–	–	[illegible]4,266	24,459	-0.8
Renault	45,087	60,856	488	–	–	448	106,879	–	1,201	339	4[illegible]4,889	353,286	14.6
Shanghai AIC	–	–	371	–	24	–	395	–	–	–	1[illegible]2,885	298,120	-48.7
Subaru	–	932	–	–	–	–	932	–	849	–	2[illegible]6,979	272,942	5.1
Suzuki	1	5,569	412	–	–	5	5,987	5,274	258	–	4[illegible]2,733	405,864	18.9
Tata Motors	–	8,180	394	–	–	–	8,574	–	207	3,894	6[illegible]9,334	645,086	8.4
Toyota	33,579	50,150	8,102	1,627	–	11,608	105,066	8,291	1,866	54,260	2,5[illegible]1,782	2,093,135	21.4
Volkswagen	20,919	118,113	2,489	–	–	1,565	143,086	508	5,863	8,505	8[illegible]7,425	750,888	15.5
Volvo Truck	–	17,610	2,480	5,313	158	1,316	26,877	–	1,672	4,912	1[illegible]5,152	174,641	0.3
Zhejiang Geely	–	2,396	–	–	–	–	2,396	–	196	–	1[illegible]0,702	82,173	46.9
Other	19,122	72,131	12,100	18,211	1,037	294	122,895	7,799	1,718	5,390	1,5[illegible]3,368	1,652,443	-4.8
TOTAL CV 2012	**233,661**	**959,923**	**105,318**	**34,711**	**3,604**	**69,777**	**1,406,994**	**56,129**	**45,164**	**174,635**	**26,5[illegible]8,032**	**25,114,272**	**5.6**
TOTAL CV 2011	**209,497**	**983,974**	**111,067**	**28,268**	**3,561**	**55,350**	**1,391,717**	**42,992**	**48,021**	**166,983**	**25,1[illegible]4,272**		
Vol. % Chg.	**11.5**	**-2.4**	**-5.2**	**22.8**	**1.2**	**26.1**	**1.1**	**30.6**	**-5.9**	**4.6**	**5.6**		

Note: Total World represents countries for which data is presented.
Source: WardsAuto InfoBank.

WORLD

TOTAL VEHICLE SALES BY COMPANY GROUP AND COUNTRY, 2012

Company	Australia	China	India	Indonesia	Japan	Malaysia	New Zealand	Pakistan	Philippines	Singapore	S. Korea	Taiwan
Auto Vaz	–	–	–	–	–	–	–	–	–	–	–	–
Beijing AIC	–	225,708	–	–	–	–	–	–	–	–	–	–
BMW	20,829	147,374	9,016	2,188	–	6,659	2,069	–	839	5,373	34,106	12,642
Brilliance	–	490,605	–	–	–	–	–	–	–	1	–	–
BYD	–	456,318	–	–	–	–	–	–	–	–	–	–
Chana	–	1,294,265	–	–	–	–	–	–	–	2	–	–
Chery	1,133	563,305	–	–	–	1,636	226	–	–	5	–	–
China National	–	137,802	–	–	–	–	–	–	–	–	–	–
Chrysler	21,401	–	–	762	–	–	1,408	–	–	38	4,123	–
Daimler	27,839	125,400	6,106	68,385	34,715	8,085	2,180	–	1,444	4,325	20,389	24,716
Dongfeng	–	981,419	–	–	–	–	–	–	–	–	–	–
FAW	–	739,318	–	680	–	–	–	–	–	–	–	–
Fiat	4,646	28,296	11,145	–	–	–	433	–	–	238	–	–
Ford	90,408	276,322	87,600	11,958	–	7,108	11,051	–	9,398	354	5,126	18,459
Foton	111	512,150	–	–	–	–	–	–	26	9	–	–
GAZ	–	–	–	–	–	–	–	–	–	–	–	–
GM	115,206	2,821,720	92,059	5,643	1	2,026	9,426	–	3,328	537	[illegible]46,177	954
Great Wall Zhongxing	11,006	624,602	–	–	–	–	996	–	–	–	–	–
Honda	35,812	627,610	73,134	69,320	745,018	34,950	3,003	13,551	12,653	515	3,944	18,794
Hyundai Group	122,359	1,367,303	391,276	20,035	–	24,265	10,545	–	29,523	1,755	1[illegible]49,837	17,063
Isuzu	14,972	1,865	–	33,165	59,805	10,673	885	458	11,759	–	–	–
Jianghuai Auto Co	–	486,760	–	–	–	–	27	–	–	–	–	–
Lifan Motor Co	–	270,303	–	–	–	–	–	–	–	–	–	–
Mahindra & Mahindra	–	–	449,149	–	–	1	–	–	–	–	–	–
Mazda	103,886	336,084	12,169	12,391	218,361	6,332	6,298	–	1,504	348	–	13,222
Mitsubishi	58,868	107,793	–	85,376	127,726	11,652	5,174	–	34,272	61	81	41,269
Nissan	79,832	778,793	45,239	67,179	617,433	34,966	6,359	–	6,404	759	3,501	40,901
PACCAR	2,678	–	–	–	–	–	209	–	–	–	–	391
Perodua	–	–	–	–	–	189,137	–	–	–	8	–	–
Peugeot Citroen	6,773	440,028	–	333	–	6,114	1,266	–	–	917	2,662	–
Proton	1,074	45,421	–	1,017	–	141,121	1	–	–	62	–	–
Renault	4,581	–	35,157	–	–	90	107	–	–	114	59,926	–
Shanghai AIC	–	345,298	–	–	–	–	–	–	114	1	–	–
Subaru	40,189	–	–	204	177,721	53	1,686	–	1,752	135	627	1,611
Suzuki	24,057	170,037	1,063,599	126,577	672,110	8,087	5,412	95,763	4,501	144	–	4,008
Tata Motors	8,715	–	838,045	331	–	643	568	–	–	767	8,535	–
Toyota	229,231	751,317	172,241	602,984	2,391,675	114,507	22,815	46,672	65,876	4,309	15,771	131,054
Volkswagen	72,140	2,608,896	110,412	1,449	–	14,417	6,000	–	123	5,459	33,656	19,622
Volvo Truck	3,473	–	1,274	2,925	9,104	1,445	340	254	270	–	–	141
Zhejiang Geely	5,375	547,508	–	1,040	–	797	251	–	212	1,066	1,768	4,181
Other	5,438	996,815	195,595	2,288	316,052	2,989	1,849	883	250	10,330	51,486	16,843
TOTAL VEHICLE 2012	**1,112,032**	**19,306,435**	**3,593,216**	**1,116,230**	**5,369,721**	**627,753**	**100,584**	**157,581**	**184,248**	**37,632**	**1,[illegible]41,715**	**365,871**
TOTAL VEHICLE 2011	**1,008,437**	**18,505,352**	**3,299,439**	**894,183**	**4,210,220**	**599,760**	**84,484**	**163,271**	**164,921**	**39,809**	**1,[illegible]79,674**	**378,287**
Vol. % Chg.	**10.3**	**4.3**	**8.9**	**24.8**	**27.5**	**4.7**	**19.1**	**-3.5**	**11.7**	**-5.5**	**-2.4**	**-3.3**

Note: Total World represents countries for which data is presented.
Source: WardsAuto InfoBank.

TOTAL VEHICLE SALES BY COMPANY GROUP AND COUNTRY, 2012

Company	Thailand	Vietnam	Total Asia Pacific	Austria	Belgium	Czech Republic	Denmark	Finland	France	Germany	Greece	Hungary
Auto Vaz	–	–	–	136	40	93	–	–	414	1,575	–	11
Beijing AIC	–	–	225,708	–	–	–	–	–	–	–	–	–
BMW	6,256	–	247,351	17,961	35,905	4,058	3,329	4,475	69,667	284,650	1,996	1,515
Brilliance	–	460	491,066	–	–	–	–	–	–	–	–	–
BYD	–	–	456,318	–	–	–	–	–	–	–	–	–
Chana	–	–	1,294,267	–	–	–	–	–	–	–	–	–
Chery	35	157	566,497	–	–	–	–	–	–	–	–	–
China National	–	–	137,802	–	–	–	–	–	–	–	–	–
Chrysler	–	–	27,732	1,007	754	380	7	257	3,939	7,958	76	69
Daimler	8,372	2,383	334,339	14,884	29,124	5,735	7,075	7,342	80,523	393,155	1,846	2,196
Dongfeng	2,158	–	983,577	–	–	–	–	–	–	–	–	–
FAW	–	–	739,998	–	–	–	–	–	–	–	–	–
Fiat	–	–	44,758	20,357	24,626	5,935	11,015	1,605	112,201	115,111	5,399	4,794
Ford	55,901	4,790	578,475	25,527	32,578	14,290	16,044	11,002	110,947	230,882	3,348	6,941
Foton	–	–	512,296	–	–	–	–	–	–	–	–	–
GAZ	–	–	–	–	–	–	–	–	1	–	–	–
GM	75,457	5,613	3,278,147	26,054	44,408	6,766	15,648	5,528	103,717	254,413	8,235	9,028
Great Wall Zhongxing	–	–	636,604	–	–	–	–	–	–	–	–	–
Honda	171,208	1,804	1,811,316	2,200	2,439	1,659	2,129	3,144	8,429	26,306	657	1,065
Hyundai Group	6,280	18,865	3,159,106	32,078	26,831	23,700	20,172	10,777	62,304	156,213	4,134	3,971
Isuzu	213,380	1,493	348,455	216	485	44	1	138	1,876	768	235	–
Jianghuai Auto Co	–	–	486,787	–	–	–	–	–	–	–	–	–
Lifan Motor Co	–	–	270,303	–	–	–	–	–	–	–	8	–
Mahindra & Mahindra	–	–	449,150	2	–	–	–	–	1	–	13	–
Mazda	72,728	–	783,323	8,474	3,186	1,171	2,808	1,346	5,267	40,730	259	621
Mitsubishi	129,460	1,141	602,873	4,243	3,782	1,135	492	1,377	5,375	25,831	768	483
Nissan	123,850	313	1,805,529	9,477	19,007	3,288	5,578	7,631	79,209	68,017	3,846	2,746
PACCAR	–	–	3,278	1,001	1,670	1,348	253	102	5,549	6,635	–	414
Perodua	–	–	189,145	–	–	–	–	–	–	–	–	–
Peugeot Citroen	331	–	458,424	28,847	91,874	14,424	31,810	7,055	700,889	153,835	7,076	4,290
Proton	1,789	–	190,485	8	37	–	1	–	74	56	–	–
Renault	–	–	99,975	26,911	68,314	15,692	10,524	2,477	549,093	169,471	1,489	7,055
Shanghai AIC	–	–	345,413	–	–	–	–	–	–	–	–	–
Subaru	118	–	224,096	1,638	528	815	64	1,031	985	9,824	41	100
Suzuki	24,685	3,409	2,202,389	6,079	5,100	2,310	6,560	814	16,350	30,843	2,278	3,300
Tata Motors	4,899	–	862,503	2,541	5,291	666	152	477	10,145	16,065	15	110
Toyota	533,145	25,559	5,107,156	8,954	18,869	4,286	17,125	15,130	75,322	86,859	6,641	2,918
Volkswagen	899	–	2,873,073	127,310	112,926	76,967	43,505	33,421	279,510	1,216,420	12,253	13,766
Volvo Truck	959	–	20,185	873	2,139	1,546	1,179	1,063	20,811	8,040	–	879
Zhejiang Geely	1,854	–	564,052	3,128	15,402	1,458	1,758	7,573	13,555	32,750	1,380	1,137
Other	2,571	14,785	1,618,174	4,923	5,206	6,029	1,917	2,738	15,578	57,595	525	985
TOTAL VEHICLE 2012	**1,436,335**	**80,772**	**35,030,125**	**374,829**	**550,521**	**193,795**	**199,146**	**126,503**	**2,331,731**	**3,394,002**	**62,518**	**68,394**
TOTAL VEHICLE 2011	**794,944**	**111,116**	**31,833,897**	**396,655**	**644,238**	**195,350**	**198,445**	**144,418**	**2,687,053**	**3,508,456**	**104,658**	**61,026**
Vol. % Chg.	**80.7**	**-27.3**	**10.0**	**-5.5**	**-14.5**	**-0.8**	**0.4**	**-12.4**	**-13.2**	**-3.3**	**-40.3**	**12.1**

Note: Total World represents countries for which data is presented.
Source: WardsAuto InfoBank.

TOTAL VEHICLE SALES BY COMPANY GROUP AND COUNTRY, 2012

Company	Ireland	Italy	Lithuania	Luxem-bourg	Nether-lands	Norway	Poland	Portugal	Romania	Russia	Serbia	Slovakia
Auto Vaz	–	69	19	8	3	–	–	–	–	537,625	509	87
Beijing AIC	–	–	–	–	–	–	–	–	–	1,845	–	–
BMW	3,769	60,302	474	5,497	22,688	8,181	5,555	7,422	2,329	40,144	510	2,418
Brilliance	–	–	–	–	–	–	–	–	–	–	–	–
BYD	–	–	–	–	1	–	–	–	–	2,630	–	–
Chana	–	–	–	–	26	–	–	–	–	–	–	–
Chery	–	–	–	–	–	–	–	–	–	19,004	92	–
China National	–	–	–	–	–	–	–	–	–	–	–	–
Chrysler	117	6,561	41	263	869	467	–	183	183	5,104	59	184
Daimler	2,269	70,754	236	4,135	21,358	8,980	11,625	8,368	3,302	46,098	552	2,569
Dongfeng	–	–	–	–	5	–	–	–	–	–	–	1
FAW	–	–	–	–	–	–	–	–	–	1,151	–	–
Fiat	1,237	468,638	1,149	2,478	28,417	2,508	25,545	8,008	3,669	9,653	3,197	3,831
Ford	10,841	105,046	556	2,836	40,468	13,801	19,835	5,589	6,342	138,479	1,763	2,389
Foton	–	–	–	–	–	–	–	–	–	141	–	–
GAZ	–	–	23	–	3	–	–	–	–	122,627	–	–
GM	5,621	114,020	534	2,704	50,106	4,767	31,040	9,524	4,355	376,540	2,251	4,629
Great Wall Zhongxing	–	980	–	–	–	–	–	–	–	14,373	–	–
Honda	1,102	5,314	215	233	2,969	1,731	6,896	1,487	553	21,512	391	1,100
Hyundai Group	8,518	70,573	1,078	2,240	49,165	10,833	30,747	3,094	4,310	362,536	2,679	10,084
Isuzu	81	1,587	17	61	37	672	–	183	38	1,716	–	3
Jianghuai Auto Co	–	–	–	–	–	–	–	–	–	–	–	–
Lifan Motor Co	–	–	–	–	–	–	–	–	–	20,544	–	–
Mahindra & Mahindra	–	279	–	–	–	–	–	–	–	–	–	5
Mazda	1,433	4,396	208	124	6,150	2,934	–	430	436	44,443	243	741
Mitsubishi	499	4,309	200	186	3,662	5,457	4,301	1,517	1,020	74,294	88	497
Nissan	6,544	54,123	891	1,501	10,233	7,661	14,280	4,235	1,848	162,956	429	1,917
PACCAR	117	1,228	–	167	3,762	122	3,545	82	420	–	–	415
Perodua	8	–	–	–	–	–	–	–	–	–	–	–
Peugeot Citroen	4,449	155,536	1,119	7,083	74,773	15,042	29,276	17,939	4,393	77,279	295	7,865
Proton	–	–	–	10	1	–	–	1	–	–	–	–
Renault	6,521	96,349	858	6,361	53,243	1,184	28,455	14,087	28,213	189,852	2,638	6,197
Shanghai AIC	–	–	–	–	–	–	–	–	–	–	–	–
Subaru	157	3,719	315	110	967	3,559	–	3	175	14,296	23	237
Suzuki	630	14,793	81	589	12,522	2,060	5,988	169	1,638	32,684	460	2,431
Tata Motors	1,140	15,789	103	957	1,863	954	–	572	794	20,549	107	311
Toyota	11,236	58,397	1,386	1,239	38,296	20,209	20,680	5,040	2,616	168,700	804	2,152
Volkswagen	22,461	192,249	4,059	13,874	124,253	46,807	65,084	21,903	18,458	315,937	6,325	25,367
Volvo Truck	346	2,469	–	238	2,546	1,967	3,542	779	680	6,250	–	1,084
Zhejiang Geely	1,154	14,918	166	1,201	16,963	11,550	4,480	1,977	455	37,930	79	456
Other	1,478	10,938	2,946	954	6,544	5,463	18,929	819	1,278	212,779	16	1,219
TOTAL VEHICLE 2012	**91,728**	**1,533,336**	**16,674**	**55,049**	**571,893**	**176,909**	**329,803**	**113,411**	**87,505**	**3,079,671**	**23,510**	**78,189**
TOTAL VEHICLE 2011	**102,448**	**1,941,221**	**17,913**	**55,015**	**627,865**	**180,313**	**338,766**	**191,316**	**110,668**	**2,787,208**	**31,780**	**77,904**
Vol. % Chg.	**-10.5**	**-21.0**	**-6.9**	**0.1**	**-8.9**	**-1.9**	**-2.6**	**-40.7**	**-20.9**	**10.5**	**-26.0**	**0.4**

Note: Total World represents countries for which data is presented.
Source: WardsAuto InfoBank.

TOTAL VEHICLE SALES BY COMPANY GROUP AND COUNTRY, 2012

Company	Slovenia	Spain	Sweden	Switzer-land	Turkey	U.K.	Ukraine	Total Europe	Canada	Mexico	U.S.	Total North America
Auto Vaz	–	37	24	–	–	–	19,718	560,368	–	–	–	–
Beijing AIC	–	–	–	–	–	–	–	1,845	–	–	–	–
BMW	1,492	32,991	17,527	23,561	17,005	178,940	2,674	857,035	37,707	11,827	347,583	397,117
Brilliance	–	–	–	–	–	–	–	–	–	–	–	–
BYD	–	2	–	–	–	–	–	2,633	–	–	–	–
Chana	–	–	–	–	–	–	–	26	–	–	–	–
Chery	–	–	–	–	810	–	3,519	23,425	–	–	–	–
China National	–	–	–	–	–	–	–	–	–	–	–	–
Chrysler	37	1,674	784	3,088	799	5,639	–	40,499	243,845	88,023	,651,787	1,983,655
Daimler	968	30,123	15,567	23,883	36,797	127,934	3,696	961,094	47,114	20,788	408,225	476,127
Dongfeng	–	–	–	–	921	–	–	927	–	–	–	–
FAW	–	–	–	–	–	–	130	1,281	–	–	–	–
Fiat	2,297	33,607	7,845	19,823	119,454	74,295	5,359	1,122,053	–	5,131	–	5,131
Ford	2,823	59,588	17,396	20,991	112,200	347,168	11,714	1,371,384	276,068	84,382	2,250,165	2,610,615
Foton	–	–	–	–	–	–	–	141	–	–	–	–
GAZ	–	2	–	–	–	–	3,563	126,219	–	–	–	–
GM	5,416	67,871	10,120	21,460	67,882	272,328	16,823	1,541,788	226,825	186,383	2,595,717	3,008,925
Great Wall Zhongxing	–	–	–	–	–	476	–	15,829	–	–	–	–
Honda	300	5,917	2,821	5,540	16,516	54,208	3,752	180,585	148,712	54,515	,422,785	1,626,012
Hyundai Group	5,003	47,827	27,851	16,292	58,834	141,293	39,475	1,232,612	214,083	–	,260,606	1,474,689
Isuzu	–	223	355	–	5,294	3,904	517	18,451	528	3,443	12,778	16,749
Jianghuai Auto Co	–	–	–	–	–	–	–	–	–	–	–	–
Lifan Motor Co	–	–	–	–	–	–	–	20,552	–	–	–	–
Mahindra & Mahindra	–	22	–	–	–	–	–	322	–	–	–	–
Mazda	481	4,600	2,187	7,211	1,013	26,183	1,714	168,789	71,638	25,424	277,046	374,108
Mitsubishi	112	4,319	4,294	4,902	5,142	11,402	5,501	175,188	19,671	8,753	57,790	86,214
Nissan	1,517	40,923	9,305	10,656	20,161	116,507	12,609	677,095	82,194	245,698	1,141,656	1,469,548
PACCAR	–	1,189	129	161	–	11,153	150	39,612	11,123	11,314	63,957	86,394
Perodua	–	–	–	–	–	416	–	424	–	–	–	–
Peugeot Citroen	9,472	133,182	22,431	28,052	50,315	213,156	9,200	1,900,957	–	5,204	–	5,204
Proton	–	2	2	–	760	345	–	1,297	–	–	–	–
Renault	9,902	82,766	14,275	27,380	118,169	55,685	13,567	1,606,728	–	25,030	–	25,030
Shanghai AIC	–	–	–	–	–	782	–	782	–	–	–	–
Subaru	75	995	4,305	8,483	640	2,023	1,142	56,250	31,003	283	336,441	367,727
Suzuki	204	3,717	2,691	7,702	1,780	24,893	1,687	190,353	5,458	10,733	25,358	41,549
Tata Motors	134	7,593	956	4,567	2,718	68,652	1,698	164,919	4,912	1,044	55,675	61,631
Toyota	1,087	39,638	18,987	16,298	34,576	100,912	17,597	795,954	193,784	57,358	2,088,809	2,339,951
Volkswagen	13,454	174,828	86,218	99,945	122,509	431,594	32,486	3,733,889	79,132	164,560	577,443	821,135
Volvo Truck	4	4,626	2,551	761	–	7,968	2,420	74,761	6,044	1,449	39,252	46,745
Zhejiang Geely	123	6,318	52,755	8,059	6,266	31,790	13,144	287,925	5,578	1,513	68,117	75,208
Other	1,348	6,381	4,644	7,458	17,059	25,692	39,749	461,190	10,759	9,838	104,746	125,343
TOTAL VEHICLE 2012	**56,249**	**790,961**	**326,020**	**366,273**	**817,620**	**2,335,338**	**263,604**	**18,415,182**	**1,716,178**	**1,022,693**	**1,785,936**	**17,524,807**
TOTAL VEHICLE 2011	**66,404**	**931,389**	**359,066**	**355,256**	**913,367**	**2,249,483**	**230,890**	**19,508,571**	**1,620,221**	**935,093**	**3,040,613**	**15,595,927**
Vol. % Chg.	**-15.3**	**-15.1**	**-9.2**	**3.1**	**-10.5**	**3.8**	**14.2**	**-5.6**	**5.9**	**9.4**	**13.4**	**12.4**

Note: Total World represents countries for which data is presented.
Source: WardsAuto InfoBank.

TOTAL VEHICLES SALES BY COMPANY GROUP AND COUNTRY, 2012

Company	Argentina	Brazil	Chile	Colombia	Uruguay	Venezuela	Total South America	Egypt	Israel	S. Africa	2012	2011	Vol. % Chg.
Auto Vaz	–	–	–	–	–	–	–	–	–	–	560,368	610,291	-8.2
Beijing AIC	–	–	–	–	–	–	–	–	–	–	227,553	170,578	33.4
BMW	–	10,833	3,286	–	292	18	14,429	2,729	1,679	27,538	1,547,378	1,441,980	7.3
Brilliance	–	1,287	800	148	8	–	2,243	1,305	–	–	494,614	473,907	4.4
BYD	–	–	2,360	301	898	–	3,559	–	–	–	462,510	454,676	1.7
Chana	–	11,403	4,137	1,116	1,455	–	18,111	1,524	–	249	1,314,177	1,187,386	10.7
Chery	–	14,216	5,390	2,238	2,556	–	24,400	6,297	–	–	620,619	702,900	-11.7
China National	–	–	–	–	47	–	47	–	–	–	137,349	159,162	-13.4
Chrysler	–	7,134	8,187	4,178	299	9,423	29,221	1,982	1,776	9,586	2,094,451	1,770,669	18.3
Daimler	16,194	58,569	10,872	3,131	749	1,121	90,636	945	2,906	30,339	1,896,386	1,844,240	2.8
Dongfeng	–	–	1,746	2,065	867	3,126	7,804	–	–	–	992,308	1,028,454	-3.5
FAW	–	2	1,744	–	1,834	352	3,932	–	–	–	745,211	963,601	-22.7
Fiat	85,662	854,977	4,838	1,846	3,783	3,986	955,092	641	5,265	5,986	2,138,926	2,336,021	-8.4
Ford	105,192	345,461	12,045	11,212	1,904	23,447	499,261	863	20,521	42,436	5,123,555	5,231,309	-2.1
Foton	–	2	912	–	384	–	1,298	–	–	–	513,735	541,092	-5.1
GAZ	–	–	–	–	–	–	–	–	–	–	126,219	125,708	0.4
GM	131,515	642,553	59,043	86,242	10,033	39,277	968,663	54,378	13,987	53,442	8,919,330	8,645,694	3.2
Great Wall Zhongxing	–	–	8,202	1,112	855	–	10,169	192	205	–	662,399	514,265	28.9
Honda	15,962	134,938	3,038	2,670	527	–	157,135	691	6,112	9,093	3,790,944	3,059,946	23.9
Hyundai Group	–	149,643	64,816	56,641	3,606	10,618	285,324	62,228	50,752	–	6,264,711	5,820,776	7.6
Isuzu	–	–	–	–	–	4,426	4,426	–	1,834	16,515	406,430	294,234	38.1
Jianghuai Auto Co	–	18,037	5,589	–	1,024	–	24,650	970	–	–	512,407	524,677	-2.3
Lifan Motor Co	–	1,219	1,674	127	302	–	3,322	–	–	–	294,177	224,861	30.8
Mahindra & Mahindra	–	327	4,377	–	123	–	4,827	242	–	3,741	458,282	368,340	24.4
Mazda	–	–	7,853	10,747	155	3,102	21,857	9	11,092	6,788	1,365,366	1,188,247	15.0
Mitsubishi	–	60,572	8,599	1,001	161	7,308	77,641	7,193	2,552	3,577	955,238	981,688	-2.7
Nissan	–	104,791	34,157	25,110	3,828	11	167,897	400	10,071	49,927	4,180,467	3,951,521	5.8
PACCAR	–	3	140	–	–	–	143	–	898	86	130,411	117,299	11.2
Perodua	–	–	–	–	–	–	–	–	–	–	189,569	180,571	5.0
Peugeot Citroen	113,886	146,650	13,821	1,415	3,683	2,322	281,777	1,728	12,697	5,944	2,666,731	2,922,365	-8.7
Proton	–	–	5	–	–	–	5	1,544	–	–	193,331	204,193	-5.3
Renault	124,289	241,556	9,301	39,411	2,354	1,093	418,004	11,895	6,645	10,555	2,178,332	2,409,727	-9.6
Shanghai AIC	–	37	1,831	–	24	–	1,892	–	168	–	348,255	457,627	-23.9
Subaru	–	1,274	3,926	401	29	–	5,630	270	1,746	861	656,580	554,594	18.4
Suzuki	6	7,115	17,389	1,255	3,290	5	29,060	7,281	9,540	4,724	2,484,396	2,328,473	6.7
Tata Motors	–	8,221	1,038	–	7	–	9,266	202	281	14,284	1,113,086	1,030,759	8.0
Toyota	49,301	113,748	28,720	17,545	1,507	17,742	228,563	16,601	20,556	121,276	8,630,057	7,055,353	22.3
Volkswagen	157,379	773,299	9,592	8,642	7,197	1,566	957,675	8,912	29,142	107,611	8,531,137	7,872,895	8.4
Volvo Truck	–	17,610	2,480	5,313	158	1,316	26,877	–	1,672	4,912	175,152	174,641	0.3
Zhejiang Geely	–	2,960	4,378	797	959	–	9,094	1,431	712	2,984	941,106	894,795	5.2
Other	30,672	73,422	16,045	31,304	1,939	294	153,676	7,799	1,735	6,905	2,374,322	2,402,450	-1.1
TOTAL VEHICLE 2012	**830,058**	**3,801,859**	**362,331**	**315,968**	**56,837**	**130,553**	**5,497,606**	**200,252**	**214,544**	**539,359**	**77,421,375**	**73,221,965**	**5.7**
TOTAL VEHICLE 2011	**883,350**	**3,633,006**	**356,183**	**324,570**	**55,004**	**120,689**	**5,372,802**	**176,157**	**236,925**	**497,686**	**73,221,965**		
Vol. % Chg.	**-6.0**	**4.6**	**1.7**	**-2.7**	**3.3**	**8.2**	**2.3**	**13.7**	**-9.4**	**8.4**	**5.7**		

Note: Total World represents countries for which data is presented.
Source: WardsAuto InfoBank.

WORLD VEHICLES IN OPERATION BY VEHICLE TYPE

Year	Cars	Commercial Vehicles	Total
2012	773,322,945	341,234,690	1,114,557,635
2011	753,157,957	321,273,313	1,074,431,270
2010	726,893,302	308,030,199	1,034,923,501
2009	704,882,018	293,142,601	998,024,619
2008	687,209,647	281,796,333	969,005,980
2007	662,876,351	273,321,017	936,197,368
2006	643,470,372	261,814,156	905,284,528
2005	627,606,128	249,931,437	877,537,565
2004	610,643,250	238,035,402	848,678,652
2003	595,860,625	227,617,691	823,478,316
2002	581,867,870	214,323,466	796,191,336
2001	567,392,271	210,298,749	777,691,020
2000	553,943,527	204,176,435	758,119,962
1999	496,058,947	200,308,471	696,367,418
1998	483,621,920	199,384,765	683,006,685
1997	481,755,182	196,055,598	677,810,780
1996	485,954,009	185,404,036	671,358,045
1995	477,010,289	169,748,819	646,759,108
1994	479,532,564	149,544,553	629,077,117
1993	469,460,221	147,626,840	617,087,061
1992	469,942,881	143,587,318	613,530,199
1991	456,032,819	139,273,829	595,306,648
1990	444,899,624	138,082,153	582,981,777
1989	424,365,795	132,565,691	556,931,486
1988	412,907,178	126,882,443	539,789,621
1987	394,029,867	121,176,363	515,206,230
1986	386,349,579	113,436,144	499,785,723
1985	374,483,098	113,023,510	487,506,608
1984	365,104,996	107,924,511	473,029,507
1983	352,031,754	103,887,834	455,919,588
1982	340,265,535	98,787,169	439,052,704

VEHICLES IN OPERATION BY COUNTRY

Country	2012 Cars	2012 Commercial Vehicles	2012 Total	2011 Cars	2011 Commercial Vehicles	2011 Total
AFRICA						
Algeria	2,958,000	1,374,000	4,332,000	2,803,000	1,333,000	4,136,000
Angola	55,200	71,600	126,800	52,700	70,100	122,800
Benin	12,500	19,300	31,800	11,900	18,800	30,700
Botswana	221,209	135,971	357,180	195,000	140,000	335,000
Burkina Faso	131,700	57,200	188,900	125,000	55,000	180,000
Burundi	33,000	27,200	60,200	31,500	26,600	58,100
Cameroon	243,500	74,900	318,400	230,000	73,000	303,000
Central African Republic	2,500	2,150	4,650	2,400	2,100	4,500
Congo	47,700	22,100	69,800	45,600	21,300	66,900
Congo, Democratic Republic of	981,800	740,500	1,722,300	943,000	726,000	1,669,000
Egypt	3,231,513	1,136,388	4,367,901	3,074,862	1,089,244	4,164,106
Ethiopia	82,300	60,200	142,500	79,000	59,000	138,000
Ghana	123,000	63,000	186,000	118,000	62,000	180,000
Ivory Coast	396,700	107,800	504,500	376,000	104,000	480,000
Kenya	652,100	441,700	1,093,800	598,622	427,741	1,026,363
Liberia	16,200	39,800	56,000	15,000	39,000	54,000
Libya	1,831,000	505,700	2,336,700	1,730,000	487,000	2,217,000
Madagascar	137,900	122,700	260,600	130,000	118,000	248,000
Malawi	13,400	18,700	32,100	13,100	18,000	31,100
Mali	11,900	12,500	24,400	11,200	12,000	23,200
Mauritania	14,800	9,600	24,400	14,100	9,400	23,500
Mauritius	147,733	43,152	190,885	136,225	43,771	179,996
Morocco	2,222,000	1,025,000	3,247,000	2,083,710	836,598	2,920,308
Mozambique	46,800	26,800	73,600	44,300	25,800	70,100
Namibia	104,700	133,800	238,500	101,000	130,000	231,000
Niger	117,500	32,900	150,400	109,000	31,000	140,000
Nigeria	825,000	496,700	1,321,700	791,000	487,000	1,278,000
Reunion	254,900	100,900	355,800	245,000	99,000	344,000
Sierra Leone	7,500	16,000	23,500	6,500	13,190	19,690
South Africa	6,112,000	2,832,000	8,944,000	5,832,000	2,741,000	8,573,000
Sudan	43,600	55,400	99,000	42,000	54,000	96,000
Tanzania	26,200	56,400	82,600	25,000	55,000	80,000
Togo	116,300	52,400	168,700	111,000	51,000	162,000
Tunisia	889,000	422,000	1,311,000	848,000	411,000	1,259,000
Uganda	123,600	205,200	328,800	110,000	190,000	300,000
Zambia	186,900	106,800	293,700	179,000	104,000	283,000
Zimbabwe	430,700	72,900	503,600	412,000	71,000	483,000
Total Africa	**22,852,355**	**10,721,361**	**33,573,716**	**21,675,719**	**10,234,644**	**31,910,363**
AMERICA, Caribbean						
Bahamas	94,500	32,400	126,900	93,000	32,000	125,000
Barbados	83,200	19,900	103,100	82,000	19,000	101,000

VEHICLES IN OPERATION BY COUNTRY

Country	2012 Cars	2012 Commercial Vehicles	2012 Total	2011 Cars	2011 Commercial Vehicles	2011 Total
Bermuda	22,500	5,400	27,900	22,000	5,000	27,000
Cuba	218,900	233,900	452,800	214,000	229,000	443,000
Dominican Republic	663,700	748,300	1,412,000	662,633	706,447	1,369,080
Haiti	39,900	38,200	78,100	38,000	37,000	75,000
Jamaica	141,400	36,500	177,900	139,000	36,000	175,000
Puerto Rico	1,101,600	482,800	1,584,400	1,087,000	473,000	1,560,000
Trinidad and Tobago	328,900	42,900	371,800	324,000	42,000	366,000
Virgin Islands (US)	39,200	21,600	60,800	38,000	21,000	59,000
Total Caribbean	**2,733,800**	**1,661,900**	**4,395,700**	**2,699,633**	**1,600,447**	**4,300,080**
AMERICA, Central & South						
Argentina	9,100,000	3,000,000	12,100,000	8,274,108	2,684,976	10,959,084
Belize	14,700	17,500	32,200	13,900	16,600	30,500
Bolivia	248,000	278,000	526,000	218,736	295,010	513,746
Brazil	29,566,116	7,705,144	37,271,260	27,490,694	7,164,275	34,654,969
Chile	2,383,638	1,208,390	3,592,028	2,169,280	1,138,252	3,307,532
Colombia	2,526,000	974,000	3,500,000	2,289,740	883,814	3,173,554
Costa Rica	634,000	201,500	835,500	611,175	199,696	810,871
Ecuador	509,800	706,700	1,216,500	455,502	630,094	1,085,596
El Salvador	97,300	126,400	223,700	92,000	120,000	212,000
Guatemala	604,000	1,069,000	1,673,000	563,000	998,000	1,561,000
Guyana	35,800	17,100	52,900	33,700	16,200	49,900
Honduras	36,400	102,800	139,200	34,200	97,800	132,000
Nicaragua	96,400	341,400	437,800	94,300	305,000	399,300
Panama	390,300	125,600	515,900	380,401	105,201	485,602
Paraguay	400,300	104,000	504,300	379,000	95,000	474,000
Peru	927,698	809,016	1,736,714	860,366	771,460	1,631,826
Suriname	85,200	37,000	122,200	80,800	35,100	115,900
Uruguay	656,500	89,900	746,400	625,000	82,000	707,000
Venezuela	2,531,529	807,867	3,339,396	2,480,000	789,000	3,269,000
Total Central & South America	**50,843,681**	**17,721,317**	**68,564,998**	**47,145,902**	**16,427,478**	**63,573,380**
AMERICA, North						
Canada	20,750,000	995,000	21,745,000	20,450,000	970,000	21,420,000
Mexico	22,961,571	9,881,728	32,843,299	22,368,598	9,584,003	31,952,601
United States	120,901,628	130,595,459	251,497,087	127,576,670	121,354,963	248,931,633
Total North America	**164,613,199**	**141,472,187**	**306,085,386**	**170,395,268**	**131,908,966**	**302,304,234**
ASIA, Far East						
Afghanistan	745,875	292,402	1,038,277	691,573	259,633	951,206
Bangladesh	242,004	344,035	586,039	219,976	276,505	496,481
Brunei	168,000	16,000	184,000	132,743	11,455	144,198
Burma	11,000	20,000	31,000	10,000	19,400	29,400

VEHICLES IN OPERATION BY COUNTRY

Country	2012 Cars	2012 Commercial Vehicles	2012 Total	2011 Cars	2011 Commercial Vehicles	2011 Total
China	52,165,000	57,275,000	109,440,000	43,220,000	50,280,000	93,500,000
Hong Kong	455,000	158,000	613,000	435,000	156,000	591,000
India	18,796,000	10,558,000	29,354,000	16,771,943	9,834,025	26,605,968
Indonesia	10,494,000	7,510,000	18,004,000	9,459,597	7,214,711	16,674,308
Japan	59,421,009	15,061,199	74,482,208	58,670,314	15,196,370	73,866,684
Korea, South	14,577,182	4,293,351	18,870,533	14,136,465	4,300,908	18,437,373
Malaysia	9,833,000	1,171,000	11,004,000	9,721,447	1,178,647	10,900,094
Pakistan	1,997,000	588,000	2,585,000	1,849,229	560,000	2,409,229
Philippines	488,600	1,424,800	1,913,400	828,600	2,396,400	3,225,000
Singapore	605,149	205,789	810,938	592,361	203,865	796,226
Sri Lanka	533,000	423,000	956,000	468,168	400,038	868,206
Taiwan	6,091,324	1,054,584	7,145,908	5,960,088	1,042,944	7,003,032
Thailand	6,273,983	6,591,061	12,865,044	5,404,763	6,237,480	11,642,243
Vietnam	178,000	192,000	370,000	169,000	186,000	355,000
Total Far East	**183,075,126**	**107,178,221**	**290,253,347**	**168,741,267**	**99,754,381**	**268,495,648**
ASIA, Middle East						
Azerbaijan	958,594	159,666	1,118,260	871,449	151,371	1,022,820
Bahrain	360,000	89,000	449,000	342,218	84,455	426,673
Cyprus	551,000	148,000	699,000	530,258	147,550	677,808
Iran	3,332,000	859,000	4,191,000	3,195,000	830,000	4,025,000
Iraq	1,126,000	528,000	1,654,000	1,052,000	511,000	1,563,000
Israel	2,246,053	391,198	2,637,251	2,164,385	397,230	2,561,615
Jordan	891,000	131,000	1,022,000	819,706	129,706	949,412
Kuwait	1,326,000	336,000	1,662,000	1,260,000	325,000	1,585,000
Lebanon	490,000	101,000	591,000	467,000	98,000	565,000
Oman	409,000	133,000	542,000	383,000	129,000	512,000
Qatar	217,000	106,000	323,000	205,000	103,000	308,000
Saudi Arabia	3,803,000	2,044,000	5,847,000	3,614,000	1,994,000	5,608,000
Syria	918,000	839,000	1,757,000	825,000	792,000	1,617,000
Turkey	8,648,875	4,178,324	12,827,199	8,113,111	3,948,903	12,062,014
United Arab Emirates	1,945,000	120,000	2,065,000	1,812,000	115,000	1,927,000
Yemen	331,000	359,000	690,000	316,000	351,000	667,000
Total Middle East	**27,552,522**	**10,522,188**	**38,074,710**	**25,970,127**	**10,107,215**	**36,077,342**
EUROPE, East						
Belarus	2,797,000	175,000	2,972,000	2,646,500	164,211	2,810,711
Bulgaria	2,799,000	383,000	3,182,000	2,696,000	370,000	3,066,000
Croatia	1,532,000	158,000	1,690,000	1,518,278	159,725	1,678,003
Czech Republic	4,638,372	707,784	5,346,156	4,582,903	702,499	5,285,402
Hungary	2,986,028	431,706	3,417,734	2,967,808	432,790	3,400,598
Lithuania	1,745,000	142,000	1,887,000	1,713,277	141,280	1,854,557
Poland	18,744,000	3,278,000	22,022,000	18,125,000	3,231,000	21,356,000

VEHICLES IN OPERATION BY COUNTRY

Country	2012 Cars	2012 Commercial Vehicles	2012 Total	2011 Cars	2011 Commercial Vehicles	2011 Total
Romania	4,487,000	762,000	5,249,000	4,335,000	737,000	5,072,000
Russian Federation	38,482,000	6,901,000	45,383,000	36,415,100	6,446,678	42,861,778
Serbia & Montenegro	1,792,000	190,000	1,982,000	1,677,510	180,703	1,858,213
Slovakia	1,771,000	315,000	2,086,000	1,709,000	308,300	2,017,300
Slovenia	1,066,028	85,445	1,151,473	1,066,495	85,544	1,152,039
Ukraine	8,301,000	1,642,000	9,943,000	8,058,000	1,603,000	9,661,000
Uzbekistan	1,453,000	510,000	1,963,000	1,375,000	487,000	1,862,000
Total Eastern Europe	**92,593,428**	**15,680,935**	**108,274,363**	**88,885,871**	**15,049,730**	**103,935,601**
EUROPE, West						
Austria	4,584,202	426,081	5,010,283	4,513,421	417,054	4,930,475
Belgium	5,392,908	846,083	6,238,991	5,359,014	831,836	6,190,850
Denmark	2,237,122	473,042	2,710,164	2,197,831	483,262	2,681,093
Finland	3,036,618	518,024	3,554,642	2,958,568	498,357	3,456,925
France	31,600,000	6,538,000	38,138,000	31,425,000	6,516,000	37,941,000
Germany	43,431,124	3,107,000	46,538,124	42,927,647	3,055,708	45,983,355
Greece	5,167,557	1,345,880	6,513,437	5,203,591	1,348,417	6,552,008
Iceland	208,000	32,000	240,000	206,112	32,181	238,293
Ireland	1,895,000	409,000	2,304,000	1,886,421	399,935	2,286,356
Italy	37,078,274	4,921,712	41,999,986	37,113,300	4,953,778	42,067,078
Latvia	618,274	81,347	699,621	612,321	77,808	690,129
Luxembourg	355,358	42,303	397,661	344,951	41,811	386,762
Malta	249,612	47,390	297,002	239,987	49,785	289,772
Netherlands	8,142,387	1,071,783	9,214,170	8,135,102	1,099,802	9,234,904
Norway	2,442,964	561,420	3,004,384	2,376,426	553,501	2,929,927
Portugal	4,497,000	1,310,000	5,807,000	4,522,000	1,351,000	5,873,000
Spain	22,247,528	5,232,813	27,480,341	22,277,244	5,319,109	27,596,353
Sweden	4,457,145	572,416	5,029,561	4,408,749	563,178	4,971,927
Switzerland	4,254,725	420,204	4,674,929	4,163,003	403,975	4,566,978
United Kingdom	31,481,823	4,279,078	35,760,901	31,362,716	4,269,641	35,632,357
Total Western Europe	**213,377,621**	**32,235,576**	**245,613,197**	**212,233,404**	**32,266,138**	**244,499,542**
PACIFIC						
Australia	12,714,235	3,265,521	15,979,756	12,474,044	3,164,896	15,638,940
Fiji	90,400	64,800	155,200	89,756	61,613	151,369
French Polynesia	46,100	24,900	71,000	45,500	24,400	69,900
Guam	73,800	35,900	109,700	72,400	35,000	107,400
New Caledonia	67,100	30,500	97,600	66,300	30,000	96,300
New Zealand	2,633,378	510,284	3,143,662	2,607,266	501,205	3,108,471
Papua New Guinea	39,200	95,800	135,000	38,700	94,200	132,900
Samoa (American)	7,600	7,800	15,400	7,500	7,700	15,200
Vanuatu	9,400	5,500	14,900	9,300	5,300	14,600
Total Pacific	**15,681,213**	**4,041,005**	**19,722,218**	**15,410,766**	**3,924,314**	**19,335,080**
WORLD TOTAL	**773,322,945**	**341,234,690**	**1,114,557,635**	**753,157,957**	**321,273,313**	**1,074,431,270**

NOTE: Some country figures estimated by *Ward's*.

ASSEMBLY PLANTS BY MANUFACTURER IN AFRICA

Manufacturer/ Country	Plant Site	Products
Arab American Vehicle		
Egypt	Cairo	Jeep Cherokee, J8, Wrangler; Toyota Fortuner
Associated Vehicle Assemblers		
Kenya	Mombasa	Toyota Land Cruiser
BMW		
Egypt	Sixth of October City	BMW Vehicles
South Africa	Rosslyn	BMW 3-Series
Egyptian-German Automotive		
Egypt	Sixth of October City	Mercedes C-Class, E-Class, GLK, S-Class
FAW Xiali		
Ethiopia	Adama	Weizhi Cars; Xiali Cars
Ford Southern Africa		
South Africa	Pretoria	Ford Ranger; Mazda BT-50
Foton		
Kenya	Nairobi	Foton Commercial Vehicles
General Motors		
South Africa	Kempston Road	Isuzu Commercial Vehicles
General Motors East Africa		
Kenya	Nairobi	Isuzu Commercial Vehicles
General Motors Egypt		
Egypt	Sixth of October City	Chevrolet Light Vehicles, Move; Opel Light Vehicles
General Motors South Africa		
South Africa	Struandale	Chevrolet Montana, Spark
Giad Motors		
Sudan	Khartoum	Hyundai Cars; Nissan Light Trucks
Great Wall		
Ethiopia	Addis Ababa	Great Wall Haval, Wingle
Kenya Vehicle Manufacturers		
Kenya	Thika	Foton Vehicles; Iveco Trucks; Land Rover Vehicles; MAN Commercial Vehicles; Mercedes-Benz Trucks; Nissan Urvan; UD Trucks Commercial Vehicles
MAN		
South Africa	Olifantsfontein	MAN Buses
	Pinetown	MAN Buses, Trucks
Mercedes-Benz South Africa		
South Africa	East London	Mercedes-Benz C-Class
Mesfin Industrial		
Ethiopia	Mekelle	Geely Cars
NEOPLAN		
Ghana	Kumasi	NEOPLAN Buses
Nissan Egypt		
Egypt	Giza	Nissan Pickup, Sunny, X-Trail

ASSEMBLY PLANTS BY MANUFACTURER IN AFRICA

Manufacturer/ Country	Plant Site	Products
Nissan South Africa		
South Africa	Rosslyn	Nissan Grand Livina, Livina, NP200, NP300, Tiida; Renault Sandero
Quest Motor		
Zimbabwe	Mutare	Chery QQ, Tiggo; Jiangling Pickup
Somaca		
Morocco	Casablanca	Dacia Logan, Sandero; Renault Kangoo
Suzuki Egypt		
Egypt	Sixth of October City	Suzuki Cars, SUVs
Tata Hispano Carrocera		
Morocco	Casablanca	Tata Hispano Buses
Tata Motors South Africa		
South Africa	Rosslyn	Tata Hispano Commercial Vehicles
Temsa		
Egypt	Cairo	Temsa Buses
Toyota South Africa		
South Africa	Durban	Toyota Corolla, Dyna, Fortuner, Hilux
Volkswagen South Africa		
South Africa	Uitenhage	Volkswagen Polo
Volvo Truck		
South Africa	Durban	Volvo Truck Medium- and Heavy-duty Trucks
Willowvale Mazda		
Zimbabwe	Harare	Mazda BT-50, Mazda3
Yangfan Motors		
Ethiopia	Addis Ababa	Lifan 520, Breez

ASSEMBLY PLANTS BY MANUFACTURER IN ASIA

Manufacturer/ Country	Plant Site	Products
Aftab Automobiles		
Bangladesh	Dhaka	Hino Trucks and Buses; Toyota Land Cruiser
Anhui Ankai		
China	Anhui	Ankai Buses
Ashok Leyland		
India	Alwar	Ashok Leyland Buses
	Bhandara	Ashok Leyland Commercial Vehicles
	Ennore	Ashok Leyland Commercial Vehicles
	Hosur	Ashok Leyland Dost, Medium- and Heavy-duty Trucks
	Pantnagar	Ashok Leyland Avia, Medium- and Heavy-duty Trucks
Asia Motor Works		
India	Buhj	AMW Medium- and Heavy-duty Trucks
Assembly Services		
Malaysia	Selangor	Daihatsu Delta; Hino Dutro; Toyota Camry, Fortuner, Hiace, Hilux, Innova, Vios
Astra Daihatsu		
Indonesia	Sunter	Daihatsu Gran Max, Luxio, Terios, Xenia; Toyota Avanza, Rush
Astra Honda		
Indonesia	Jakarta	Honda Beat
Astra Multi Trucks Indonesia		
Indonesia	Sunter	UD Trucks Buses, Trucks
AutoAlliance Thailand		
Thailand	Rayong	Ford Everest, Fiesta, Ranger; Mazda BT-50, Mazda2, Mazda3
Automotive Manufacturers Malaysia (AMM)		
Malaysia	Pahang	Mercedes-Benz Actros, C-Class, E-Class, S-Class; Mitsubishi Fuso Commercial Vehicles; Suzuki Swift; Volkswagen Passat
BAIC Motor		
China	Beijing	BAIC Buses, CUVs, Light trucks, Mini Trucks
	Zhuzhou	BAIC Cars
Beijing Benz		
China	Beijing	Mercedes-Benz C-Class, E-Class, GLK
Beijing Foton Daimler		
China	Huairou	Auman Medium- and Heavy-duty Trucks
Beijing Hyundai		
China	Beijing	Hyundai Accent, Avante, Elantra, i30, ix35, Moinca, Sonata, Tucson, Verna, Yuedong
BMW		
India	Chennai	BMW 3-Series, 5-Series, X1, X3
Thailand	Amata City	BMW 3-Series, 5-Series, 7-Series, X3
Brilliance BMW		
China	Shenyang	BMW 3-Series, 5-Series, X1
BYD		
China	Xi'an	BYD Buses, E6, F0, F3, F3DM, F6, G3, G6, L3, M6, S6
Changan Ford Mazda		
China	Chongqing	Ford Focus, Mondeo, S-Max
	Nanjing	Ford Fiesta; Mazda2, Mazda3
Chery Automobiles		
China	Dalian	Chery Eastar
	Kaifeng	Karry Mini-Truck

ASSEMBLY PLANTS BY MANUFACTURER IN ASIA

Manufacturer/ Country	Plant Site	Products
	Wuhu	Chery A1, A3, A5, Cowin, E5, Fulwin, QQ3, Tiggo; Karry Yousheng, Youya, Yo-Yo; Rely X5; Riich G5, M1
China Motor		
Taiwan	Hsinchu	Mitsubishi Fuso Medium- and Heavy-duty Trucks
	Yangmei	Mitsubishi Colt Plus, Delica, Grunder, Lancer Fortis, Outlander, Savrin, Veryca, Zinger
Chongqing Changan Suzuki		
China	Yudong	Suzuki Alto, Lingyang, Swift, SX4
Columbian Motors		
Philippines	Manila	UD Trucks Buses, Medium- and Heavy-duty Trucks
Daewoo Bus		
China	Guilin	Daewoo Buses
South Korea	Ulsan	Daewoo Buses
Taiwan	Pingtung	Daewoo Buses
Daihatsu		
Japan	Ikeda	Daihatsu Boon, Coo, Copen, Rush; Toyota bB, Passo
	Kyoto	Toyota Probox, Sienta, Succeed
	Ryuo	Daihatsu Move, Tanto; Subaru Stella Pixis Space
Daihatsu Kyushu		
Japan	Nakatsu	Daihatsu Atrai, Hijet, Mira, Mira Cocoa, Move Conte, Tanto Exe; Subaru Dias, Lucra, Sambar; Toyota Pixis Epoch, Pixis Space, Pixis Truck
Daimler India		
India	Chennai	BharatBenz Commercial Vehicles
DBR-Hicom		
Malaysia	Pegoh	Volkswagen Passat
Dewan Farooque		
Pakistan	Sujawal	Hyundai Santro, Trucks
Dongfeng		
China	Changzhou	Dongfeng Light Commercial Vehicles
	Xiangyang	Dongfeng Light Commercial Vehicles; Nissan Cabstar
Dongfeng Nissan		
China	Huadu	Nissan Livina, March, Murano, Qashqai, Sunny, Sylphy, Teana, Tiida, X-Trail; Venucia D50, R50
Dongfeng Peugeot Citroen		
China	Wuhan	Citroen C2, C4, C5, C-Elysee; Peugeot 207, 307, 308, 408, 508
Dongfeng Yueda-Kia		
China	Yancheng	Kia Cerato, Forte, Optima, Rio, Soul, Sportage
FAW-GM		
China	Changchun	FAW Light Trucks; GM Light Trucks
FAW-Mazda		
China	Changchun	FAW Vehicles; Mazda6, Mazda8
FAW-Volkswagen		
China	Changchun	Audi A4, A6, Q5; Volkswagen Bora, Golf, Jetta, Magotan, Sagitar
	Chengdu	Volkswagen Jetta, Sagitar

ASSEMBLY PLANTS BY MANUFACTURER IN ASIA

Manufacturer/ Country	Plant Site	Products
Fiat India		
India	Ranjangaon	Fiat Grande Punto, Linea; Tata Indica Vista, Indigo Manza
Force		
India	Pithampur	Force One, Traveller, Trax
Ford		
Thailand	Rayong	Ford Focus
Ford Australia		
Australia	Broadmeadows	Ford Falcon, Territory
Ford India		
India	Chennai	Ford Endeavour, Fiesta, Figo
Ford Lio Ho		
Taiwan	Zhongli	Ford Econovan, Escape, Focus, i-Max, Mondeo; Mazda3, Mazda5
Ford Philippines		
Philippines	Santa Rosa	Ford Escape, Focus; Mazda3, Mazda Tribute
Ford Thailand		
Thailand	Rayong	Ford Focus
Ford Vietnam		
Vietnam	Hanoi	Ford Escape, Everest, Focus, Mondeo, Ranger, Transit
Formosa Automobile		
Taiwan	Changhua	DAF Trucks
GAC Changfeng		
China	Guangzhou	Mitsubishi Pajero
GAC Fiat		
China	Changsha	Fiat Viaggio
GAC Toyota		
China	Guangzhou	Toyota Camry, Highlander, Yaris
Gaya Motor		
Indonesia	Sunter	BMW 3-Series; Isuzu Panther; UD Trucks Commercial Vehicles
Geely		
China	Jinan	Emgrand EC8
	Lanzhou	Englon SC6, SC6-RV; Gleagle GC3
	Linhai	Gleagle GC3, GX2, Panda
	Ningbo	Emgrand EC7
	Taizhou	Englon SC3, SC515-RV, SX5; Gleagle GX7
	Xiangtan	Geely Vision/Tiger; Gleagle GC7, GV5
General Motors India		
India	Halol	Chevrolet Captiva, Cruze, Enjoy, Sail, Tavera
	Talegaon	Chevrolet Beat, Spark
General Motors Thailand		
Thailand	Rayong	Chevrolet Captiva, Colorado, Cruze, Sonic, Trailblazer
Ghandhara Industries		
Pakistan	Karachi	Isuzu Buses, Trucks
Ghandhara Nissan		
Pakistan	Karachi	Nissan Sunny; UD Trucks Buses, Medium- and Heavy-duty Trucks
GM Korea		
South Korea	Changwon	Chevrolet Spark; Daewoo Damas, Labo
	Gunsan	Chevrolet Cruze, Orlando
	Incheon	Chevrolet Alpheon, Aveo, Captiva, Malibu, Trax

ASSEMBLY PLANTS BY MANUFACTURER IN ASIA

Manufacturer/ Country	Plant Site	Products
GM Vietnam		
Vietnam	Hanoi	Chevrolet Aveo, Captiva, Cruze, Lacetti, Orlando, Spark
Great Wall		
China	Baoding	Great Wall Cowry, Florid, Hover, Voleex
Guangzhou Honda		
China	Guangzhou	Honda Accord, City, Crosstour, Fit, Odyssey
Hindustan		
India	Pithampur	Hindustan Winner
	Thiruvallur	Mitsubishi Cedia, Lancer, Outlander, Pajero
	Uttarpara	Hindustan Ambassador, Veer, Winner
Hino		
Indonesia	Pulo Gadung	Hino Buses, Trucks; Toyota Dyna
Japan	Hamura	Hino Trucks; Toyota Dyna, FJ Cruiser, Land Cruiser Prado
Thailand	Samutprakarn	Hino Buses, Medium- and Heavy-duty Trucks
Hinopak		
Pakistan	Karachi	Hino Buses, Trucks
Holden		
Australia	Elizabeth	Holden Calais, Caprice, Commodore, Cruze, Ute
Honda		
Japan	Sayama	Acura RLX, TSX; Honda Accord, CR-V, Fit, Freed, Inspire, Legend, Odyssey, Step Wagon, Stream
	Suzuka	Honda CR-Z, Fit, Insight
Honda Atlas		
Pakistan	Lahore	Honda City, Civic
Honda Malaysia		
Malaysia	Pegoh	Honda Accord, City, Civic, CR-V
Honda Philippines		
Philippines	Santa Rosa	Honda City, Civic
Honda Prospect		
Indonesia	Karawang	Honda Brio, CR-V, Freed, Jazz
Honda Siel Cars India		
India	Noida	Honda Accord, Brio, City, Jazz
Honda Taiwan		
Taiwan	Pingtung	Honda Accord, Civic, CR-V, Fit
Honda Thailand		
Thailand	Ayutthaya	Honda Accord, Brio, City, Civic, CR-V, Jazz
Honda Vietnam		
Vietnam	Hanoi	Honda Civic, CR-V
Hyundai		
India	Sriperumbudur	Hyundai Accent, Elantra, Eon, i10, i20, Santro, Sonata, Verna
South Korea	Asan	Hyundai Grandeur, Sonata
	Jeonju	Hyundai Buses, Medium- and Heavy-duty Trucks
	Ulsan	Hyundai Accent, Avante, Equus, Genesis, Grand Starex, i30, i40, Light Trucks, Santa Fe, Tucson, Veloster, Veracruz
Hyundai Indonesia		
Indonesia	Bekasi	Hyundai Accent, Grand Avega, Sonata, Starex, Tucson
Indus Motors		
Pakistan	Karachi	Toyota Corolla, Hilux

ASSEMBLY PLANTS BY MANUFACTURER IN ASIA

Manufacturer/ Country	Plant Site	Products
Inokom		
Malaysia	Padang Meha	BMW 3-Series, 5-Series; Hyundai Elantra, i10, Santa Fe; Land Rover Vehicles; Mazda3
International Cars & Motors		
India	Amb	International Extreme
Irizar-TVS		
India	Viralimalai	Irizar Buses
Isuzu		
Japan	Fujisawa	Isuzu Commercial Vehicles
Thailand	Gateway	Isuzu Buses, D-Max, Trucks
	Samrong	Isuzu D-Max
Isuzu Astra Indonesia		
Indonesia	Pondok Ungu	Isuzu Medium- and Heavy-duty Trucks
Isuzu Hicom Malaysia		
Malaysia	Pahang	Isuzu D-Max, Medium- and Heavy-duty Trucks
Isuzu Philippines		
Philippines	Binan	Isuzu Buses, Light- and Medium-duty Trucks, Pickups, SUVs
Isuzu Taiwan		
Taiwan	Dadu	Isuzu Buses, Elf, Light- and Medium-duty Trucks
Isuzu Vietnam		
Vietnam	Ho Chi Minh City	Isuzu Buses, Light Trucks, Pickups, SUVs
JBCL		
India	Punjab	JBCL Buses
Jianghuai Auto		
China	Hefei	JAC Buses, Refine, Rein, Tojoy, Tongyue, Trucks, Yueyue
Jiangling Motors		
China	Jiangxi/Nanchang	Ford Transit; JMC Buses, Light Commercial Vehicles
Kenworth		
Australia	Melbourne	Kenworth Medium- and Heavy-duty Trucks
Kia		
South Korea	Gwangju	Kia Buses, Carens, Light Trucks, Soul, Sportage
	Hwasung	Kia Carens, Cerato, Forte, K3, K5, K7, Mohave, Sorento
	Sohari	Kia Carnival, K9, Pride
Korindo Heavy Industry		
Indonesia	Tangerang	Hyundai Medium- and Heavy-duty Trucks
Krama Yudha Ratu		
Indonesia	Jakarta	Mitsubishi Outlander Sports; Mitsubishi Fuso Trucks
Kuozui		
Taiwan	Guanyin	Hino Commercial Vehicles; Toyota Camry, Innova, Vios, Wish, Yaris
	Zhongli	Toyota Corolla Altis
Land Rover		
India	Pune	Land Rover Freelander
Mack		
Australia	Wacol	Mack Medium- and Heavy-duty Trucks

ASSEMBLY PLANTS BY MANUFACTURER IN ASIA

Manufacturer/ Country	Plant Site	Products
Mahindra and Mahindra		
India	Chakan	Mahindra Genio, Maxximo, XUV500; Mahindra Navistar Medium- and Heavy-duty Trucks; Ssangyong Rexton
	Haridwar	Mahindra Gio
	Kandivali	Mahindra Bolero, Maxi Truck
	Nashik	Mahindra Bolero, Scorpio, Verito, Xylo
	Zaheerabad	Mahindra Buses, Maxx, Pickup
Mahindra Reva		
India	Bangalore	Reva Electric Cars
MAN Trucks India		
India	Pithampur	MAN Medium- and Heavy-duty Trucks
	Pune	Force Commercial Vehicles
Maruti Suzuki		
India	Gurgaon	Maruti Suzuki 800, Alto, Eeco, Ertiga, Estilo, Omni, Ritz, Wagon R
	Manesar	Maruti Suzuki A-Star, Swift, SX4
Master Motor		
Pakistan	Karachi	Master Buses, Pickup, Trucks
Mazda		
Japan	Hofu	Mazda3, Mazda6
	Ujina	Mazda Biante, Bongo, CX-5, CX-7, CX-9, Flairwagon, Mazda2, Mazda5, Mazda8, MX-5, RX-8, Verisa
MCV India		
India	Chakan	Mercedes-Benz Buses
Mekong Auto		
Vietnam	Hanoi	PMC Light Commercial Vehicles
Mercedes-Benz India		
India	Chakan	Mercedes-Benz C-Class, E-Class, M-Class, S-Class
Mercedes-Benz Indonesia		
Indonesia	Wanaherang	Mercedes-Benz Buses, C-Class, E-Class
Mercedes-Benz Vietnam		
Vietnam	Ho Chi Minh City	Mercedes-Benz C-Class, E-Class, GLK, Sprinter
Mithra JBCL		
India	Vijiayawada	JBCL Buses
Mitsubishi		
Japan	Mizushima	Mitsubishi eK, Galant, i, i-Miev, Lancer, Minicab, Outlander, Toppo
	Nagoya	Mitsubishi ASX, Outlander, RVR
	Pajero	Mitsubishi Delica D5, Montero, Pajero
Mitsubishi Fuso		
Japan	Fuchu-machi	Mitsubishi Fuso Buses
	Kawasaki	Mitsubishi Fuso Light-, Medium-, and Heavy-duty Trucks
Mitsubishi Philippines		
Philippines	Cainta	Mitsubishi Adventure, Delica, Lancer
Mitsubishi Thailand		
Thailand	Laem Chabang	Mitsubishi Lancer, Mirage, Pajero Sport, Space Wagon, Triton; Nissan Navara
Nanjing Iveco		
China	Nanjing	Iveco Daily

ASSEMBLY PLANTS BY MANUFACTURER IN ASIA

Manufacturer/ Country	Plant Site	Products
National Assembler		
Indonesia	Jakarta	Audi A4, A6; Chery QQ, Tiggo; Volkswagen Golf, Multivan, Touran, Transporter
Naza		
Malaysia	Gurun	Kia Forte, Pregio, Sportage; Peugeot 207, 308, 508, 3008
Nissan		
Japan	Kyushu	Nissan Dualis, Murano, Note, Rogue, Serena, Teana, X-Trail
	Oppama	Nissan Cube, Juke, Leaf
	Tochigi	Infiniti EX, FX, G, M; Nissan 370Z, Cima, Fairlady Z, Fuga, GT-R, Skyline
Thailand	Bangna Trad	Nissan Almera, March, Sylphy, Teana
Nissan Indonesia		
Indonesia	Purwakarta	Nissan Evalia, Grand Livina, Juke, Livina, March, Serena, X-Trail
Nissan Philippines		
Philippines	Santa Rosa	Nissan Grand Livina, Sentra, X-Trail
Nissan Shatai		
Japan	Kanagawa	Infiniti QX56; Nissan AD Expert, Atlas, Caravan, Cedric, Elgrand, NV200, NV350, Patrol, Pickup, Quest, Wingroad
Nita		
Bangladesh	Dhaka	Tata Commercial Vehicles
Oriental Assemblers		
Malaysia	Johor Bahru	Changan Trucks; Chery B14, Eastar, Tiggo
Pak Suzuki		
Pakistan	Bin Qasim	Suzuki Bolan, Cultus, Liana, Mehran, Ravi, Swift
Perodua		
Malaysia	Selangor	Perodua Alza, Myvi, Viva
Piaggio		
India	Baramati	Piaggio Ape
Premier		
India	Chinchwad	Pimpri Rio, Roadstar, Sigma
Prince Motors		
Taiwan	Tucheng	Chery Apola, Fresh
Progoti		
Bangladesh	Chittagong	Mitsubishi Pajero
Proton		
Malaysia	Shah Alam	Proton Exora, Inspira, Saga, Waja
	Tanjung Malim	Proton Gen-2, Persona, Preve, Satria Neo
Qingling Motors		
China	Chongqing	Isuzu Rodeo; Qingling Trucks
Renault Samsung		
South Korea	Busan	Renault Fluence, Koleos, Latitude; Samsung QM5, SM3, SM5, SM7
Renault-Nissan India		
India	Orgadam	Nissan Evalia, Micra, Sunny; Renault Duster, Fluence, Koleos, Pulse, Scala
SAIC Motor		
China	Nanjing	MG Cars; Roewe Cars
	Shanghai	MG Cars; Roewe Cars

ASSEMBLY PLANTS BY MANUFACTURER IN ASIA

Manufacturer/ Country	Plant Site	Products
SAIC-GM-Wuling		
China	Liuzhou	Baojun 630, Lechi; Wuling Hongguang, Sunshine
	Qingdao	GM Spark; Wuling 1025 Pickup, PN Series
Sanyang Industries		
Taiwan	Hsinchu	Hyundai Elantra, i30, Lativa, Matrix, Porter, Santa Fe, Tucson
Scania		
India	Bangalore	Scania Buses, Medium- and Heavy-duty Trucks
Scania Siam		
Thailand	Bangpakong	Scania Buses, Heavy-duty Trucks
Shanghai GM		
China	Shanghai	Buick Excelle, LaCrosse, Regal; Cadillac SLS; Chevrolet Aveo, Captiva Sport, Epica, Malibu, Sail
Shanghai GM Dong Yue		
China	Yantai	Buick Encore, Excelle GT, Excelle XT; Chevrolet Aveo, Captiva, Epica, Sail
Shanghai GM Norsom		
China	Liaoning	Buick GL8; Chevrolet Cruze
Shanghai Sunwin		
China	Shanghai	Volvo Truck Buses
Shanghai-Volkswagen		
China	Anting	Skoda Fabia, Octavia, Superb; Volkswagen Lavida, Passat, Polo, Santana, Tiguan, Touran
	Nanjing	Volkswagen Passat, Santana
Shenyang Brilliance Jinbei		
China	Shenyang	Jinbei Minibus
Sichuan FAW Toyota		
China	Changchun	Toyota Corolla, Land Cruiser, Prius
	Chengdu	Toyota Coaster, Land Cruiser Prado
Sigma		
Pakistan	Islamabad	Land Rover Defender
Skoda India		
India	Aurangabad	Audi A4, A6, Q5; Skoda Laura, Superb, Yeti; Volkswagen Jetta, Passat
SML Isuzu		
India	Asron	Isuzu Medium- and Heavy-duty Trucks; SML Medium- and Heavy-duty Trucks
Southeast (Fujian)		
China	Fuzhou	Mitsubishi Galant, Lancer, Zinger; Soueast Vehicles
Ssangyong		
South Korea	Pyongtaek	Ssangyong Actyon, Chairman, Korando, Kyron, Rexton, Rodius
Subaru		
Japan	Gunma Main	Scion FR-S; Subaru BRZ; Toyota 86/GT86
	Gunma Yajima	Subaru Exiga, Forester, Impreza, Legacy, XV/Crosstrek
Sutlej Motors		
India	Jalandhar	Mercedes-Benz Buses
Suzuki		
Japan	Iwata	Mazda Scrum; Suzuki Carry Pickup, Escudo, Every, Grand Vitara, Jimny

ASSEMBLY PLANTS BY MANUFACTURER IN ASIA

Manufacturer/ Country	Plant Site	Products
	Kosai	Mazda AZ Wagon, Carol, Flairwagon; Mitsubishi Delica D2; Nissan Moco, Pino, ROOX; Suzuki Alto, Lapin, MR Wagon, Palette, Solio, Wagon R
	Sagara	Suzuki Escudo/Grand Vitara, Kizashi, Swift, SX4
Thailand	Rayong	Suzuki Swift
Vietnam	Bien Hoa	Suzuki Carry
Suzuki Indomobil		
Indonesia	Tambun	Suzuki APV, Carry, Ertiga, Grand Vitara, Swift, SX4
Swedish Motor Assemblies		
Malaysia	Shah Alam	King Long Placer; Volvo S40, S60, S80, V50, V60, XC60, XC90
Tan Chong Motor		
Malaysia	Segambut	Nissan Sentra, Serena, Urvan, X-Trail; UD Trucks Medium- and Heavy-duty Trucks
	Serendah	Nissan Almera, Latio, Livina, Sylphy, Teana
Tata		
Burma	Magway	Tata Commercial Vehicles
India	Chikhli	Land Rover Freelander 2
	Dharwad	Tata Ace Zip, Magic Iris
	Jamshedpur	Tata Medium and Heavy Commercial Vehicles
	Lucknow	Tata Buses
	Pantnagar	Tata Ace, Magic
	Pune	Tata Aria, Commercial Vehicles, Indica, Indigo, Safari, Sumo, Venture, Winger
	Sanand	Tata Nano
Thailand	Samutprakarn	Tata Xenon
Tata Daewoo		
South Korea	Gunsan	Tata Daewoo Medium- and Heavy-duty Trucks
Tata Marcopolo		
India	Dharwad	Tata Marcopolo Buses
Thai Auto Works		
Thailand	Samutprakarn	Toyota Hiace
Thai Swedish Assembly		
Thailand	Bangna Trad	Volvo S60, S70, V70; Volvo Truck Buses, Medium- and Heavy-duty Trucks
Thonburi		
Thailand	Samutprakarn	Hyundai Sonata; Mercedes-Benz Buses, C-Class, E-Class, M-Class
Tianjin FAW Toyota		
China	Tianjin	Toyota Corolla, Crown, RAV4, Reiz, Vios
Tianjin FAW Xiali		
China	Tianjin	FAW Vita, Xiali
Toyota		
Australia	Altona	Toyota Camry
Japan	Motomachi	Lexus LFA; Toyota Crown, Estima, Mark X
	Tahara	Lexus GS, GX, IS, LS; Toyota 4Runner, Land Cruiser, RAV4, Vanguard, Wish
	Takaoka	Scion iQ; Toyota Corolla
	Tsutsumi	Scion tC; Toyota Allion, Camry, Premio, Prius
Toyota Auto Body		
Japan	Fujimatsu	Toyota Alphard, Coaster, Estima, Hiace, Land Cruiser, Noah, Prius, Vellfire, Voxy

ASSEMBLY PLANTS BY MANUFACTURER IN ASIA

Manufacturer/ Country	Plant Site	Products
Toyota Indonesia		
Indonesia	Karawang	Toyota Fortuner, Innova
Toyota Kirloskar		
India	Bangalore	Toyota Corolla, Etios, Fortuner, Innova, Liva
Toyota Kyushu		
Japan	Miyawaka	Lexus CT, ES, HS, IS, RX; Toyota Harrier, Highlander, Sai
Toyota Motor East Japan		
Japan	Higashi Fuji	Toyota Century, Comfort, Crown, Isis, Porte, Ractis, Spade
	Iwate	Toyota Aqua, Ist, Ractis
	Ohira	Toyota Corolla
Toyota Philippines		
Philippines	Santa Rosa	Toyota Innova, Vios
Toyota Thailand		
Thailand	Ban Pho	Toyota Fortuner, Hilux
	Gateway	Toyota Camry, Corolla, Prius, Vios, Yaris
	Samrong	Toyota Hilux
Toyota Vietnam		
Vietnam	Me Linh	Toyota Camry, Corolla, Fortuner, Hiace, Innova, Vios
UD Trucks		
Japan	Ageo	UD Trucks Condor, Medium- and Heavy-duty Trucks
Universal Motors		
Philippines	Manila	Nissan Frontier, Navara, Patrol, Urvan
Vietnam Motors		
Vietnam	Hanoi	Nissan Grand Livina
VinaStar		
Vietnam	Binh Duong	Mitsubishi Pajero Sport, Zinger
Volkswagen India		
India	Chakan	Skoda Fabia, Rapid; Volkswagen Polo, Vento
Volvo		
China	Chongqing	Volvo S40, S80L
Volvo Eicher		
India	Pithampur	Eicher Commercial Vehicles; Volvo Truck Commercial Vehicles
Volvo India		
India	Bangalore	Volvo Truck Buses, Medium- and Heavy-duty Trucks
Volvo Truck		
Australia	Brisbane	Volvo Truck Medium- and Heavy-duty Trucks
China	Jinan	Volvo Truck Medium- and Heavy-duty Trucks
Yachiyo Industry		
Japan	Yokkaichi	Honda Acty, Life, Vamos, Zest
Yulon		
Taiwan	Sanyi	Nissan Grand Livina, March, Serena, Sylphy, Teana, Tiida; UD Trucks Commercial Vehicles
Zhengzhou Nissan		
China	Zhengzhou	Dongfeng Light Commercial Vehicles, Oting; Nissan NV200, Paladin, Pickup

ASSEMBLY PLANTS BY MANUFACTURER IN EUROPE

Manufacturer/ Country	Plant Site	Products
AIOS Isuzu		
Turkey	Gebze	Isuzu Light-duty Commercial Vehicles
Alexander Dennis		
United Kingdom	Guildford	Alexander Dennis Buses
Aston Martin		
United Kingdom	Gaydon	Aston Martin Cygnet, DB9, Rapide, Vanquish, Vantage, Virage
Audi		
Belgium	Brussels	Audi A1
Germany	Ingolstadt	Audi A3, A4, A5, Q5
	Neckarsulm	Audi A4, A5, A6, A7, A8, R8
Audi Hungaria		
Hungary	Gyor	Audi A3, TT
Avia		
Czech Republic	Letnany	Avia Medium- and Heavy-duty Trucks
Avtoframos		
Russia	Moscow	Dacia Duster, Logan, Sandero; Renault Koleos, Latitude, Megane
AvtoKraz		
Ukraine	Kremenchug	AvtoKraz Trucks
Avtotor		
Russia	Kaliningrad	BMW 3-Series, 5-Series, 7-Series, X1, X3, X5, X6; Cadillac Escalade, SRX; Chevrolet Antara, Aveo, Captiva Sport, Lacetti, Malibu, Orlando, Tahoe; Kia Ceed, Cerato, Mohave, Sorento, Soul, Sportage, Venga; Opel Astra, Insignia, Meriva, Zafira
AvtoVAZ		
Russia	Togliatti	IzhAvto Light Trucks; Lada Granta, Kalina, Largus, Niva, Priora, Samara
BAW		
Russia	Ulyanovsk	BAW Trucks
BelAZ		
Belarus	Zhodino	BelAZ Medium- and Heavy-duty Trucks
Bentley		
United Kingdom	Crewe	Bentley Continental, Mulsanne
BMC		
Turkey	Pinarbasi	BMC Buses, Trucks
BMW		
Germany	Dingolfing	BMW 5-Series, 6-Series, 7-Series
	Leipzig	BMW 1-Series, X1
	Munich	BMW 3-Series
	Regensburg	BMW 1-Series, 3-Series, M3, Z4
Bogdan		
Ukraine	Cherkassy	Bogdan 2310; Hyundai Accent, Elantra, Trucks, Tucson; Isuzu Trucks; Lada 2110
	Lutsk	Bogdan Buses
Bugatti		
France	Molsheim	Bugatti Veyron
BZKT		
Russia	Byansk	BAZ Medium- and Heavy-Duty Trucks
Caterham		
United Kingdom	Dartford	Caterham Super 7

ASSEMBLY PLANTS BY MANUFACTURER IN EUROPE

Manufacturer/ Country	Plant Site	Products
Craiova Assembly		
Romania	Craiova	Ford B-Max
DAF		
Belgium	Westerlo	DAF Medium- and Heavy-duty Trucks
Netherlands	Eindhoven	DAF Medium- and Heavy-duty Trucks
Daimler		
France	Hambach	Smart Fortwo
Dennis Eagle		
United Kingdom	Warwick	Dennis Eagle Medium- and Heavy-duty Trucks
Derways		
Russia	Cherkessk	Geely MK; Lifan Breez
Eurocar		
Ukraine	Uzhhorod	Skoda Vehicles
Evobus		
France	Ligny	Mercedes-Benz Buses
Germany	Dortmund	Mercedes-Benz Buses
	Mannheim	Mercedes-Benz Buses
	Ulm	Mercedes-Benz Buses
FAAM		
Italy	Monterubbiano	FAAM Light-duty Trucks
Ferrari		
Italy	Maranello	Ferrari 458 Italia, California, FF
Fiat		
Italy	Cassino	Alfa Romeo Giulietta; Fiat Bravo; Lancia Delta
	Melfi	Fiat Punto Evo
	Mirafiori	Alfa Romeo MiTo; Lancia Musa
	Pomigliano d'Arco	Fiat Panda
Fiat Poland		
Poland	Tychy	Fiat 500, Panda; Ford Ka; Lancia Ypsilon
Fiat Serbia		
Serbia	Kragujevac	Fiat 500, Punto
Ford		
Belgium	Genk	Ford Galaxy, Mondeo, S-Max
Germany	Cologne	Ford Fiesta
	Saarlouis	Ford Focus, Kuga
Romania	Craiova	Ford B-Max, Transit Connect
Russia	Vsevolozhsk	Ford Focus
Spain	Valencia	Ford C-Max
Turkey	Kocaeli	Ford Transit, Transit Connect
United Kingdom	Southampton	Ford Transit
Ford Sollers		
Russia	Elabuga	Ford Explorer, Galaxy, Kuga, S-Max, Transit
	St. Petersburg	Ford Focus, Mondeo
Fuso Kamaz Trucks		
Russia	Tatarstan	Mitsubishi Fuso Canter
GAZ		
Russia	Nizhny Novgorod	GAZ Buses, Gazelle, Medium-duty Trucks, Sobol
General Motors		
Russia	St. Petersburg	Chevrolet Astra, Captiva, Cruze; Opel Antara
GM Avtovaz		
Russia	Togliatti	Chevrolet Niva

ASSEMBLY PLANTS BY MANUFACTURER IN EUROPE

Manufacturer/ Country	Plant Site	Products
GM Poland		
Poland	Gliwice	Opel Astra
GolAZ		
Russia	Odintsovo	GolAZ Buses
Great Wall		
Russia	Moscow	Great Wall Hover
Honda		
Turkey	Gebze	Honda Civic
United Kingdom	Swindon	Honda Civic, CR-V, Jazz
Hyundai		
Czech Republic	Nosovice	Hyundai i30, i30, ix35
Russia	St. Petersburg	Hyundai Solaris; Kia Rio
Turkey	Izmit	Hyundai Accent, i20, Matrix
Ikarus		
Hungary	Szekesfehervar	Ikarus Buses
Irisbus		
Czech Republic	Vysoke Myto	Karosa Buses
France	Annonay	Irisbus Buses
	Rorthais	Heuliez Buses
Irizar		
Spain	Ormaiztegi	Irizar Buses
Iveco		
Germany	Ulm	Iveco Medium- and Heavy-duty Trucks
Italy	Brescia	Iveco Medium- and Heavy-duty Trucks
Russia	Chelyabinsk	Iveco Heavy-duty Trucks
Spain	Madrid	Iveco Medium- and Heavy-duty Trucks
	Valladolid	Iveco Daily
Jaguar Land Rover		
United Kingdom	Castle Bromwich	Jaguar XF, XJ, XK
	Halewood	Land Rover Freelander
	Solihull	Land Rover Defender, Discovery, Range Rover, Range Rover Evoque, Range Rover Sport
KamAZ		
Russia	Naberezhnye Chelny	KamAZ Trucks
Karsan		
Turkey	Bursa	Citroen Berlingo First; Hyundai Trucks; Peugeot Partner Origin
KAVZ		
Russia	Kurgan	KAVZ Buses
Kia Slovakia		
Slovakia	Zilina	Hyundai ix35; Kia Cee'd, Sportage, Venga
KrASZ		
Ukraine	Kremenchug	Geely Vehicles; Ssangyong Vehicles
Lamborghini		
Italy	Sant'Agata Bolognese	Lamborghini Gallardo, Murcielago
Leyland Trucks		
United Kingdom	Leyland	DAF Medium- and Heavy-duty Trucks
LiAZ		
Russia	Likino	LiAZ Buses
Litex		
Bulgaria	Bahovitsa	Great Wall Steed, Voleex
Lotus		
United Kingdom	Hethel	Lotus Elise, Evora, Exige

ASSEMBLY PLANTS BY MANUFACTURER IN EUROPE

Manufacturer/ Country	Plant Site	Products
LTI		
United Kingdom	Coventry	LTI TX Taxi
Magna-Steyr		
Austria	Graz	Aston Martin Rapide; Mercedes-Benz G-Class; Mini Countryman; Peugeot 308 RCZ
Magyar Suzuki		
Hungary	Esztergom	Fiat Sedici; Opel Agila; Suzuki Splash, Swift, SX4
MAN		
Austria	Steyr	MAN Medium- and Heavy-duty Trucks
Germany	Munich	MAN Heavy-duty Trucks
	Plauen	MAN Buses
	Salzgitter	MAN Buses, Heavy-duty Trucks
Poland	Poznan	MAN Buses
Turkey	Ankara	MAN Buses
Maserati		
Italy	Modena	Maserati GranCabrio, GranTurismo, Quattroporte
MAZ		
Belarus	Minsk	MAZ Buses, Trucks
Mazda Sollers		
Russia	Vladivostok	Mazda CX-5, Mazda6
McLaren		
United Kingdom	Woking	McLaren MP4-12C
Mercedes-Benz		
Germany	Bremen	Mercedes-Benz C-Class, E-Class, GLK, SL, SLK
	Dusseldorf	Mercedes-Benz Sprinter; Volkswagen Crafter
	Ludwigsfelde	Mercedes-Benz Sprinter, Vario; Volkswagen Crafter
	Rastatt	Mercedes-Benz A-Class, B-Class
	Sindelfingen	Maybach Cars; Mercedes-Benz C-Class, CL, CLS, E-Class, S-Class, SLS AMG
	Worth	Mercedes-Benz Actros, Atego, Axor, Econic, Unimog, Zetros
Hungary	Kecskemet	Mercedes-Benz B-Class
Russia	Tatarstan	Mercedes-Benz Actros, Atego, Axor
Spain	Vitoria	Mercedes-Benz Viano, Vito
Turkey	Aksaray	Mercedes-Benz Actros, Atego, Axor, Unimog
	Hosdere	Mercedes-Benz Buses
MG UK		
United Kingdom	Longbridge	MG MG6
Mini		
United Kingdom	Oxford	Mini Cooper
Minsk		
Belarus	Minsk	Minsk Trucks
Mitsubishi Fuso		
Portugal	Tramagal	Mitsubishi Canter
MOAZ		
Belarus	Minsk	MOAZ Trucks
Morgan		
United Kingdom	Malvern	Morgan 4/4, Aero, Plus4, Plus8, Roadster
NedCar		
Netherlands	Born	Mitsubishi Colt, Outlander

ASSEMBLY PLANTS BY MANUFACTURER IN EUROPE

Manufacturer/ Country	Plant Site	Products
NefAZ		
Russia	Bashkortostan	NefAZ Buses
Nissan Iberica		
Spain	Barcelona	Nissan Atleon, Cabstar, Navara, NV200, Pathfinder, Primastar
Nissan Russia		
Russia	St. Petersburg	Infiniti FX; Nissan Murano, Teana, X-Trail
Nissan UK		
United Kingdom	Sunderland	Nissan Juke, Leaf, Note, Qashqai
Opel		
Germany	Bochum	Opel Astra, Zafira
	Eisenach	Opel Corsa
	Russelsheim	Buick Regal; Opel Astra, Insignia
Spain	Zaragoza	Opel Combo, Corsa, Meriva
Optare		
United Kingdom	Leeds	Optare Buses
Otokar		
Turkey	Karasu	Land Rover Defender; Otokar Commercial Vehicles
Oyak-Renault		
Turkey	Bursa	Renault Clio, Fluence, Megane, Symbol
PCMA Rus		
Russia	Kaluga	Citroen C4, C-Crosser; Mitsubishi Outlander; Peugeot 308, 4007
Peugeot-Citroen		
France	Aulnay	Citroen C3
	Mulhouse	Citroen C4, DS4; Peugeot 206, 308
	Poissy	Citroen C3, DS3; Peugeot 207
	Rennes	Citroen C5, C6; Peugeot 407, 508
	Sochaux	Citroen DS5; Peugeot 308, 3008, 5008
Portugal	Mangualde	Citroen Berlingo; Peugeot Partner
Slovakia	Trnava	Citroen C3 Picasso; Peugeot 207
Spain	Madrid	Peugeot 207
	Vigo	Peugeot Partner
	Vigo	Citroen Berlingo, C4 Picasso
Pininfarina		
Italy	San Giorgio	Pininfarina Bluecar
Porsche		
Germany	Leipzig	Porsche Cayenne, Panamera
	Zuffenhausen (Stuttgart)	Porsche 911, Boxster, Cayman
Renault		
France	Batilly	Renault Master
	Dieppe	Renault Clio, Megane
	Douai	Renault Megane, Scenic
	Flins	Renault Clio
	Maubeuge	Renault Kangoo
	Sandouville	Renault Espace, Laguna
Romania	Pitesti	Dacia Duster, Logan, Sandero
Slovenia	Novo Mesto	Renault Clio, Twingo, Wind
Spain	Palencia	Renault Megane
	Valladolid	Renault Clio, Modus
Renault Trucks		
France	Blainville	Renault Trucks Medium- and Heavy-duty Trucks

ASSEMBLY PLANTS BY MANUFACTURER IN EUROPE

Manufacturer/ Country	Plant Site	Products
	Bourg-en-Bresse	Renault Trucks Heavy-duty Trucks
	Limoges	Renault Trucks Medium- and Heavy-duty Trucks
	Venissieux	Renault Trucks Medium- and Heavy-duty Trucks
Rolls-Royce		
United Kingdom	Goodwood	Rolls-Royce Ghost, Phantom
Roman		
Romania	Brasov	Roman Trucks
Scania		
France	Angers	Scania Trucks
Netherlands	Zwolle	Scania Trucks
Poland	Slupsk	Scania Buses
SEAT		
Spain	Martorell	Audi Q3; SEAT Altea, Exeo, Ibiza, Leon
Sevelnord		
France	Lieu Saint Amand	Citroen C8, Jumpy; Fiat Scudo; Peugeot 807, Expert
Sevelsud		
Italy	Val di Sangro	Citroen Jumper; Fiat Ducato; Peugeot Boxer
Sisu		
Finland	Karjaa	Sisu Trucks
Skoda		
Czech Republic	Kvasiny	Skoda Superb, Yeti
	Mlada Boleslav	Skoda Fabia, Octavia
	Vrchlabi	Skoda Octavia, Praktik, Roomster
Sollers Far East		
Russia	Vladivostok	Ssangyong Actyon, Kyron, Rexton
Sollers-Isuzu		
Russia	Ulyanovsk	Isuzu Light Trucks
TagAZ		
Russia	Taganrog	BYD F3; Hyundai Elantra, Porter; TagAZ C10, C190, LC-100 Master, Road Partner, Tager; Vortex Corda, Elena, Estina, Estina, Yuna
Tata Hispano Carrocera		
Spain	Zaragoza	Tata Hispano Buses
Tatra		
Czech Republic	Koprivnice	Tatra Trucks
Temsa		
Turkey	Adana	Mitsubishi Canter; Temsa Buses
Tofas		
Turkey	Bursa	Citroen Nemo; Fiat Albea, Doblo, Fiorino, Linea, Palio; Opel Combo; Peugeot Bipper
Toyota		
France	Valenciennes	Toyota Yaris
Russia	St. Petersburg	Toyota Camry
Turkey	Gebze/Sakarya	Toyota Auris, Verso
United Kingdom	Burnaston	Toyota Auris, Avensis
Toyota Caetano		
Portugal	Ovar	Toyota Dyna
Toyota Peugeot Citroen		
Czech Republic	Kolin	Citroen C1; Peugeot 107; Toyota Aygo
UAZ		
Russia	Ulyanovsk	UAZ Commercial Vehicles, Pickup

ASSEMBLY PLANTS BY MANUFACTURER IN EUROPE

Manufacturer/ Country	Plant Site	Products
Unison		
Belarus	Minsk	IKCO Samand
Valmet		
Finland	Uusikaupunki	Fisker Karma; Mercedes-Benz A-Class
Van Hool		
Belgium	Lier-Koningshooikt	Van Hool Buses, Trucks
Vauxhall Opel		
United Kingdom	Ellesmere Port	Opel Astra
	Luton	Opel Vivaro; Renault Trafic
Volkswagen		
Germany	Dresden	Volkswagen Phaeton
	Emden	Volkswagen CC, Passat
	Hanover	Volkswagen T5
	Osnabruck	Volkswagen Golf
	Wolfsburg	Volkswagen Golf, Tiguan, Touran
	Zwickau	Volkswagen Golf, Passat
Poland	Poznan	Volkswagen Caddy, T5
Portugal	Palmela	SEAT Alhambra; Volkswagen Eos, Scirocco, Sharan
Russia	Kaluga	Skoda Fabia, Octavia; Volkswagen Polo, Tiguan
Spain	Pamplona	Volkswagen Polo
Volkswagen Slovakia		
Slovakia	Bratislava	Audi Q7; Porsche Cayenne; Seat Mii; Skoda Citigo; Volkswagen Touareg, Up!
Volvo Cars		
Belgium	Ghent	Volvo C30, S60, V40, XC60
Sweden	Torslanda	Volvo S80, V60, V70, XC70, XC90
	Uddevalla	Volvo C70
Volvo Truck		
Belgium	Ghent	Volvo Truck Medium- and Heavy-duty Trucks
Poland	Wroclaw	Volvo Truck Buses
Russia	Kaluga	Volvo Truck Commercial Vehicles
Sweden	Boras	Volvo Truck Buses
	Gothenburg	Volvo Truck Medium- and Heavy-duty Trucks
	Saffle	Volvo Truck Buses
	Uddevalla	Volvo Truck Buses
	Umea	Volvo Truck Medium- and Heavy-duty Trucks
Westfalia Werke		
Germany	Rheda-Wiedenbruck	Westfalia Conversion Vans
ZAZ		
Ukraine	Zaporizhya	Chery A13; Forza Cars; Kia Mohave, Sorento; Tata Trucks; ZAZ Chance, Lanos, Sens, Vida
ZIL		
Russia	Moscow	ZIL Commercial Vehicles

ASSEMBLY PLANTS BY MANUFACTURER IN THE MIDDLE EAST

Manufacturer/ Country	Plant Site	Products
Agromashholding Kazakhstan		
Kazakhstan	Kostanay	Ssangyong Actyon, Actyon Sports, Korando C, Kyron, Rexton II
Asia Auto		
Kazakhstan	Ust-Kamenogorsk	Chevrolet Aveo, Captiva, Cruze, Malibu, Orlando; Kia Cadenza, Cerato, Mojave, Optima, Sorento, Soul, Sportage; Lada Niva 4x4, Samara; Skoda Fabia, Octavia, Yeti; UAZ Hunter, Patriot
Automotive Industries (AIL)		
Israel	Nazareth Illit	AIL Abir II, Storm 3
Bahman		
Iran	Tehran	Isuzu Light Trucks; Mazda2, Mazda3, Mazda Pickups
Daewoo Bus		
Kazakhstan	Ust-Kamenogorsk	Daewoo Buses
General Motors Uzbekistan		
Uzbekistan	Asaka	Chevrolet Damas, Lacetti, Matiz, Nexia, Spark
	Tashkent	Chevrolet Captiva, Epica, Malibu
Hyundai Auto Trans		
Kazakhstan	Almaty	Hyundai Commercial Vehicles
Iran Khodro		
Iran	Khorasan	Peugeot 405, Pars; Suzuki Grand Vitara
	Mazandaran	IKCO Sarir
	Tabriz	IKCO Bardo
	Tehran	IKCO Dena, Runna, Samand, Soren, Tondar 90; Peugeot 206, 207, 405, Roa
Iran Khodro Diesel		
Iran	Tehran	Hyundai Buses, Light Trucks; Khawar Buses, Trucks; Mercedes-Benz Trucks
Kerman		
Iran	Tehran	Lifan 520, 620; Mitsubishi Tojoy
Kish Khodro		
Iran	Tehran	IKCO Sinad
Morattab Khodro		
Iran	Tehran	Land Rover Pazhan; Ssangyong Musso
Oghab-Scania		
Iran	Semnan	Scania Buses
Pars Khodro		
Iran	Tehran	Nissan Vehicles; Renault Vehicles
Saipa		
Iran	Tehran	Saipa 131, 132, 141
Syria	Hassia	Saipa 131, 132
Shahab Khodro		
Iran	Tehran	Shahab Khodro Commercial Vehicles
Siamco		
Syria	Damascus	IKCO Samand
State Company for Automotive Industry (SCAI)		
Iraq	Iskandariyah	Scania Commercial Vehicles
Zagross Khodro		
Iran	Borujerd	Proton Gen-2, Impian, Wira

ASSEMBLY PLANTS BY MANUFACTURER IN NORTH AMERICA

Manufacturer/ Country	Plant Site	Products
Blue Diamond		
Mexico	Escobedo 2	Ford Medium-duty Trucks; International Medium-duty Trucks
BMW		
U.S.	Spartanburg	BMW X3, X5, X6
Chrysler		
Canada	Brampton	Chrysler 300 Series; Dodge Challenger, Charger; Lancia Thema
	Windsor	Chrysler Town & Country; Dodge Caravan; Lancia Voyager; Ram Cargo Van; Volkswagen Routan
Mexico	Saltillo	Ram Pickup
	Toluca	Dodge Journey; Fiat 500, Freemont
U.S.	Belvidere	Dodge Dart; Jeep Compass, Patriot
	Detroit (Jefferson North)	Dodge Durango; Jeep Grand Cherokee
	Sterling Heights	Chrysler 200 Series; Dodge Avenger; Lancia Flavia
	Warren (Dodge City)	Ram Pickup
	Toledo Supplier Park	Jeep Wrangler
Dina Camiones		
Mexico	Tepeapulco	Dina Medium- and Heavy-duty Trucks
Flat Rock Assembly		
U.S.	Flat Rock	Ford Mustang
Ford		
Canada	Oakville	Ford Edge, Flex; Lincoln MKT, MKX
Mexico	Cuautitlan	Ford Fiesta
	Hermosillo	Ford Fusion; Lincoln MKZ
U.S.	Chicago	Ford Explorer, Taurus; Lincoln MKS
	Louisville	Ford Escape
	Louisville (Kentucky Truck)	Ford Expedition, F-Series; Lincoln Navigator
	Dearborn Truck	Ford F-Series; Lincoln Mark LT
	Detroit	Ford Chassis
	Michigan Assembly	Ford C-Max, Focus
	Kansas City 2	Ford F-Series
	Avon Lake (Ohio Assembly)	Ford Econoline
Freightliner		
Mexico	Santiago Tianquistenco	Freightliner Medium- and Heavy-duty Trucks; Mercedes-Benz Medium- and Heavy-duty Trucks
U.S.	Cleveland	Freightliner Heavy-duty Trucks
	High Point	Thomas Built Buses
	Mt. Holly	Freightliner Medium-duty Trucks
	Portland	Freightliner Heavy-duty Trucks; Western Star Heavy-duty Trucks
General Motors		
Canada	Ingersoll	Chevrolet Equinox; GMC Terrain
	Oshawa 1	Buick Regal; Cadillac XTS; Chevrolet Camaro
	Oshawa 2	Chevrolet Equinox, Impala
Mexico	Ramos Arizpe	Cadillac SRX; Chevrolet Captiva, Sonic
	Silao	Cadillac Escalade EXT; Chevrolet Avalanche, Silverado; GMC Sierra
	San Luis Potosi	Chevrolet Aveo, Trax
U.S.	Fort Wayne	Chevrolet Silverado; GMC Sierra
	Fairfax	Buick LaCrosse; Chevrolet Malibu

ASSEMBLY PLANTS BY MANUFACTURER IN NORTH AMERICA

Manufacturer/ Country	Plant Site	Products
	Bowling Green	Chevrolet Corvette
	Flint 1	Chevrolet Silverado; GMC Sierra
	Hamtramck	Chevrolet Malibu, Volt; Opel Ampera
	Lansing Delta	Buick Enclave; Chevrolet Traverse; GMC Acadia
	Lansing Grand River	Cadillac ATS, CTS
	Orion Township	Buick Verano; Chevrolet Sonic
	Wentzville	Chevrolet Express; GMC Savana
	Lordstown	Chevrolet Cruze
	Spring Hill	Chevrolet Equinox
	Arlington	Cadillac Escalade, Escalade ESV; Chevrolet Suburban, Tahoe; GMC Yukon, Yukon XL
Hino		
Canada	Woodstock	Hino Medium-duty Trucks
Mexico	Silao	Hino Medium-duty Trucks
U.S.	Williamstown	Hino Medium-duty Trucks
Honda		
Canada	Alliston 1	Honda Civic
	Alliston 2	Acura MDX, ZDX; Honda CR-V
Mexico	El Salto	Honda CR-V
U.S.	Lincoln	Honda Odyssey, Pilot, Ridgeline
	Greensburg	Acura ILX; Honda Civic
	East Liberty	Acura RDX; Honda Crosstour, CR-V
	Marysville	Acura TL; Honda Accord
Hyundai		
U.S.	Montgomery	Hyundai Elantra, Sonata
International		
Mexico	Escobedo	International Medium- and Heavy-duty Trucks
U.S.	Conway	International Buses
	Springfield	International Buses, Medium-duty Trucks
	Tulsa	International Buses
	Garland	International Medium- and Heavy-duty Trucks
Isuzu		
U.S.	Charlotte	Isuzu Light and Medium-duty trucks
Kenworth		
Canada	St. Therese	Kenworth Medium-duty Trucks; Peterbilt Medium-duty Trucks
Mexico	Mexicali	Kenworth Medium- and Heavy-duty Trucks
U.S.	Chillicothe	Kenworth Heavy-duty Trucks
	Denton	Peterbilt Heavy-duty Trucks
	Renton	Kenworth Heavy-duty Trucks
Kia		
U.S.	West Point	Hyundai Santa Fe; Kia Optima, Sorento
Mack		
U.S.	Macungie	Mack Heavy-duty Trucks
MAN		
Mexico	Puebla	Volkswagen Medium-duty Trucks
Mercedes-Benz		
U.S.	Vance	Mercedes GL, M-Class, R-Class
	Gaffney	Freightliner Medium-duty Trucks
	Ladson	Mercedes Sprinter Van
Mitsubishi		
U.S.	Normal	Mitsubishi Outlander Sport

ASSEMBLY PLANTS BY MANUFACTURER IN NORTH AMERICA

Manufacturer/ Country	Plant Site	Products
Nissan		
Mexico	Aguascalientes	Nissan March, Sentra, Versa
	Cuernavaca	Nissan Chassis, Frontier, Pickup, Tiida, Tsuru
U.S.	Canton	Nissan Altima, Armada, Frontier, NV, Titan, Xterra
	Smyrna	Infiniti JX; Nissan Altima, Leaf, Maxima, Pathfinder
Subaru		
U.S.	Lafayette	Subaru Legacy, Outback, Tribeca; Toyota Camry
Toyota		
Canada	Cambridge	Lexus RX; Toyota Corolla
	Woodstock	Toyota RAV4
Mexico	Tijuana	Toyota Tacoma
U.S.	Princeton	Toyota Highlander, Sequoia, Sienna
	Georgetown	Toyota Avalon, Camry, Venza
	Blue Springs	Toyota Corolla
	San Antonio	Toyota Tacoma, Tundra
Volkswagen		
Mexico	Puebla	Volkswagen Beetle, Bora, Jetta
U.S.	Chattanooga	Volkswagen Passat
Volvo Truck		
Mexico	Mexico City	Volvo Heavy-duty Trucks and Buses
U.S.	Dublin	Mack Heavy-duty Trucks; Volvo Heavy-duty Trucks

ASSEMBLY PLANTS BY MANUFACTURER IN SOUTH AMERICA

Manufacturer/ Country	Plant Site	Products
Agrale		
Brazil	Caxias do Sul	Agrale Trucks
Aymesa		
Ecuador	Quito	Chevrolet Vitara; Kia Grand Pregio, Pride, Rio, Sportage
Bramont-Montadora		
Brazil	Manaus	Mahindra Scorpio
Caoa Montadora de Veiculos		
Brazil	Anapolis	Hyundai H100, HR, ix35, Porter, Tucson
Chery		
Venezuela	Aragua	Chery A1/Kimino, A3
Chrysler de Venezuela		
Venezuela	Carabobo	Dodge Caliber; Jeep Cherokee, Grand Cherokee
Compania Colombiana Automotriz (CCA)		
Colombia	Santa Fe de Bogota	Mazda BT-50, Mazda2, Mazda3
Daewoo Bus		
Costa Rica	Santa Ana	Daewoo Buses
Fiat		
Argentina	Cordoba	Fiat Palio, Siena
Brazil	Betim	Fiat Bravo, Doblo, Fiorino, Idea, Linea, Palio, Punto, Strada, Uno
Fiat Industrial		
Brazil	Sete Lagoas	Fiat Ducato; Iveco Daily
Ford		
Argentina	Pacheco	Ford Focus, Ranger
Brazil	Camacari	Ford EcoSport, Fiesta
	Sao Bernardo	Ford Courier, Ka
Venezuela	Valencia	Ford Explorer, Fiesta, F-Series, Trucks
General Motors Colmotores		
Colombia	Santafe de Bogota	Chevrolet Aveo, Cobalt, Optra, Sail, Spark; Isuzu Trucks
General Motors de Argentina		
Argentina	Rosario	Chevrolet Agile, Classic, Corsa
General Motors del Ecuador		
Ecuador	Quito	Chevrolet Aveo, Colorado, Grand Vitara, LUV, Sail
General Motors do Brasil		
Brazil	Gravatai	Chevrolet Celta, Prisma
	Sao Caetano do Sul	Chevrolet Cobalt, Corsa, Cruze, Montana, Spin
	Sao Jose dos Campos	Chevrolet Blazer, Corsa, S-10
General Motors Venezolana		
Venezuela	Mariara	Chevrolet LUV; Isuzu Trucks
	Valencia	Chevrolet Aveo, Cruze, Optra, Orlando, Silverado, Spark, Trucks
Hino		
Colombia	Cota	Hino Trucks
Honda Automoveis do Brasil		
Brazil	Sumare	Honda City, Civic, Fit
Iveco Argentina		
Argentina	Cordoba	Iveco Buses, Medium- and Heavy- duty Trucks
Iveco Venezuela		
Venezuela	La Victoria	Iveco Daily, Light-, Medium-, and Heavy-duty Trucks

ASSEMBLY PLANTS BY MANUFACTURER IN SOUTH AMERICA

Manufacturer/ Country	Plant Site	Products
Mack de Venezuela		
Venezuela	Las Tejerias	Mack Trucks
MAN		
Brazil	Resende	Volkswagen Buses, Medium- and Heavy-duty Trucks
Maresa		
Ecuador	Quito	Mazda BT-50
Mercedes-Benz Argentina		
Argentina	Buenos Aires	Mercedes-Benz Buses, Sprinter, Trucks
Mercedes-Benz do Brazil		
Brazil	Juiz de Fora	Mercedes-Benz Light Trucks
	Sao Bernardo do Campo	Mercedes-Benz Buses, Trucks
Mitsubishi Brazil		
Brazil	Catalao	Mitsubishi Challenger, L200, Pajero, Triton
MMC Automotriz		
Venezuela	Barcelona	Hyundai Elantra, Getz; Mitsubishi Canter, L300, Lancer
PSA Peugeot Citroen		
Argentina	Villa Bosch	Citroen Berlingo, C4; Peugeot 206, 207, 308, 408, Partner
Brazil	Porto Real	Citroen AirCross, C3; Peugeot 207, Amlat, Hoggar
Renault Argentina		
Argentina	Santa Isabel	Renault Clio, Fluence, Kangoo, Megane, Symbol
Renault do Brasil		
Brazil	Curitiba	Dacia Duster; Nissan Frontier, Livina; Renault Logan, Master, Megane, Sandero
Scania Latin America		
Brazil	Sao Paulo	Scania Buses, Medium- and Heavy-duty Trucks
Sofasa		
Colombia	Envigado	Renault Clio, Logan, Megane, Sandero, Twingo
SVB Automotive		
Brazil	Itumbiara	Suzuki Jimny
Toyota Argentina		
Argentina	Zarate	Toyota Hilux
Toyota de Venezuela		
Venezuela	Cumana	Toyota Corolla, Fortuner, Hilux, Terios
Toyota do Brasil		
Brazil	Indaiatuba	Toyota Corolla
	Sorocaba	Toyota Etios
Volkswagen Argentina		
Argentina	Pacheco	Volkswagen Amarok, Suran
Volkswagen do Brasil		
Brazil	Anchieta	Volkswagen Gol, Parati, Polo, Saveiro, T2
	Curitiba	Volkswagen Fox, Golf, Suran
	Taubate	Volkswagen Gol, Voyage
Volvo Truck		
Brazil	Curitiba	Volvo Buses, Heavy-duty Trucks